Computational Simulation in Architectural and Environmental Acoustics

Tetsuya Sakuma · Shinichi Sakamoto
Toru Otsuru
Editors

Computational Simulation in Architectural and Environmental Acoustics

Methods and Applications of Wave-Based Computation

Springer

Editors
Tetsuya Sakuma
Shinichi Sakamoto
The University of Tokyo
Tokyo
Japan

Toru Otsuru
Oita University
Oita
Japan

ISBN 978-4-431-54453-1 ISBN 978-4-431-54454-8 (eBook)
DOI 10.1007/978-4-431-54454-8
Springer Tokyo Heidelberg New York Dordrecht London

Library of Congress Control Number: 2014942712

Printed on acid-free paper

Springer is part of Springer Science+Business Media (www.springer.com)

Preface

With the consistent progress of computer technology, computational acoustic simulation is now becoming a popular, indispensable, and powerful tool for sound environmental design of architectural and urban spaces. Indeed, there are high expectations for a wide variety of applications for prediction of room acoustics and noise propagation, development of building materials/components and audio equipment, and visualization and auralization of sound environment, among other effects. In the first decade of this century, remarkable advances have been made in wave-based acoustic simulation techniques, which are steadily increasing their practicality and applicability.

In the meantime, the Subcommittee of Computational Acoustics on Built Environment of the Architectural Institute of Japan proposed a book in Japanese to review a variety of numerical methods for wave-based acoustic simulation and recent applications to architectural and environmental acoustics, and it was published by the institute in 2011. Following the original concept and enriching the contents, this present book was composed in English for international publication by the editorial board and new authors were added.

This book has two main parts, following an introduction providing an overview of computational simulation of sound environment. Part I explains the fundamentals and advanced techniques for three popular methods, namely, the finite-difference time-domain method, the finite element method, and the boundary element method, as well as alternative time-domain methods. Part II demonstrates various applications to room acoustics simulation, noise propagation simulation, and acoustic property simulation for building components, and auralization.

All authors willingly contributed the latest fruits of their own research, which led to the successful completion of this edition. The editors have tried to make terminology and mathematical notation consistent to some extent; however, we beg the readers' pardon if incomplete descriptions remain. We hope that this book will be helpful to researchers, engineers, and students in deepening their interests and knowledge of computational acoustic simulation.

The editors sincerely thank Dr. Yuko Sumino of Springer Japan for her advice and assistance, and we are grateful to all authors for their cooperation. As a final word, we deeply regret that one of the authors, Dr. Tomonao Okubo, passed away during the pre-publication editorial process. This book is one of his posthumous works and also serves as our tribute to the memory of a distinguished acoustician and great friend.

Tokyo, March 2014

Tetsuya Sakuma
Shinichi Sakamoto
Toru Otsuru

Contents

Part II Applications to Architectural and Environmental Acoustic Problems

Contributors

Takumi Asakura Institute of Technology, Shimizu Corporation, Koto-ku, Tokyo, Japan

Yoshinari Horinouchi Faculty of Engineering, Kyoto University, Nishikyo-ku, Kyoto, Japan

Kazuma Hoshi Junior college, Nihon University, Funabashi-shi, Chiba, Japan

Naohisa Inoue Graduate School of Frontier Sciences, The University of Tokyo, Kashiwa-shi, Chiba, Japan

Takashi Ishizuka Institute of Technology, Shimizu Corporation, Koto-ku, Tokyo, Japan

Yasuhito Kawai Faculty of Environmental and Urban Engineering, Kansai University, Suita-shi, Osaka, Japan

Takayuki Masumoto Cybernet Systems Co., Ltd., Chiyoda-ku, Tokyo, Japan

Yusuke Naka Institute of Aeronautical Technology, Japan Aerospace Exploration Agency, Mitaka-shi, Tokyo, Japan

Noriko Okamoto Department of Architecture, Ariake National College of Technology, Omuta-shi, Fukuoka, Japan

Kan Okubo Graduate School of System Design, Tokyo Metropolitan University, Hino-shi, Tokyo, Japan

Tomonao Okubo Kobayasi Institute of Physical Research, Kokubunji-shi, Tokyo, Japan

Takeshi Okuzono Department of Architecture, Graduate School of Engineering, Kobe University, Kobe, Japan

Hirofumi Onitsuka Yamaha Corporation Research and Development Division, Iwata-shi, Shizuoka, Japan

Takuya Oshima Faculty of Engineering, Niigata University, Nishi-ku, Niigata-shi, Niigata, Japan

Toru Otsuru Faculty of Engineering, Department of Architecture and Mechatronics, Oita University, Oita-shi, Japan

Shinichi Sakamoto Institute of Industrial Science, The University of Tokyo, Meguro-ku, Tokyo, Japan

Tetsuya Sakuma Graduate School of Frontier Sciences, The University of Tokyo, Kashiwa-shi, Chiba, Japan

Hisaharu Suzuki Evixar Japan, Inc., Chuou-ku, Tokyo, Japan

Reiji Tomiku Faculty of Engineering, Oita University, Oita-shi, Oita, Japan

Masahiro Toyoda Faculty of Environmental and Urban Engineering, Kansai University, Suita-shi, Osaka, Japan

Hideo Tsuru Nittobo Acoustic Engineering Co., Ltd., Sumida-ku, Tokyo, Japan

Yosuke Yasuda Faculty of Engineering, Kanagawa University, Kanagawa-ku, Yokohama, Japan

Takatoshi Yokota Kobayasi Institute of Physical Research, Kokubunji-shi, Tokyo, Japan

Symbols

The time convention $e^{-j\omega t}$ is used.

c	Sound speed
E	Young's modulus
f	Frequency
G	Green's function
H	Conjugate transpose
H_n, $H_n^{(1)}$, $H_n^{(2)}$	Hankel function,—of the first kind,—of the second kind
h_n, $h_n^{(1)}$, $h_n^{(2)}$	Spherical Hankel function,—of the first kind,—of the second kind
J_n	Bessel function
j_n	Spherical Bessel function
j	$\sqrt{-1}$
K_n	Order-n Krylov subspace
k, k	Wavenumber, wavenumber vector
N	Degrees of freedom, number of unknowns
n	Normal unit vector
$\partial/\partial n$	Normal derivative
$O(\cdot)$	Landau symbol
p	Sound pressure
T	Transpose
t	Time
α	Absorption coefficient
Γ	Boundary
δ	Dirac delta function
δ_{ij}	Kronecker delta
∇	Hamiltonian
∇^2	Laplacian
η	Loss factor
λ	Wavelength
ν	Poisson ratio
ρ	Density
Ω	Region
ω	Angular frequency
$\int_\Omega d\Omega$	Region integral
$\int_\Gamma d\Gamma$	Boundary integral

$\int\int\int_V dV$	Volume integral
$\int\int_S dS$	Surface integral
$\int\int_l dl$	Line integral
$*$	Complex conjugate

Chapter 1
Introduction

Tetsuya Sakuma and Toru Otsuru

Abstract This chapter introduces the significance of computational simulation for sound environmental design, and overviews the progress of computational acoustics for the last 5 decades. At present and in the future, computational acoustic simulation will be more and more widely used in the practice of room acoustics, building acoustics, noise control engineering, and so on. In the state of the art where a variety of numerical methods have been developed, it is important to select a suitable method for one's application, with understanding the features of each method.

Keywords Sound environment · Computational acoustics · Architectural acoustics · Environmental acoustics · Room acoustics · Noise propagation · Wave-based acoustic simulation

1.1 Computational Simulation for Sound Environment Design

In the late nineteenth century, W. C. Sabine started his study on the improvement of acoustics in a lecture hall of the old Fogg Art Museum in Harvard University, USA. After confirming that optical instruments, e.g., dancing flame, were of no help for measuring sound pressure in ordinary room acoustics cases, Sabine used his own ear and an organ pipe of 512 Hz and showed that the duration of the sound (known today as "reverberation time") decreases smoothly with the addition of lengths of seat cushions. Then, in the fall of 1898, he could have found the "Sabine formula

T. Sakuma (✉)
Graduate School of Frontier Sciences, The University of Tokyo, 5-1-5 Kashiwanoha, Kashiwa-shi, Chiba 277-8563, Japan
e-mail: sakuma@k.u-tokyo.ac.jp

T. Otsuru
Faculty of Engineering, Oita University, 700 Dannoharu, Oita-shi, Oita 870-1192, Japan
e-mail: otsuru@oita-u.ac.jp

T. Sakuma et al. (eds.), *Computational Simulation in Architectural and Environmental Acoustics*, DOI: 10.1007/978-4-431-54454-8_1,

for reverberation time." The formula marked the beginning of the science of architectural acoustics and its application to Boston Symphony Hall achieved satisfactory results [1, 2].

Still now, Sabine formula is used widely. Sabine formula is simple and easy to use especially for practical applications. However, it is also well-known that the formula contains a contradiction in an extreme case: if the mean absorption coefficient equals unity, reverberation time should equal zero because there is no reflections, physically; but the formula does not become zero, mathematically.

To overcome the contradiction, in 1930, Eyring published a paper on reverberation time in "DEAD" rooms [3]. At that time, radio broadcasting and sound pictures (records or CDs at this time) were one of the most important and popular appliances for people's daily life world widely. With the poor microphones and other electrical instruments at that time, better acoustics with short reverberation time, i.e., very "dead," in studios helped them to improve the quality of the radio broadcasting as well as of sound picture recording, and the prediction of their reverberation times were of significant importance for the modern society. In such cases, Sabine formula results overestimations for absorptive materials which might cause economical losses.

Eyring introduced a discretely decaying absorption model with the concept of "mean free path p," and he developed Eyring formula. When a room is acoustically "live," the values calculated by Eyring formula come close to those by Sabine formula. While, if the room is acoustically "dead," the former gives shorter reverberation times; even if in the extreme case with the mean absorption coefficient equals unity, Eyring formula results reverberation time of zero correctly.

Though Eyring could have established the general formula for reverberation time, the formula contains a constant k the magnitude of which is controlled by the mean free path p. The formula loses its effectiveness if one needs to change the value k depending on the rooms' shapes. If a room's shape is simple like cubical, cylindrical, or spherical, the value of p can be calculated without much difficulty using the method of images. However, if the shape is complicated to some extent, the problem would become "unmanageable."

To handle such unmanageable problems with unmanageably complicated shapes and/or with unmanageably huge degrees-of-freedom, S. Ulman and von Neumann had an idea to use random number and named it "Monte Carlo method." In 1946, the first electronic general-purpose computer Electronic Numerical Integrator and Computer (ENIAC) was announced, and von Neumann was also one of the members of its development team.

Thus, it is around 1950 when effective environments enough for solving such unmanageable problems have been set up. Then, in 1958, utilizing Monte Carlo method with a digital computer firstly on acoustics, Allred and Newhouse calculated sound reflections up to 9,000 times in various rectangular rooms, i.e., unmanageably complicated problems, and they examined the relationship between the mean free paths and the proportions of rectangular rooms [4].

The wave equation governs any sound field and a room's acoustics is given by solving the equation with appropriate boundary conditions. If the sound field is too complicated to be dealt with by any analytical approach, a numerical method would be effective to solve the wave equation. The effectiveness generally depends on the balance of the complexity and the specification of a computer, speed, and memory. The complexity and the degrees-of-freedom of an ordinary sound field on architectural/environmental acoustics frequently arise so huge that we needed to wait for next 50 years for computers with appropriate specifications to come.

Meanwhile, using computers with less specifications, various methods, like image methods, ray methods, energy-based methods, and so on, have been developed and applied to solve rooms' sound fields. Although they are helpful especially for practical application stages, their physical modelings of sound wave are not always appropriate. Then, scale model techniques have also been developed and successfully applied to many acoustic design of music halls, noise barriers, etc. Still now, the techniques are helpful tools, but there remains some difficulties like finding materials that satisfy similitude within required frequency ranges.

In these days of the 2nd decade in the twenty-first century, acoustic designs of rooms in music halls, lecture rooms, hospitals, schools, public spaces, etc., are important and the social demand for better acoustics is increasing; acoustic designs of outside environments are also of growing importance. Whereas, the specifications of computers have almost come to satisfy the requirements for wave-based acoustic simulations. Therefore, various applications of wave-based simulations are expected to be executed effectively. Such simulations can result more than thousands and millions of values of sound pressure and/or particle velocity. The values enable us to investigate the sound fields in far more details than what Sabine or Eyring formula gives us.

Note that the boundary conditions of the sound fields on architectural and environmental acoustics are generally not only complicated but also miscellaneous. In ordinary situations, atmospheric conditions, e.g., temperature and humidity, might be unstable. One needs to be careful enough to bring such unmanageably complicated and miscellaneous conditions to stay within the manageable ranges for meaningful simulation results. The following chapters present useful explanations of fundamentals as well as practical examples of computer simulations for sound environment design.

1.2 Progress of Computational Acoustics

1.2.1 Advent of Wave-Based Acoustic Analysis

Just after the age of computers began in the 1950s, early studies on geometrical acoustic analysis, based on the ray tracing method and the image source method, had gone ahead for room acoustics. Geometrical acoustic analysis is that follows energy propagation disregarding the wave nature of sound. With less computational resources, it can apparently predict transient responses in large rooms such as concert halls. Through various improvements in the methods, up to now commercial software

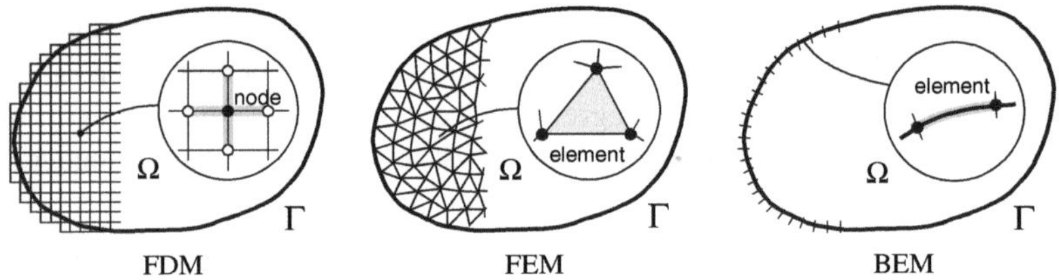

Fig. 1.1 Schematics of FDM, FEM, and BEM

for geometrical analysis has been widely used in practical acoustic design. Nevertheless, it is no more than energy-based approximation, and in principle, different from wave-based computation.

Researches on wave-based acoustic analysis appeared in the middle 1960s, and then a variety of methods have been developed with the rapid progress of computational mechanics. Wave-based acoustic analysis is that models a sound field as a discrete system in the time or frequency domain on the basis of the wave equation. Since it has the potential to accurately simulate wave phenomena, high expectations have been placed in the field of acoustics from the early stage. However, at the time, its application to indoor and outdoor living spaces in the audible frequency range was far from realization, due to the requirement of enormous computational resources.

As numerical methods, the three major techniques, the finite difference method (FDM), the finite element method (FEM) and the boundary element method (BEM) are well known. In the field of acoustics, following the pioneers' work for those methods [5–7], the fundamental algorithms have been mostly established in the 1970s. The three methods have respective features, which led to somewhat different expansion of techniques and applications with the exponential increase of computational resources.

1.2.2 Features of Numerical Methods

The three numerical methods, FDM, FEM, and BEM (Fig. 1.1), respectively have advantages and disadvantages, depending on the object and purpose of simulation such as for room acoustics, noise propagation, acoustic materials and components, visualization, auralization, and so on [8]. In the following, the features of the three methods are briefly described from the viewpoints of target space, frequency and time domains, computational load, program coding, and applicability.

The FDM is a means of discretizing partial differential equations by using a nodal grid system for a finite space. Although frequency domain analysis is possible based on the Helmholtz equation [9], time-domain analysis is quite popular with explicit schemes based on the Euler equations and the equation of continuity, called the finite-difference time-domain (FDTD) method. In the FDM, orthogonal grids are widely used for simple and easy coding, however requiring staircase approximation for boundaries. It can be applied to vibro-acoustic coupled systems and inhomogeneous sound fields, but not directly for modal analysis. A usual way of computing frequency domain responses is to apply the Fourier transform to the responses calculated in the time domain.

The FEM is a means of discretizing domain integral equations that originate in the weighted residual method, by subdividing a finite space into elements. The resulting system is composed of mass, stiffness, and damping matrices, which enables both frequency and time-domain analysis, and also modal analysis in the eigenvalue problem of the system. For computing time-domain responses, a step-by-step integration scheme is used in the time-domain analysis; otherwise, a superposition technique is applied through the modal analysis. The FEM uses a similar number of nodes to the FDM, however requiring more computational resources to solve a band matrix system. There have been developed many general-purpose FEM codes, and its applicability to vibro-acoustic coupled systems and inhomogeneous sound fields is superior.

The BEM is a means of discretizing boundary integral equations by subdividing boundaries into elements, regardless of exterior or interior space. Although time-domain analysis is possible based on the boundary integral equation in the corresponding domain, frequency domain analysis is more widely used. Nodes are placed only on the boundaries, so that the number of nodes is considerably smaller than the other two methods. However, the BEM is not always advantageous in computational resources, because it requires solving a dense matrix system. In numerical computation, special treatments are required for singular integrals and the non-uniqueness problem, and it is not feasible for analysis of inhomogeneous sound fields. Nevertheless, the strongest point is that exterior problems, such as scattering and radiation problems in the free field, can be efficiently analyzed.

1.2.3 Application to Sound Environmental Simulation

The FDTD method was proposed in the field of electromagnetics in the late 1960s [5], however until the 1990s, it had not been actively applied to acoustics due to the limitation of computational resources. After that, the simplicity of coding was again recognized with the need of time-domain analysis, and its application to room acoustics simulation was started [10, 11]. In the early 2000s, simulation of concert halls was realized up to the middle frequency range [12], and its practical use in room acoustics design will be achievable in the near future. Recently, it is also applied to a variety of noise propagation simulations, extending to vibro-acoustic coupling analysis of sound insulation and floor impact noise [13, 14].

In the 1970s, the FEM has been intensively developed in various fields such as structural mechanics, thermodynamics and fluid dynamics, and then commercial software has widely spread as a general-purpose method. In the filed of acoustics, it was used for analysis of frequency responses and normal modes in ducts, cavities, and rooms [6, 15], and moreover for optimization of room shape regarding modal distribution in the low frequency range [16]. From the early stage, it has been also applied to vibro-acoustic coupled systems and inhomogeneous sound fields [17, 18], which is currently useful for evaluation of sound insulation, floor impact noise, acoustic materials and components. With the recent progress of iterative solvers for linear systems, its application to concert halls is being challenged [19].

In the 1960s, the boundary integral equation method [7] has begun to be applied in the field of acoustics, and later, it was called the BEM with the use of boundary elements in the late 1970s [20]. For the need of analysis of exterior problems, numerical techniques in the BEM have been developed, and it has been mainly applied to analysis of scattering and diffraction by obstacles, radiation from noise sources and loudspeakers. Recently, owing to the development of iterative solvers and the fast multipole algorithm [21, 22], acceleration of computation and reduction of memory requirements were drastically achieved, and large-scale analysis becomes more and more popular. Up to now, characteristics of diffusers and loudspeakers, and head-related transfer functions can be simulated in the full audible frequency range [23], and its practical application is spreading in detailed shape design of acoustic components.

References

1. L.L. Beranek, The notebooks on wallace c. sabine. J. Acoust. Soc. Am. **61**, 629–639 (1977)
2. W.C. Sabine, Reverberation, The American Architect, 1900, in *Collected Papers on Acoustics*. Reprinted by (Peninsula Publishing, 1993)
3. C.F. Eyring, Reverberation time in "dead" rooms. J. Acoust. Soc. Am. **1**, 217–241 (1930)
4. J.C. Allred, A. Newhouse, Application of the Monte Carlo method to architectural acoustics. J. Acoust. Soc. Am. **30**, 1–3 (1958)
5. K.S. Yee, Numerical solution of initial boundary value problems involving Maxwell's equations in isotropic media. IEEE Trans. Antennas Propag. **17**, 585–589 (1966)
6. G. Gladwell, A finite element method for acoustics, in Proceedings of *5th International Congress of Acoustics*, Liege, L33, pp. 1–4 (1965)
7. L.H. Chen, D.G. Schweikert, Sound radiation from an arbitrary body. J. Acoust. Soc. Am. **35**, 1626–1632 (1963)
8. T. Sakuma, Prospects for the application of numerical simulation techniques to room acoustic design. J. Acoust. Soc. Jpn. **57**, 463–469 (2001). (in Japanese)
9. T. Shuku, Finite difference analysis of the acoustic field in irregular rooms. J. Acoust. Soc. Jpn. **28**, 5–12 (1972). (in Japanese)
10. D. Botteldooren, Finite-difference time-domain simulation of low frequency room acoustic problems. J. Acoust. Soc. Am. **98**, 3302–3308 (1995)
11. S. Sakamoto, Y. Tokita, H. Tachibana, Calculation of impulse responses of rooms by using of the finite difference method, in *Proceedings of 3rd ASA/ASJ Joint Meeting*, pp. 1307–1310 (1996)
12. S. Sakamoto, T. Seimiya, H. Tachibana, Visualization of sound reflection and diffraction using finite difference time domain method. Acoust. Sci. Tech. **23**, 34–39 (2002)
13. T. Asakura, S. Sakamoto, Finite-difference time-domain analysis of sound insulation performance of wall systems. Build. Acoust. **16**, 267–281 (2009)
14. M. Toyoda, D. Takahashi, Prediction for architectural structure-borne sound by the finite-difference time-domain method. Acoust. Sci. Tech. **30**, 265–276 (2009)

15. A. Craggs, The use of simple three dimensional acoustic finite elements for determining the natural modes and frequencies of complex shaped enclosures. J. Sound Vib. **23**, 331–339 (1972)
16. J.R. Milner, R.J. Bernhard, An investigation of the modal characteristics of nonrectangular reverberation rooms. J. Acoust. Soc. Am. **85**, 772–779 (1989)
17. A. Craggs, The transient response of a coupled plate-acoustic system using plate and acoustic finite elements. J. Sound Vib. **15**, 509–528 (1971)
18. A. Craggs, A finite element model for rigid porous absorbing materials. J. Sound Vib. **61**, 101–111 (1978)
19. N. Okamoto, Y. Yasuda, T. Otsuru, R. Tomiku, Numerical analysis of large-scale sound fields using iterative methods, part II: application of Krylov subspace methods to finite element analysis. J. Comput. Acoust. **15**, 473–493 (2007)
20. C.A. Brebbia, *The Boundary Element Method for Engineers* (Pentech Press, London, 1978)
21. Y. Yasuda, Y. Kosaka, N. Okamoto, T. Oshima, T. Sakuma, S. Sakamoto, Numerical analysis of large-scale sound fields using iterative methods, part I: application of Krylov subspace methods to boundary element analysis. J. Comput. Acoust. **15**, 449–471 (2007)
22. T. Sakuma, Y. Yasuda, Fast multipole boundary element method for large-scale steady-state sound field analysis. part I: setup and validation. Acta Acustica United Acustica **88**(4), 513–525 (2002)
23. T. Masumoto, T. Oshima, Y. Yasuda, T. Sakuma, M. Kabuto, M. Akiyama, HRTF caulculation in the full audible frequency range using FMBEM, in *Proceedings of 19th International Congress of Acoustics*, Madrid, COM-06-012 (2007)

Part I
Methods of Wave-Based Acoustic Simulation

Chapter 2
Finite-Difference Time-Domain Method

Shinichi Sakamoto, Hideo Tsuru, Masahiro Toyoda and Takumi Asakura

Abstract In this chapter, analysis of sound and vibration using the Finite-Difference Time-Domain method (FDTD method) is illustrated. In Sect. 2.1, the fundamentals of the FDTD method are described. In the FDTD method, several error factors caused by discretization of sound field are pointed out. As methods to solve such problems, in Sect. 2.2, the compact finite difference is described in detail. The FDTD method can not only be applied to acoustic problem of air-borne sound, but also vibroacoustic problems such as a floor impact sound and a sound insulation problem through a wall structure. In Sect. 2.3, therefore, application of the FDTD method to vibroacoustic problems is focused on, and the theoretical background and its numerical formulation are described in detail.

Keywords Finite-difference time-domain (FDTD) method · Stability condition · FDTD(2, 4) method · Perfectly matched layer (PML) · Compact finite difference · Vibro-acoustic problem

S. Sakamoto (✉)
Institute of Industrial Science, The University of Tokyo, 4-6-1, komaba, Meguro-ku, Tokyo 153-8505, Japan
e-mail: sakamo@iis.u-tokyo.ac.jp

H. Tsuru
Nittobo Acoustic Engineering Co., Ltd., 1-21-10 Midori, Sumida-ku, Tokyo 130-0021, Japan
e-mail: tsuru@noe.co.jp

M. Toyoda
Faculty of Environmental and Urban Engineering, Kansai University, 3-3-35 Yamate-cho, Suita-shi, Osaka 564-8680, Japan
e-mail: toyoda@kansai-u.ac.jp

T. Asakura
Institute of Technology, Shimizu Corporation, 3-4-17 Etchujima, Koto-ku, Tokyo 135-8530, Japan
e-mail: t_asakura@shimz.co.jp

T. Sakuma et al. (eds.), *Computational Simulation in Architectural and Environmental Acoustics*, DOI: 10.1007/978-4-431-54454-8_2,

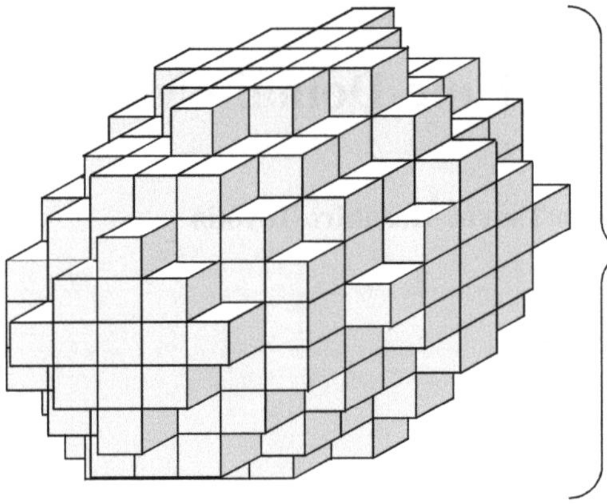

Fig. 2.1 Discretization of sound field

2.1 Fundamentals

2.1.1 Basic Equations

Sound propagation in the air is described by two kinds of differential equations, Euler's equation and the equation of continuity.

$$\rho\frac{\partial u}{\partial t}+\frac{\partial p}{\partial x}=0, \tag{2.1}$$

$$\rho\frac{\partial v}{\partial t}+\frac{\partial p}{\partial y}=0, \tag{2.2}$$

$$\rho\frac{\partial w}{\partial t}+\frac{\partial p}{\partial z}=0, \tag{2.3}$$

$$\frac{\partial p}{\partial t}+\kappa\left(\frac{\partial u}{\partial x}+\frac{\partial v}{\partial y}+\frac{\partial w}{\partial z}\right)=0, \tag{2.4}$$

where p, u, v, w are the sound pressure [Pa] and the particle velocities [m/s] in x, y, z directions, respectively, κ, ρ are volume elastic ratio [N/m^2] and density [kg/m^3] of the air, respectively.

By the Finite-Difference Time-Domain (FDTD) method, discrete physical acoustic quantities, sound pressure, and particle velocity are approximately updated based on finite difference schemes to which differential terms included in governed Euler's and continuity equations are substituted. For this purpose, in the first step, as shown in Fig. 2.1, sound field under test should be discretized in rectangular grids. Definition points of the sound pressure and the particle velocities are set at the

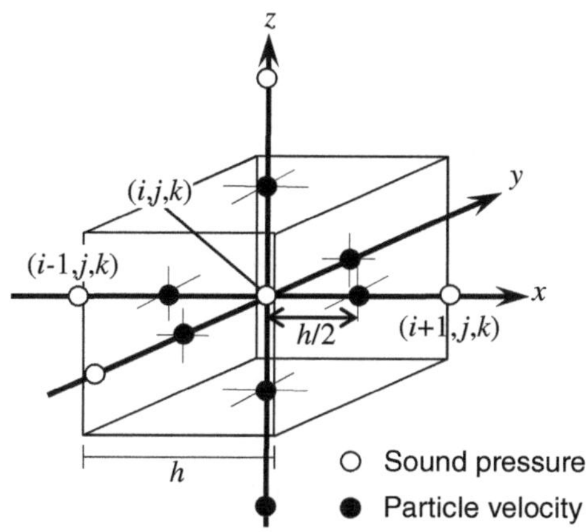

Fig. 2.2 Spatial definition points of sound pressure and particle velocities

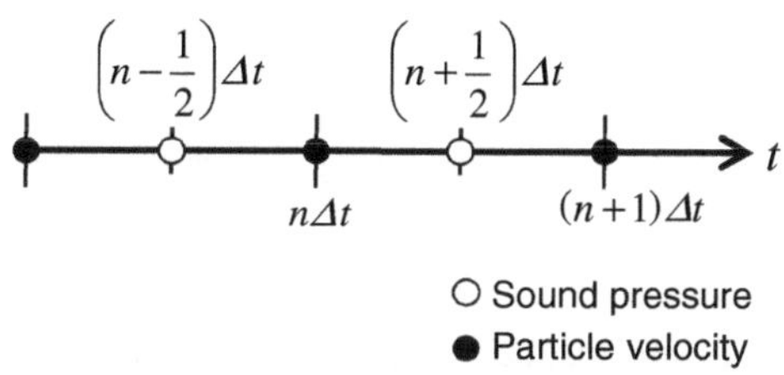

Fig. 2.3 Transient definition points of sound pressure and particle velocities

appropriate positions in the discretized grids, and the difference equations regarding the physical quantities are constructed. The difference equations based on the governed differential equations are called as "difference scheme". The most representative difference scheme is "Yee algorithm" which is described in the following section [1].

In order to approximate a value of first-order derivative of Eqs. (2.1)–(2.4) by a central difference, definition points of sound pressure and particle velocities are set at a half grid apart from each other as shown in Fig. 2.2. Such a kind of grid system where different two kinds of physical quantities are defined at different points of which distance is equal to a half of the grid size is called the staggered grid system. In the same manner as the spatial grid their temporal definition points are also a half time step apart from each other, as shown in Fig. 2.3.

Suppose that a spatial grid size and a discrete time interval are h and Δt, and that a physical quantity q at spatial grid point $(x, y, z) = (ih, jh, kh)$ at a time $n\Delta t$ is described as $q^n_{i,j,k}$, then a sound pressure and particle velocities are expressed as, $p^{n+1/2}_{i,j,k}$, $u^n_{i+1/2,j,k}$, $v^n_{i,j+1/2,k}$, $w^n_{i,j,k+1/2}$. Using these expressions, differential terms in space and time appearing in Eqs. (2.1)–(2.4) become

$$\left.\frac{\partial u}{\partial t}\right|^{n+1/2}_{i+1/2,j,k} = \frac{u^{n+1}_{i+1/2,j,k} - u^n_{i+1/2,j,k}}{\Delta t}, \quad \left.\frac{\partial u}{\partial x}\right|^{n}_{i,j,k} = \frac{u^n_{i+1/2,j,k} - u^n_{i-1/2,j,k}}{h},$$

$$\left.\frac{\partial p}{\partial t}\right|_{i,j,k}^{n} = \frac{p_{i,j,k}^{n+1/2} - p_{i,j,k}^{n-1/2}}{\Delta t}, \quad \left.\frac{\partial p}{\partial x}\right|_{i+1/2,j,k}^{n+1/2} = \frac{p_{i+1,j,k}^{n+1/2} - p_{i,j,k}^{n+1/2}}{h}.$$

Substituting these into Eqs. (2.1)–(2.4) leads to following difference scheme:

$$u_{i+1/2,j,k}^{n+1} = u_{i+1/2,j,k}^{n} - \frac{\Delta t}{\rho h}\left(p_{i+1,j,k}^{n+1/2} - p_{i,j,k}^{n+1/2}\right), \tag{2.5}$$

$$v_{i,j+1/2,k}^{n+1} = v_{i,j+1/2,k}^{n} - \frac{\Delta t}{\rho h}\left(p_{i,j+1,k}^{n+1/2} - p_{i,j,k}^{n+1/2}\right), \tag{2.6}$$

$$w_{i,j,k+1/2}^{n+1} = w_{i,j,k+1/2}^{n} - \frac{\Delta t}{\rho h}\left(p_{i,j,k+1}^{n+1/2} - p_{i,j,k}^{n+1/2}\right), \tag{2.7}$$

$$p_{i,j,k}^{n+1/2} = p_{i,j,k}^{n-1/2} - \frac{\kappa \Delta t}{h}\Big[\left(u_{i+1/2,j,k}^{n} - u_{i-1/2,j,k}^{n}\right) + \left(v_{i,j+1/2,k}^{n} - v_{i,j-1/2,k}^{n}\right) + \left(w_{i,j,k+1/2}^{n} - w_{i,j,k-1/2}^{n}\right)\Big]. \tag{2.8}$$

From Eqs. (2.5)–(2.8), we can find features of the FDTD method.

- A future value of particle velocity in each direction (at a time step $n + 1$) is calculated by a known value of that (at a time step n) and a present values of sound pressure (at a time step $n + 1/2$).
- A future value of sound pressure (at a time step $n + 1/2$) is calculated by a known value of that (at a time step $n - 1/2$) and a present values of particle velocities in all directions (at a time step n).

Therefore, after an initial distribution of sound pressure and particle velocities, the distribution of sound pressure and particle velocity at the following time steps can be calculated successively.

2.1.2 Boundary Conditions

Reflection or absorption characteristics of boundaries are generally provided by surface acoustic impedance,

$$z_n = \frac{p}{u_n}, \tag{2.9}$$

where z_n is normal acoustic impedance and u_n is normal particle velocity.

In the simplest case where surface is perfectly reflective (rigid), the surface impedance becomes infinite and it is easy to adapt this condition to the FDTD scheme by making the particle velocity on the boundary nodes zero. On the other hand, for perfectly absorbing condition, the situation is not so simple. In a special case where

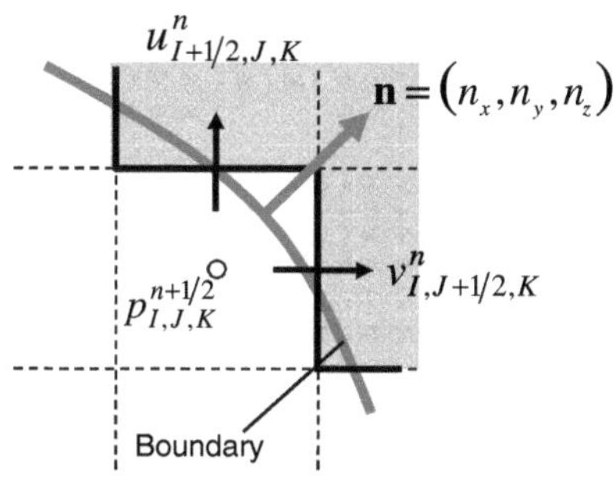

Fig. 2.4 Sound pressure and particle velocity near a boundary

a plane wave is incident perpendicularly to the absorbing boundary, the surface acoustic impedance becomes equal to the specific impedance of air. In such a case, $z_n = \rho c$ (c is speed of sound). However, if this absorbing condition is applied to a general case where the input sound is not plane wave but a cylindrical or spherical wave and its incidence angle is not perpendicular, theoretically there exists some reflection. More accurate perfectly absorbing boundary condition will be discussed in a latter part (Fig. 2.4).

In the Yee algorithm, both physical quantities of sound pressure p and particle velocity u, v, w are obtained successively in the calculation steps. Therefore, the surface acoustic impedance can be roughly approximated as [2],

$$u^{n+1}_{I+1/2,J,K} = \frac{p^{n+1/2}_{I,J,K}}{z_n} n_x, \tag{2.10}$$

$$v^{n+1}_{I,J+1/2,K} = \frac{p^{n+1/2}_{I,J,K}}{z_n} n_y, \tag{2.11}$$

$$w^{n+1}_{I,J,K+1/2} = \frac{p^{n+1/2}_{I,J,K}}{z_n} n_z, \tag{2.12}$$

where $\mathbf{n} = (n_x, n_y, n_z)$ is the normal vector of surface. Because the definition points of sound pressure and particle velocity are a half grid size apart from each other, note that this impedance is a rough approximation.

One of the most representative indeces describing absorption characteristics of materials is the absorption coefficient α. The absorption coefficient, which is obtained on energy-base, is generally used in building acoustics, room acoustics, and environmental noise. The absorption coefficient is categorized into three kinds; normal incidence absorption coefficient α_0, reverberation absorption coefficient $\overline{\alpha}$, and oblique incidence absorption coefficient α_θ. Among them, the normal incidence absorption coefficient α_0 is related to the normal acoustic impedance as

$$\alpha_0 = 1 - \left| \frac{z_n - \rho c}{z_n + \rho c} \right|. \tag{2.13}$$

Therefore, if normal incidence absorption coefficient of the material is known, normal acoustic impedance can be estimated as

$$z_n = \rho c \frac{1 + \sqrt{1 - \alpha_0}}{1 - \sqrt{1 - \alpha_0}}. \tag{2.14}$$

Acoustic impedance is generally treated in frequency domain and is expressed by complex values. By the above method, however, only real values can be treated for expressing the acoustic impedance because the FDTD method deals with physical quantities in real number, and such a treatment corresponds to a situation that acoustic impedance is a constant real value for all the frequency range under the calculation. In order to treat more complicated characteristics such that the acoustic impedance is dependent on frequency, another physical model is necessary. For such a complicated impedance model, some calculation models have been proposed. Sakamoto has proposed the mechanical substitution model in which the acoustic boundary surface has been substituted by an equivalent mechanical model being composed of mass, spring, and resistance [3]. In order to treat more general impedance characteristics, D. M. Sullivan has used Z transform to deal with a linear system in sound reflection by boundary surface [4]. Escolano has used digital signal processing, in which an acoustic admittance was modeled with IIR or FIR filters [5]. For a case where homogeneous porous materials such as glass fibrous board or urethane form are used as a surface absorption material, Suzuki has proposed a calculation method by which the inner space of the material was digitized and sound propagation was obtained based on the Rayleigh model [6].

2.1.3 FDTD(2, 4) Method

In the Yee algorithm, central difference using two reference points secures second-order accuracy. In order to raise the accuracy, various efforts have being made. In this section, as the simplest method for the implementation, FDTD(2, 4) method [7, 8] is introduced. FDTD(2, 4) means the second-order accuracy in time and the fourth-order accuracy in space. The spatial accuracy is raised by adopting the fourth-order scheme using four points central difference. For temporal derivative, however, the method remains to use two points central difference with the second-order accuracy in order to avoid increase of computational memory usage.

The difference scheme with fourth-order accuracy using four reference points is deduced from Taylor expansion of a function. Values of a function f at points $x \pm h/2$ and $x \pm 3h/2$, $f(x \pm h/2)$ and $f(x \pm 3h/2)$, are expressed as follows, by using the Taylor expansion around x.

$$f\left(x + \frac{h}{2}\right) = f(x) + \frac{1}{2} f^{(1)}(x)h + \frac{1}{8} f^{(2)}(x)h^2 + \frac{1}{48} f^{(3)}(x)h^3 + \frac{1}{384} f^{(4)}(x)h^4 + \cdots, \tag{2.15}$$

$$f\left(x-\frac{h}{2}\right)=f(x)-\frac{1}{2}f^{(1)}(x)h+\frac{1}{8}f^{(2)}(x)h^2-\frac{1}{48}f^{(3)}(x)h^3+\frac{1}{384}f^{(4)}(x)h^4+\cdots, \tag{2.16}$$

$$f\left(x+\frac{3h}{2}\right)=f(x)+\frac{3}{2}f^{(1)}(x)h+\frac{9}{8}f^{(2)}(x)h^2+\frac{9}{16}f^{(3)}(x)h^3+\frac{27}{128}f^{(4)}(x)h^4+\cdots, \tag{2.17}$$

$$f\left(x-\frac{3h}{2}\right)=f(x)-\frac{3}{2}f^{(1)}(x)h+\frac{9}{8}f^{(2)}(x)h^2-\frac{9}{16}f^{(3)}(x)h^3+\frac{27}{128}f^{(4)}(x)h^4+\cdots. \tag{2.18}$$

In order to deduce $\frac{\partial f}{\partial x}$ from the above four equations, let us calculate {(2.15) − (2.16)} × 27− {(2.17) − (2.18)}, then

$$f^{(1)}(x)=\frac{-f\,(x+3h/2)+27f\,(x+h/2)-27f\,(x-h/2)+f\,(x-3h/2)}{24h}+\epsilon, \tag{2.19}$$

where ϵ is an error term and it becomes fourth-order accurate as

$$\epsilon=\frac{3}{640}f^{(5)}(x)h^4. \tag{2.20}$$

An FDTD scheme where temporal and spatial derivatives are approximated in the second-order and fourth-order accuracy, respectively, is called the FDTD(2, 4) method, and the schemes are expressed as

$$u^{n+1}_{i+1/2,j,k}=u^{n}_{i+1/2,j,k}-\frac{\Delta t}{\rho h}\sum_{m=0}^{1}C_m\left(p^{n+1/2}_{i+m,j,k}-p^{n+1/2}_{i-m,j,k}\right), \tag{2.21}$$

$$v^{n+1}_{i,j+1/2,k}=v^{n}_{i,j+1/2,k}-\frac{\Delta t}{\rho h}\sum_{m=0}^{1}C_m\left(p^{n+1/2}_{i,j+m,k}-p^{n+1/2}_{i,j-m,k}\right), \tag{2.22}$$

$$w^{n+1}_{i,j,k+1/2}=w^{n}_{i,j,k+1/2}-\frac{\Delta t}{\rho h}\sum_{m=0}^{1}C_m\left(p^{n+1/2}_{i,j,k+m}-p^{n+1/2}_{i,j,k-m}\right), \tag{2.23}$$

$$\begin{aligned}p^{n+1/2}_{i,j,k}=p^{n-1/2}_{i,j,k}-\frac{\kappa\Delta t}{h}\Bigg[&\sum_{m=0}^{1}C_m\left(u^{n}_{i+1/2,j,k}-u^{n}_{i-1/2,j,k}\right)\\&+\sum_{m=0}^{1}C_m\left(v^{n}_{i,j+1/2,k}-v^{n}_{i,j-1/2,k}\right)\\&+\sum_{m=0}^{1}C_m\left(w^{n}_{i,j,k+1/2}-w^{n}_{i,j,k-1/2}\right)\Bigg].\end{aligned} \tag{2.24}$$

where $C_0=\frac{9}{8}$ and $C_1=\frac{1}{24}$.

2.1.4 Stability and Dispersion Error

Whereas the FDTD method is easy to be implemented, the method has a sensitive nature in which a solution unexpectedly diverges or fluctuates depending on a calculation condition. Such natures as unexpected divergence and fluctuation are caused by violation of stability condition and dispersion error, respectively. Among them, the stability condition is simpler and it is provided by relationship between discrete time step Δt and spatial grid size h as,

$$\Delta t \leq \frac{h}{c} \quad \text{(1 dimension)}, \tag{2.25}$$

$$\Delta t \leq \frac{h}{\sqrt{2}c} \quad \text{(2 dimension)}, \tag{2.26}$$

$$\Delta t \leq \frac{h}{\sqrt{3}c}. \quad \text{(3 dimension)}, \tag{2.27}$$

for the Yee algorithm using temporal second-order. This condition is called the Courant condition. As shown in the above equation, for appropriate parameter setting, a discrete time step is proportional to a spatial grid size. Therefore, more precise modeling or raising upper limit frequency by shortening a spatial grid size leads to a temporal high resolution. In this section, this stability condition is theoretically explained.

A simple technique for analyzing the stability and the dispersive phase error of a finite difference scheme has been given by Von Neumann and Richtmyer [9]. Consider a plane wave with wave number k traveling in the (θ, φ) direction in polar coordinates shown in Fig. 2.5. The sound pressure at a point (x, y, z) with a wave number k $(= \omega/c)$ and an amplitude of 1 is assumed to be

$$p_0(x, y, z, t) = e^{jk_x x} e^{jk_y y} e^{jk_z z} e^{-j\omega t}, \tag{2.28}$$

where k_x, k_y, k_z are directional component of the wave number k and they are expressed as $k_x = k \sin\varphi \cos\theta$, $k_y = k \sin\varphi \sin\theta$, $k_z = k \cos\varphi$.

Sound pressure and particle velocity at their definition points on the staggered grid system are expressed as

$$u^n_{i+1/2,j,k} = u_0 Z^n \cdot e^{jk_x(i+1/2)h} e^{jk_y(jh)} e^{jk_z(kh)}, \tag{2.29}$$

$$p^{n+1/2}_{i,j,k} = p_0 Z^{n+1/2} \cdot e^{jk_x(ih)} e^{jk_y(jh)} e^{jk_z(kh)}, \tag{2.30}$$

where u_0, v_0, w_0, p_0 are initial values, Z is a complex amplification ratio per time step. Substituting the above expressions into Eqs. (2.5)–(2.8) leads to

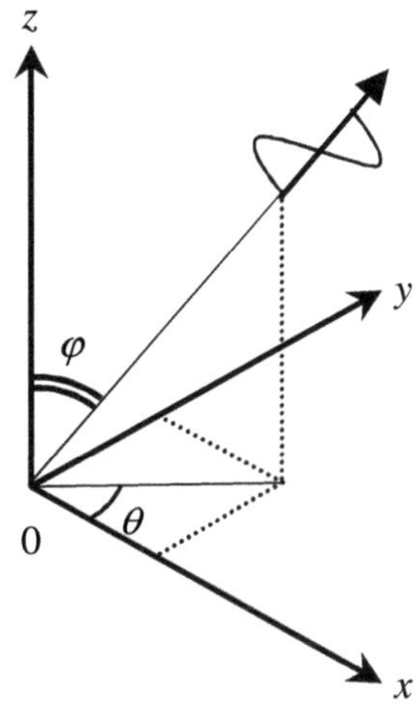

Fig. 2.5 Traveling direction of a plane wave

$$u_0\left(Z^{\frac{1}{2}}-Z^{-\frac{1}{2}}\right)-\frac{\Delta t}{\rho h}\cdot 2jp_0\sin\frac{k_xh}{2}=0, \tag{2.31}$$

$$v_0\left(Z^{\frac{1}{2}}-Z^{-\frac{1}{2}}\right)-\frac{\Delta t}{\rho h}\cdot 2jp_0\sin\frac{k_yh}{2}=0, \tag{2.32}$$

$$w_0\left(Z^{\frac{1}{2}}-Z^{-\frac{1}{2}}\right)-\frac{\Delta t}{\rho h}\cdot 2jp_0\sin\frac{k_zh}{2}=0, \tag{2.33}$$

$$p_0\left(Z^{\frac{1}{2}}-Z^{-\frac{1}{2}}\right)-\frac{\Delta t}{\rho h}\left(2j\sin\frac{k_xh}{2}u_0+2j\sin\frac{k_yh}{2}v_0+2j\sin\frac{k_zh}{2}w_0\right)=0. \tag{2.34}$$

In a matrix form of Eqs. (2.31)–(2.34),

$$\begin{bmatrix} Z^{\frac{1}{2}}-Z^{\frac{-1}{2}} & 0 & 0 & \frac{-\Delta t}{\rho h}2j\sin\frac{k_xh}{2} \\ 0 & Z^{\frac{1}{2}}-Z^{\frac{-1}{2}} & 0 & \frac{-\Delta t}{\rho h}2j\sin\frac{k_yh}{2} \\ 0 & 0 & Z^{\frac{1}{2}}-Z^{\frac{-1}{2}} & \frac{-\Delta t}{\rho h}2j\sin\frac{k_zh}{2} \\ \frac{-\kappa\Delta t}{h}2j\sin\frac{k_xh}{2} & \frac{-\kappa\Delta t}{h}2j\sin\frac{k_yh}{2} & \frac{-\kappa\Delta t}{h}2j\sin\frac{k_zh}{2} & Z^{\frac{1}{2}}-Z^{\frac{-1}{2}} \end{bmatrix}\begin{bmatrix} u_0 \\ v_0 \\ w_0 \\ p_0 \end{bmatrix}=\begin{bmatrix} 0 \\ 0 \\ 0 \\ 0 \end{bmatrix}. \tag{2.35}$$

A value of determinant of the 4×4 matrix which appears in Eq. (2.35) should be zero, in order that the vector $[u_0\ v_0\ w_0\ p_0]^T$ is non-zero vector. This condition leads the following equation regarding the complex amplification ratio Z as follows:

$$Z^2-2AZ+1=0, \tag{2.36}$$

$$A = 1 - 2\left(\frac{c\Delta t}{h}\right)^2 \left(\sin^2\frac{k_x h}{2} + \sin^2\frac{k_y h}{2} + \sin^2\frac{k_z h}{2}\right). \tag{2.37}$$

A solution of Eq. (2.36) is

$$Z = \begin{cases} A \pm \sqrt{A^2 - 1} & (A < -1) \\ A \pm j\sqrt{1 - A^2} & (-1 \le A \le 1). \end{cases} \tag{2.38}$$

If an absolute value of Z in Eq. (2.38) is greater than 1, the solution increases with advance of time steps and diverges at last. Accordingly, $-1 \le A \le 1$ is a necessary condition for the solution being stable. Actually in this case, $|Z|$ is equal to 1 and $\arg Z = \arctan A/\sqrt{1 - A^2}$. The fact means that the wave amplitude does not change and the phase of the wave shifts with the progress of time steps. Then, Eq. (2.37) leads to

$$\sin^2\frac{k_x h}{2} + \sin^2\frac{k_y h}{2} + \sin^2\frac{k_z h}{2} \le \left(\frac{h}{c\Delta t}\right)^2. \tag{2.39}$$

For arbitrary k,

$$\Delta t \le \frac{h}{\sqrt{3}c} \tag{2.40}$$

can be obtained because $\sin(k_x h/2) \le 1$, $\sin(k_y h/2) \le 1$, $\sin(k_z h/2) \le 1$. A value of A is dependent on the wave number k (see Eq. (2.37)) and therefore the degree of the numerical phase shift is also dependent on the wave number k. This error of phase shift owing to frequency is called dispersion error. Figure 2.6 shows an example of a pulse propagation with dispersion error. At a center point in a cubic room of $201\,h \times 201\,h \times 201\,h$, a pulse source is emitted and its pulse propagation is detected at a corner point. Comparing with a theoretical solution reveals that dispersion error accumulates in 10,000 time steps to raise numerical fluctuations. The influence of dispersion error on calculation results of sound propagation becomes severer as size of sound field under test is larger. Therefore, various efforts have being made to reduce such dispersion error heretofore as shown in the following section.

2.1.5 Absorbing Boundary Condition

When the FDTD method, which is categorized in a domain-type is applied to a free field, a special treatment is required to reduce reflection sound from field terminations. As a concept of absorbing boundary condition (ABC), we can choose two kinds; Differential-based Absorbing Boundary Condition (D-ABC) and Material-based Absorbing Boundary Condition (M-ABC). The former deals with progressive and regressive waves at a terminating surface, and the latter absorbs sound energy

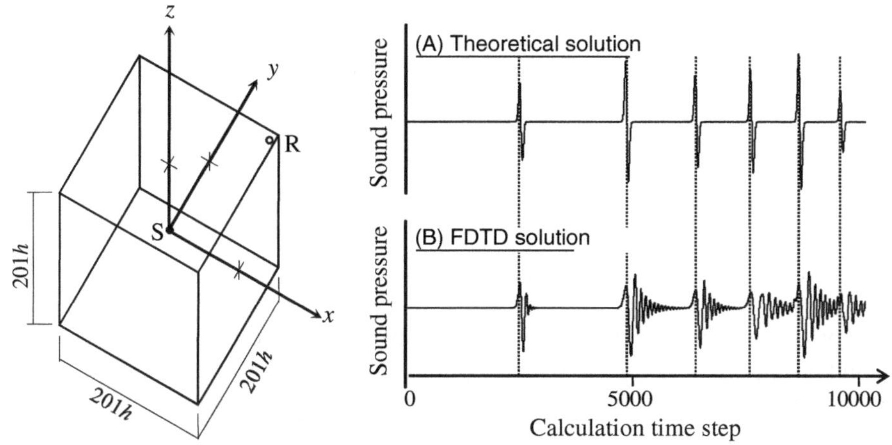

Fig. 2.6 Numerical fluctuation caused by dispersion error appearing in a numerical result

inside a numerically lossy media with sufficient thickness installed along surrounding boundary. As representative D-ABC and M-ABC, hereafter, the Mur ABC and the perfectly matched layer (PML), respectively, will be introduced.

2.1.5.1 Mur Absorbing Boundary Condition

The simplest and most commonly used grid truncation technique for open-field FDTD modeling is the Mur ABC [10]. The Mur ABC is based on a concept in which only progressive wave outgoing to an absorbing boundary exits and there exists no regressive wave inbound sound field at truncation mesh of the boundary. Now, let us consider one-dimensional wave equation in x-direction regarding x-component of particle velocity.

$$\left(\frac{\partial^2}{\partial x^2} - \frac{1}{c^2}\frac{\partial^2}{\partial t^2}\right) u = 0. \tag{2.41}$$

The above second-order differential equation is transformed as

$$\frac{\partial u}{\partial x} - \frac{1}{c}\frac{\partial u}{\partial t} = 0, \tag{2.42}$$

$$\frac{\partial u}{\partial x} + \frac{1}{c}\frac{\partial u}{\partial t} = 0, \tag{2.43}$$

then the solutions of Eqs. (**??**)–(2.43) mean a progressive and a regressive plane waves, respectively. For example, in the case of Fig. 2.7, in which absorbing boundary

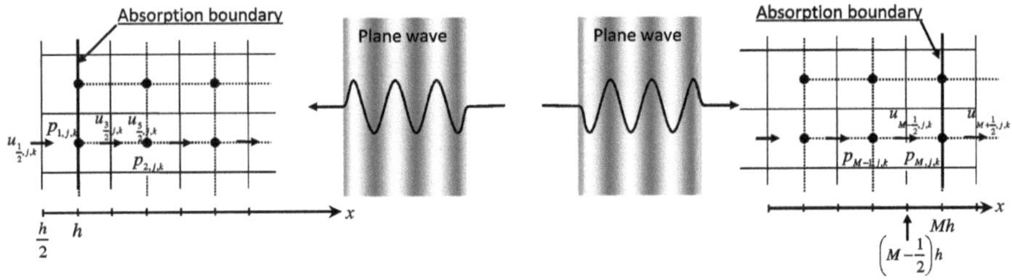

Fig. 2.7 A plane wave incidence to an absorbing boundary

is set at $x = h/2$, there exists a progressive wave in negative x-direction, Eq. (2.43) should be applied at $x = h/2$. With adopting the grid system to the Yee's staggered system, discretization of Eq. (**??**) at a time step $(n + 1/2)\Delta t$ by,

$$\left.\frac{\partial u}{\partial x}\right|_{i,j,k}^{n+\frac{1}{2}} = \left.\frac{1}{c}\frac{\partial u}{\partial t}\right|_{i,j,k}^{n+\frac{1}{2}}, \tag{2.44}$$

leads,

$$\frac{\left(u_{3/2,j,k}^{n+1} + u_{3/2,j,k}^{n}\right) - \left(u_{1/2,j,k}^{n+1} + u_{1/2,j,k}^{n}\right)}{2h} = \frac{\left(u_{3/2,j,k}^{n+1} + u_{1/2,j,k}^{n+1}\right) - \left(u_{3/2,j,k}^{n} + u_{1/2,j,k}^{n}\right)}{2c\Delta t}. \tag{2.45}$$

By arranging Eq. (2.45) regarding a particle velocity at a truncation boundary grid, the Mur ABC is obtained as

$$u_{1/2,j,k}^{n+1} = u_{3/2,j,k}^{n} - \frac{c\Delta t - h}{c\Delta t + h}\left(u_{1/2,j,k}^{n} - u_{3/2,j,k}^{n+1}\right). \tag{2.46}$$

Note that the above formulation is premised that a plane wave is normally incident on the absorbing boundary. Therefore, its accuracy deteriorates for an oblique incidence or a spherical wave incidence.

2.1.5.2 Perfectly Matched Layer

By D-ABC, a nonreflective boundary is approximated by a concept that only outward wave through a truncation boundary propagates and there exists no inward wave propagating to the problem region.

On the other hand, by M-ABC, sound wave penetrates into a lossy media with a certain thickness, which is adjacent to a problem region, and the sound energy vanishes in the lossy media, and consequently, reflecting sound from the truncation boundary is decreased. The PML developed by Berenger [11], which utilizes such a sound absorbing process, is the most flexible and efficient M-ABC.

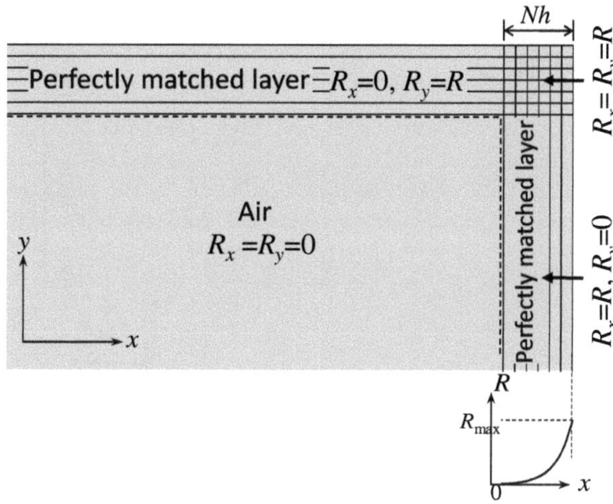

Fig. 2.8 Setting of PML

The following equations to which resistance terms specified by R_x, R_y and R_z are introduced are basic equations of the acoustic PML.

$$\rho\frac{\partial u}{\partial t} + R_x u + \frac{\partial p}{\partial x} = 0, \quad \frac{\partial p_x}{\partial t} + \frac{R_x}{\rho} p_x + \kappa\frac{\partial u}{\partial x} = 0, \tag{2.47}$$

$$\rho\frac{\partial v}{\partial t} + R_y v + \frac{\partial p}{\partial y} = 0, \quad \frac{\partial p_y}{\partial t} + \frac{R_y}{\rho} p_y + \kappa\frac{\partial v}{\partial y} = 0, \tag{2.48}$$

$$\rho\frac{\partial w}{\partial t} + R_z w + \frac{\partial p}{\partial z} = 0, \quad \frac{\partial p_z}{\partial t} + \frac{R_z}{\rho} p_z + \kappa\frac{\partial w}{\partial z} = 0, \tag{2.49}$$

$$p = p_x + p_y + p_z. \tag{2.50}$$

As shown in Eqs. (2.1)–(2.4) and Eqs. (2.47)–(2.50), governed equations in the PML are different from those in the air, and therefore, discontinuity of sound propagation media might cause sound reflection. The PML, however, contrives treatment of resistance terms so that the specific impedance of the PML media becomes equal to that of the air, and consequently, the impedance matching ensures no reflection from the interface-surface between the air and the PML media. As a special case, $R_x = R_y = R_z = 0$ leads Eqs. (2.1)–(2.4), which describe sound propagation in the air.

Figure 2.8 shows how to set PML layer for two-dimensional field. In the figure, upper PML diminishes a wave going in positive y-direction by introducing R_y as a positive value. Resistance term in x-direction does not contribute to reduction in the wave, so R_x is made to be zero. On the other hand, the right-hand side layer which weakens a wave going in positive x-direction has a positive value as a parameter of R_x and zero value as that of R_y.

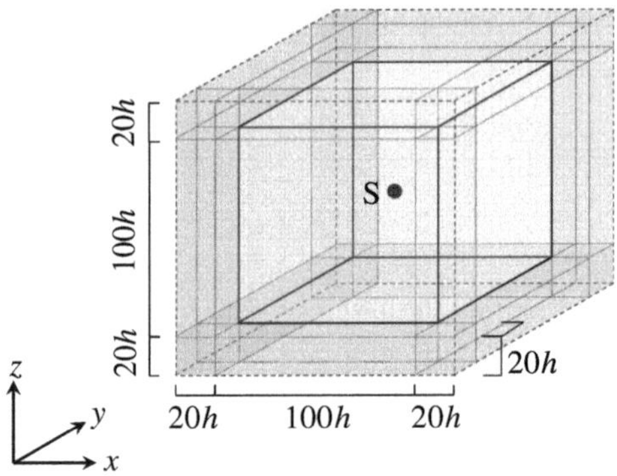

Fig. 2.9 Setting of a cubic sound field modeling a free field

When medium changes between two layers, the discontinuity of the medium results in a reflection from the interface, and the reflection may cause a numerical error. In the PML calculation, such a numerical error is usually reduced by smooth change of value of R in the PM layer. When total number of grids of a PM layer is N, for example, the value of R is given with a function of distance from the interface as

$$R(ih) = R_{\text{max}} \left(\frac{ih}{Nh} \right)^m , \tag{2.51}$$

where m is a constant determining spatial distribution pattern of R. As is shown in the equation, when $i = 0$, R is made to be zero and a sound wave smoothly penetrates through the PM layer. As i is increased, the value of R becomes rapidly larger. The PML calculation gradually decreases the amplitude of the traveling wave in the lossy media with a certain depth, and therefore the calculation method is robuster for the incidence angle than D-ABC. In order to see a difference in absorption characteristics between D-ABC and M-ABC, transition of sound pressure distribution for a cubic sound field shown in Fig. 2.9 with perfect absorption on all six walls was calculated for the following two cases—1: Mur-ABC, 2: PML with $N = 20$. Comparison of the calculation results are shown in Fig. 2.10. The figure shows every $64\Delta t$ snapshot of sound pressure distribution on a center x-y plane where a sound source exists. In a calculation result of Mur-ABC, slight reflection is seen, especially, amplitudes of the reflection from corners are larger than those from centers of boundary surfaces. On the other hand, in the case of the PML, penetrating wave gradually diminishes in the absorbing layer. As is seen in the figure, highly efficient perfectly absorbing boundary condition is realized by using the PML, but it should be noted that computational load becomes much larger than that of D-ABC from viewpoints of both of computer storage necessity and computational time.

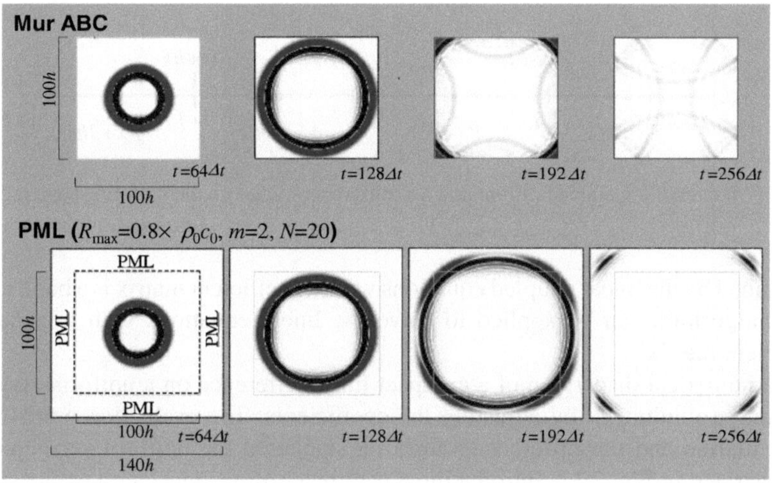

Fig. 2.10 Snapshots of sound pressure distribution in horizontal plane which includes a source

2.2 Techniques for High Accuracy

When accurate values of differentiations by finite difference are required, a grid spacing should be small compared to a typical wave length in the problem. Therefore, for a high frequency problem, the grid spacing has to be small and the degree of freedom amounts to a huge value. In order to resolve this problem, a higher order or a compact difference method can be applied. Since the time integration must be fulfilled after evaluating spatial differentiations, the accuracy of the time marching procedure must be maintained. In order to carry out the time integration accurately, a symplectic integration can be applied in the acoustic problem.

2.2.1 Compact Finite Difference

In the acoustic FDTD simulation, the variables are usually defined on a staggered grid. On the staggered grid the scalar variables (pressure, mass density, sound speed, etc.) are stored in the cell centers of the control volumes, whereas the particle velocities are located at the cell faces.

Since small grid spacing compared to the typical wavelength is required to obtain an accurate differential value when a conventional finite difference scheme is used, the grid spacing has to be small and the degree of freedom becomes huge for high frequency problems. In order to resolve this problem, a higher order explicit or a compact finite difference has been applied [12]. In the compact finite difference, differential values are coupled and fewer grid points are required for constructing the difference formula than the explicit finite difference. The differential value is

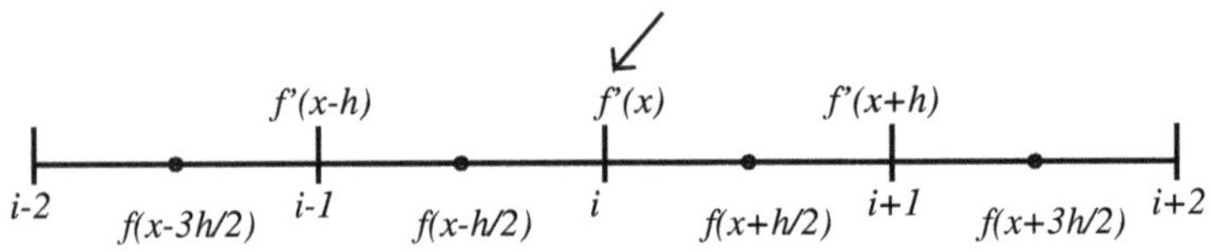

Fig. 2.11 Evalutation points of differential and difference values on the uniform staggered grid

determined by the linear coupled equations whose coefficient matrix is a band matrix. A fast algorithm can be applied to solve the linear equations with a tridiagonal coefficient matrix.

The numerical dispersion of a compact finite difference on a uniform staggered grid can be minimized by adjusting the coefficients. The evaluation points of the differentiation and the difference values on staggered the uniform staggered grid are illustrated in Fig. 2.11. Introducing a parameter α, we consider a compact finite difference on a grid with a uniform spacing h,

$$\alpha f'_{i+1} + f'_i + \alpha f'_{i-1} = b\frac{f_{i+3/2} - f_{i-3/2}}{3h} + a\frac{f_{i+1/2} - f_{i-1/2}}{h} + \epsilon. \tag{2.52}$$

Here, the coefficients a, b and the error term ϵ are related to α by

$$a = \frac{3}{8}(3 - 2\alpha), \quad b = \frac{22\alpha - 1}{8},$$
$$\epsilon = \frac{9 - 62\alpha}{1920}h^4 f^{(5)} + O(h^6). \tag{2.53}$$

The differential values are calculated by solving a linear equation with a tridiagonal coefficient matrix. When $\alpha = 1/22$, the coefficient b vanishes and the compact finite difference is represented by a least number of grid points. Thus, this case is convenient for the simulation of the wave propagation in a region of complex configuration. When $\alpha = 9/62$, the fourth-order error term in ϵ vanishes and the difference equation becomes sixth order. We evaluate effective wave number k' which is defined by the function $f(x) = \sin(kx)$ and its first-order differentiation $k' \cos(kx)$ evaluated by a finite difference. The grid wave number $w = hk$ and the effective grid wave number $w' = hk'$ are defined by the grid spacing h. For example, $w = \pi/2$ means that 4 grid points exist per one wavelength (4 PPW: point per wavelength). By the second-order explicit finite difference on the staggered grid, the effective grid wave number becomes $w' = 2\sin(w/2)$. This deviates from the exact value when the grid wave number is not very small. On the other hand, for the compact finite difference with α, we obtain

$$w'(w, \alpha) = \frac{2a\sin(\frac{w}{2}) + \frac{2}{3}b\sin(\frac{3w}{2})}{1 + 2\alpha\cos(w)}. \tag{2.54}$$

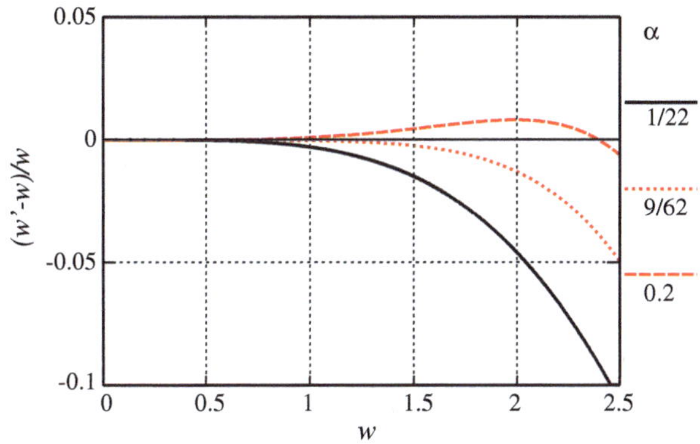

Fig. 2.12 Effective grid wave number error for various α

The exact value is $w = w'$ and the deviation from the exact value causes numerical dispersion error. Some properties of w', i.e.,

$$w'(0, \alpha) = 0,\ w'(\pi, \alpha) = \frac{7 - 10\alpha}{3(1 - 2\alpha)}, \tag{2.55}$$

$$w'(\pi, \alpha) < \pi, \ \text{when}\ \alpha < \frac{3\pi - 7}{6\pi - 10} \approx 0.27, \tag{2.56}$$

are derived. Therefore, when α is less than 0.27, there exists a point w_1 $(0 < w_1 < \pi)$ where $w = w'$, i.e.,

$$w'(w_1, \alpha) = w_1. \tag{2.57}$$

Furthermore, $w' > w$ holds in the interval $(0, w_1)$, and $w' < w$ in (w_1, π). If $\alpha > 9/62$ then $w'(w, \alpha)$ is larger than the exact grid wave number in the vicinity of $w = 0$. General tendency is that, when α is small, the effective wave number becomes less than the theoretical value even if the PPW is not so small. On the other hand, when α is large, the effective wave number can approximate the theoretical value well up to the region of short wavelength $(w \geq \pi/2)$. The effective grid wave number error is shown in Fig. 2.12. For w_1 in $(0, \pi)$, w_1 and α are related by

$$\alpha = \frac{27\sin(\frac{1}{2}w_1) - \sin(\frac{3}{2}w_1) - 12w_1}{18\sin(\frac{1}{2}w_1) - 22\sin(\frac{3}{2}w_1) + 24w_1\cos(w_1)}. \tag{2.58}$$

Therefore, adjusting α can improve the accuracy of the finite difference in a given interval of w. In order to optimize α, we have adopted the following strategy. First, we set the maximum wave number w_0 to be analyzed. Next, α is determined so that the maximum absolute value of relative error in the interval $(0, w_0)$ is minimized.

The maximum absolute value of relative error in the interval of $(0, w_0)$ is plotted in Fig. 2.13 where the abscissa means α. We see that the minimization is attained

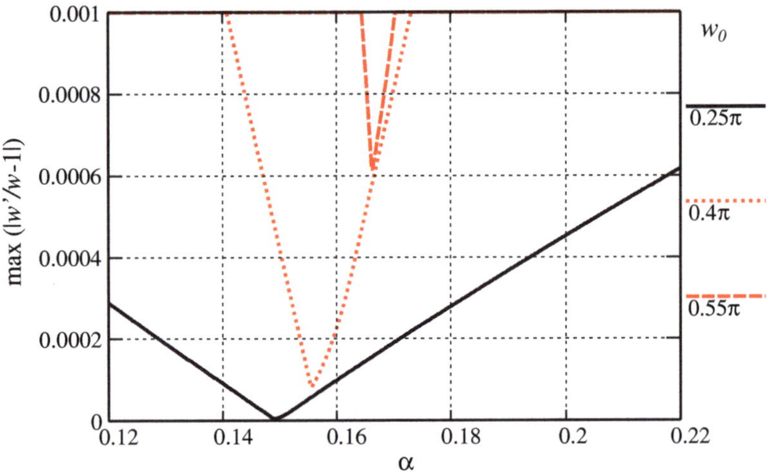

Fig. 2.13 Maximum absolute value of relative error for the various intervals of $(0, w_0)$. The abscissa represents α. The minimization of the maximum absolute value of error is realized by adjusting α

Table 2.1 Maximum wave number w_0, optimized α and maximum relative error. The optimization of α reduces the maximum absolute value of relative error below 10^{-5}

w_0	Point per wave length (PPW)	Optimized α	Maximum relative error
0.25π	8	0.14905	4.5×10^{-6}
0.3π	6.67	0.1508	1.4×10^{-5}
0.4π	5	0.15555	8.1×10^{-5}
0.5π	4	0.1621	3.3×10^{-4}
0.55π	3.64	0.16625	6.1×10^{-4}

by adjusting the parameter α. In Table 2.1, maximum wave number, optimized α, and maximum absolute value of relative error are shown. The maximum absolute value of relative error is reduced even when the PPW is not so large. This estimation has been done using a local property of coefficients, only. However, in the compact finite difference, the differential values are coupled with those of the adjacent points. Therefore, the differential value computed by the compact scheme is influenced by those of neighboring grid points. When the interval is finite and non periodic, the boundary difference scheme must be also considered.

There are two types of boundary differences in the staggered grid. The first type is that the evaluation point of the differential value is outside of the interval of difference value estimation. This type corresponds to the boundary difference of the pressure which is utilized for time evolution of the velocity. However, this differential value need not to be evaluated for a rigid boundary condition in the acoustic problem because the velocity on the rigid boundary is always zero. The evaluation points of

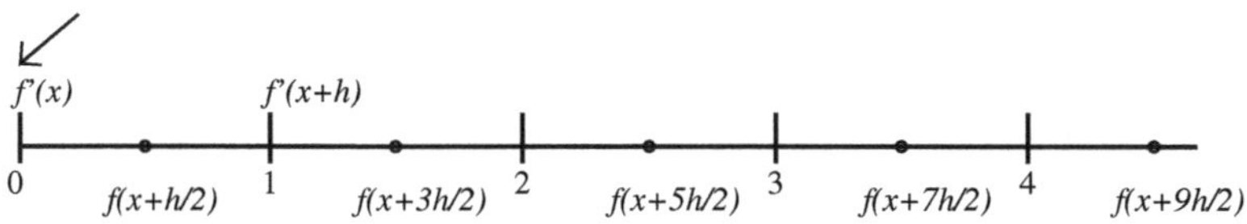

Fig. 2.14 Evaluation points of differential and difference values on boundary grid

the differentiation and the difference values on the boundary grid are illustrated in Fig. 2.14. The compact finite difference formula on the boundary is

$$f_0' + \alpha_b f_1' = \frac{1}{h}(a_b f_{1/2} + b_b f_{3/2} + c_b f_{5/2} + d_b f_{7/2} + e_b f_{9/2}) + \epsilon. \tag{2.59}$$

Here, coefficients are related to α_b by

$$\begin{aligned} a_b &= -\frac{22\alpha_b + 93}{24}, b_b = \frac{17\alpha_b + 229}{24}, c_b = \frac{3\alpha_b - 75}{8}, \\ d_b &= \frac{-5\alpha_b + 111}{24}, e_b = \frac{\alpha_b - 22}{24}, \end{aligned} \tag{2.60}$$

and ϵ is the error term,

$$\epsilon = (\frac{71\alpha_b}{1920} - \frac{563}{640})h^4 f^{(5)} + O(h^5). \tag{2.61}$$

For a function $f(x) = \cos(kx) + j\sin(kx)$, its differential value approximated by a finite difference, $f'(x) = jk'[\cos(kx) + \sin(kx)]$ is evaluated. Using grid wave number $w = hk$, w' is given by

$$jw' = \frac{a_b e^{jw/2} + b_b e^{j3w/2} + c_b e^{j5w/2} + d_b e^{j7w/2} + e_b e^{j9w/2}}{1 + \alpha_b e^{jw}}. \tag{2.62}$$

Global consideration must be executed because the differential values on the inner and boundary grid points are coupled. As an attempt to investigate the nonlocal property of the compact finite difference scheme, a spatial distribution of the amplitude error in a whole interval of a given grid is calculated by solving a system of linear equations with the tridiagonal coefficient matrix. The spatial distribution of the amplitude error in the interval with $n = 60$ grid points is demonstrated in Fig. 2.15. The wavelength is set to be 6 grid spacing (6 PPW). For the estimation of the compact difference, α of inner interval ($2 \le i \le n - 2$) is set to be the optimum value for the frequency range below 6 PPW. For the evaluation of differentiation on the grid points next to the boundary ($i = 1, n - 1$), α is set to be 1/22. At the grid points on both boundaries, α_b is set to be 21.88. The error is the largest on each boundary grid point $i = 0, n$. We evaluate the maximum absolute value of amplitude error on the boundary grid

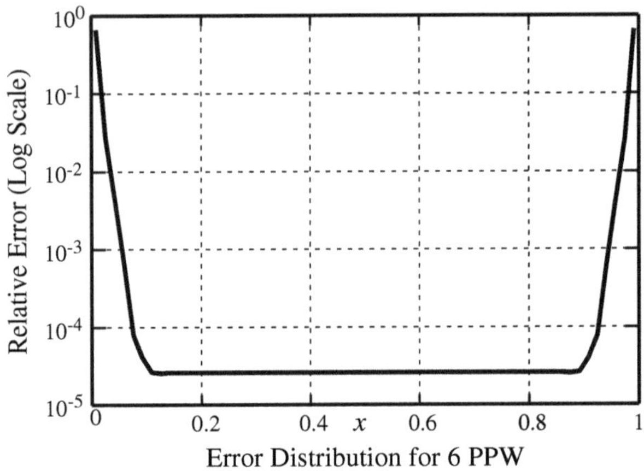

Fig. 2.15 Distribution of the absolute value of relative amplitude error on the grid with 60 points. The grid wave number is set to be $\pi/3$ (6 PPW)

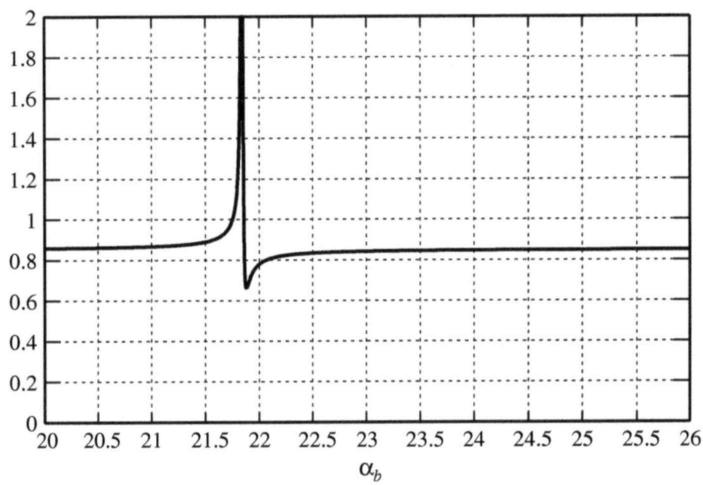

Fig. 2.16 Maximum absolute value of amplitude error versus α_b below upper bound frequency

point in the frequency range to be considered. The dependence of the error on α_b is demonstrated in Fig. 2.16. The absolute value of the relative error is not small in this case. Moreover, when α_b is near 22, the inverse of 1/22, the maximum amplitude error diverges. Thus, we can conclude that the optimization may not work well for this type of boundary difference when the value of PPW is not large.

The second type of boundary difference is that the evaluation point of differentiation lies between grid points for difference estimation. This case corresponds to the boundary difference of velocity in acoustic FDTD which is utilized

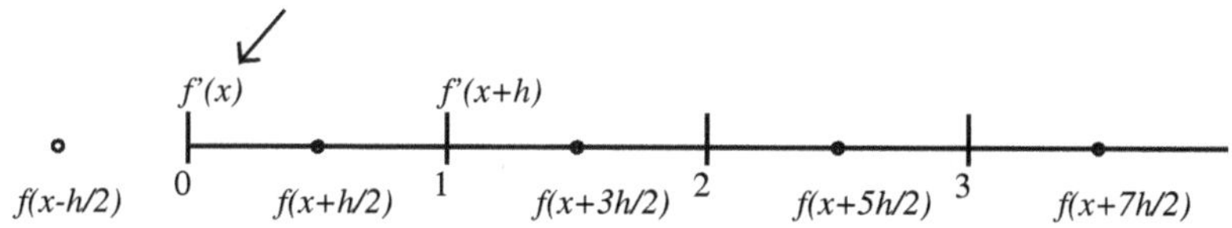

Fig. 2.17 Evaluation points of differential and difference values on boundary grid

for time evolution of the pressure at the boundary cell. The evaluation points of the differentiation and the difference values on the boundary grid is illustrated in Fig. 2.17. Its compact difference equation is

$$f_0' + \alpha_b f_1' = \frac{1}{h}(a_b f_{-1/2} + b_b f_{1/2} + c_b f_{3/2} + d_b f_{5/2} + e_b f_{7/2}) + \epsilon. \tag{2.63}$$

Relations among coefficients are

$$a_b = \frac{\alpha_b - 22}{24},\ b_b = \frac{-27\alpha_b + 17}{24},\ c_b = \frac{9\alpha_b + 3}{8},$$
$$d_b = \frac{-\alpha_b - 5}{24},\ e_b = \frac{1}{24}, \tag{2.64}$$

and

$$\epsilon = \left(-\frac{3\alpha_b}{640} + \frac{71}{1920}\right)h^4 f^{(5)} + O(h^5). \tag{2.65}$$

By the similar consideration, we obtain the effective grid wave number w' for this type as

$$jw' = \frac{a_b e^{-jw/2} + b_b e^{jw/2} + c_b e^{j3w/2} + d_b e^{j5w/2} + e_b e^{j7w/2}}{1 + \alpha_b e^{jw}}. \tag{2.66}$$

The absolute value of amplitude error of the compact difference in the interval with 60 grid points is illustrated in Fig. 2.18. Here, the grid wave number is set to be $\pi/3$ (6 PPW). The parameter α_b in Eq. (2.64) for the inner grid points also is set to be the optimum value for the frequency range below 6 PPW. At the grid point on each boundary, α_b is set to be 8.21. Though the absolute value of the error is large at the grid point on each boundary, it is less than the other case. The maximum absolute value of relative amplitude error is also estimated on the boundary point in the frequency range below a specified upper bound. By a similar procedure to the previous case, the amplitude error is minimized to 2.68×10^{-2} when α_b takes a value around 8.212. However, the efficacy of the optimization is not large for 6 PPW.

From the discussions above, the dispersion error in the finite difference at the boundary grid point cannot be greatly improved when the grid spacing is not small enough compared to the wavelength to be considered.

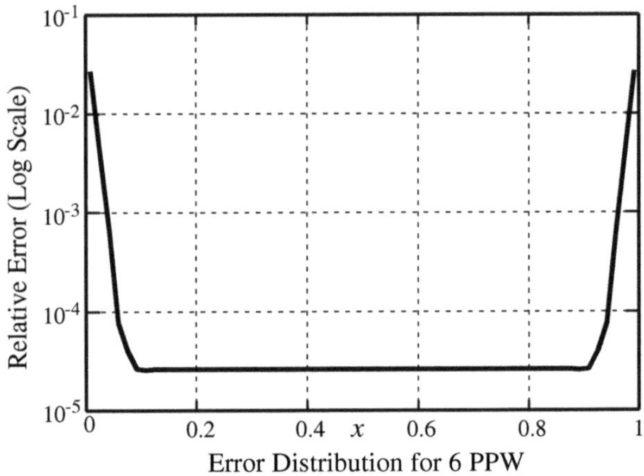

Fig. 2.18 Distribution of absolute value of relative amplitude error on the grid with 60 points. The grid wave number is set to be $\pi/3$ (6 PPW)

2.2.2 Improvement of Time Integration

Since the time integration must be fulfilled after evaluating spatial differentiations in FDTD, the next aim is to improve the time marching procedure. For the long time integrations, a symplectic integration method is an excellent scheme when the dynamics of the system possesses a Hamiltonian structure. The symplectic integration scheme is developed for the last few decades and is applied to particle dynamics and celestial mechanics. Higher order schemes of symplectic integration are considered by many authors. Extensions of the theory to a partial differential equation have been investigated recently. Here, we do not go far into theoretical details of the symplectic integration. Instead, the acoustic simulation is carried out by the symplectic integration keeping higher accuracy during long time steps.

We describe the outline of the symplectic integral method. When a set of ordinary differential equations for variables p and q are described by the following form:

$$\frac{dp}{dt} = f(q), \quad \frac{dq}{dt} = g(p), \tag{2.67}$$

the time marching by a time step Δt is carried out through m intermediate stages. The operations at i-th stage are

$$\begin{aligned} P_i &= P_{i-1} + \Delta t b_i f(Q_{i-1}), \\ Q_i &= Q_{i-1} + \Delta t \tilde{b}_i g(P_i), \end{aligned} \tag{2.68}$$

Table 2.2 Coefficients for Ruth's formula

	$i = 1$	$i = 2$	$i = 3$
b_i	7/24	3/4	$-1/24$
$\tilde{b}_i$	2/3	$-2/3$	1

where

$$P_0 = p(t), Q_0 = q(t),$$
$$P_m = p(t + \Delta t), Q_m = q(t + \Delta t). \qquad (2.69)$$

The coefficients b_i and $\tilde{b}_i$ for Ruth's formula [13] are shown in Table 2.2. For sound wave propagation, p and q are considered as acoustic pressure and velocity vector $\boldsymbol{v}$, respectively. Also, $f(q)$ and $g(p)$ are described by the compact finite differences in velocity vector $\boldsymbol{v}$ and pressure p, respectively.

Simulations of one-dimensional wave propagation have been carried out. We have chosen the initial wave form $f(x)$ as

$$f(x) = \frac{1}{2} \exp[-\ln 2(\frac{x}{3})^2] \qquad (2.70)$$

on the grid whose spacing $h = 1.0$. The sound speed c is supposed as unity. We let the initial Gaussian form travel up to 10,000 time steps by using three different schemes. The shapes of the waves at several time steps simulated by various schemes are illustrated in Fig. 2.19. In each figure, the abscissa is $x' = x - ct$. The result obtained by the combination of optimized fourth-order compact finite difference ($\alpha = 0.1475$) and Ruth's time integration scheme is excellent. Even at the time step of 10,000, the wave shape almost retains the initial form by this scheme. Therefore, the symplectic method is an effective integration technique for long time step computations

2.3 Application to Vibroacoustic Problems

In this section, application of the FDTD method to vibroacoustic problems such as a sound radiation problem and a structure-borne sound problem is presented. Herein two types of modeling methods for solid part like walls and floors, solid modeling and beam-plate modeling, are considered. The former is a straightforward method where longitudinal and shear waves are considered and solid part is discretized with many small volumes. Using the solid modeling, physical phenomena can be relatively well expressed but the computational cost will become huge. On the other hand, the latter considers solid part as a composition of plate elements. Therefore, longitudinal and bending waves are taken into account with plate theories. Although there are some limitations in applicability, this modeling has an advantage of the

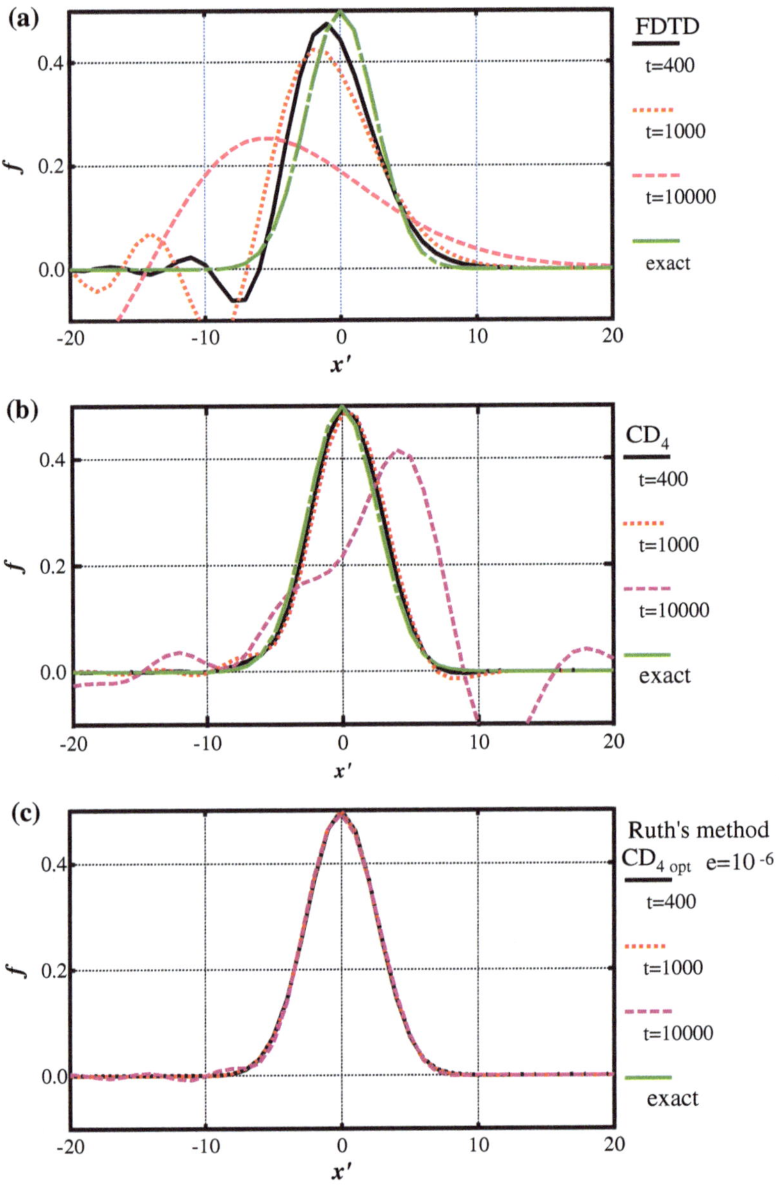

Fig. 2.19 Comparison of the wave forms at several time steps obtained by **a** the conventional FDTD scheme (explicit second-order finite difference and leap frog time integration) with CFL number 0.9, **b** the fourth-order compact finite difference and the leap frog time integration with CFL number = 0.25, and **c** the optimized fourth-order compact finite difference ($\alpha = 0.1475$) and Ruth's time integration with CFL number = 0.5

small computational load. The following introduce the formulations of vibroacoustic FDTD method according to the modeling method. Numerical examples of this section are shown in Sect. 7.6.

2.3.1 Solid Modeling

2.3.1.1 Basic Equations

To analyze vibrations of solid, consider equations of motion and constitutive equations of three-dimensional elastic bodies [14]. When probing vibration, one of the most important phenomena is damping. Many formulations and implementations have been suggested from various viewpoints, regardless if the medium is solid or fluid [15–18]. Herein two types of damping terms, which give distinct damping characteristics, are considered [19]. Then, under the conditions of small deformation, adiabatic transition, and athermic medium, the motion equations and constitutive equations with two types of damping terms can be expressed in tensor notation as [20]

$$\rho \frac{\partial v_i}{\partial t} + \zeta v_i = \frac{\partial T_{ij}}{\partial a_j}, \tag{2.71}$$

$$T_{ij} = c_{ijkl}\epsilon_{kl} + \xi_{ijkl}e_{kl}, \tag{2.72}$$

where $i, j, k, l = x, y, z$. ρ is the density, $\boldsymbol{v}$ is the velocity vector, and t is time. ζ is a constant to describe the damping force proportional only to the velocity, while $\boldsymbol{T}$ is the stress tensor. $\boldsymbol{a}$ $(= [x, y, z])$ is the position vector, $\boldsymbol{c}$ is the stiffness tensor, and $\boldsymbol{\epsilon}$ is the strain tensor. $\boldsymbol{\xi}$ is the viscosity tensor, which describes the damping force proportional to the second-order space derivative of the velocity, and $\boldsymbol{e}$ $(= \partial\boldsymbol{\epsilon}/\partial t)$ is the strain velocity tensor. It should be noted that combining the two types of damping terms ζ and $\boldsymbol{\xi}$ yields similar characteristics to Rayleigh damping [21]. Considering the reciprocities of stiffness and viscosity, stiffness tensor $\boldsymbol{c}$ and viscosity tensor $\boldsymbol{\xi}$ can be abbreviated to their six-by-six matrix forms, each of which has 21 independent constants.

Additionally, if an orthotropic medium is considered, the constants can be reduced to nine independent ones. In this case, the relationship between the stiffness matrix $\boldsymbol{C}$ and Young's modulus E, shear modulus G, and Poisson's ratios υ can be given by

$$\boldsymbol{C} = \boldsymbol{S}^{-1}, \tag{2.73}$$

$$\frac{\upsilon_{yz}}{E_y} = \frac{\upsilon_{zy}}{E_z}, \frac{\upsilon_{zx}}{E_z} = \frac{\upsilon_{xz}}{E_x}, \frac{\upsilon_{xy}}{E_x} = \frac{\upsilon_{yx}}{E_y}, \tag{2.74}$$

$$
\boldsymbol{S} = \begin{bmatrix}
\frac{1}{E_x} & -\frac{\upsilon_{xy}}{E_x} & -\frac{\upsilon_{xz}}{E_x} & 0 & 0 & 0 \\
-\frac{\upsilon_{yx}}{E_y} & \frac{1}{E_y} & -\frac{\upsilon_{yz}}{E_y} & 0 & 0 & 0 \\
-\frac{\upsilon_{zx}}{E_z} & -\frac{\upsilon_{zy}}{E_z} & \frac{1}{E_z} & 0 & 0 & 0 \\
0 & 0 & 0 & \frac{1}{G_{yz}} & 0 & 0 \\
0 & 0 & 0 & 0 & \frac{1}{G_{zx}} & 0 \\
0 & 0 & 0 & 0 & 0 & \frac{1}{G_{xy}}
\end{bmatrix}, \tag{2.75}
$$

where $\boldsymbol{S}$ is the compliance matrix. When dealing with vibroacoustic problems, behaviors of a fluid should be described with its basic equations. Interestingly, although Eqs. (2.71) and (2.72) with a stiffness matrix are derived for a solid, they can also express governing equations of a fluid with an appropriate stiffness matrix, i.e., the bulk modulus κ and zero should be substituted for $c_{11} = c_{22} = c_{33} = c_{12} = c_{21} = c_{13} = c_{31} = c_{23} = c_{32}$ and the other components of $\boldsymbol{c}$, respectively[22]. In this case, Eq. (2.72) can be interpreted as the constitutive equation of a Newtonian fluid. The linearized Navier–Stokes equation where the convection term and volumetric force are neglected can be derived by neglecting the term ζv_i of Eq. (2.71) and substituting Eq. (2.72) into Eq. (2.71). As for damping terms, $\chi \equiv \xi_{12} = \xi_{21} = \xi_{13} = \xi_{31} = \xi_{23} = \xi_{32}$ mean the second viscosity [23] and $\gamma \equiv \xi_{44} = \xi_{55} = \xi_{66}$ mean the shear viscosity, then $\xi_{11} = \xi_{22} = \xi_{33}$ can be written as $\chi + 2\gamma$. Note that the dilatational viscosity can be expressed as $\chi + (2/3)\gamma$. The sound pressure can be obtained by calculating $-\kappa \nabla \cdot \boldsymbol{u}$, where $\boldsymbol{u}$ is the displacement vector that can be calculated by integrating the velocity vector $\boldsymbol{v}$ over time.

For discretization with a difference scheme of the leap-flog algorithm, Eqs. (2.71) and the time-derivative form of (2.72) are rewritten as

$$
\rho \frac{\partial v_x}{\partial t} + \zeta v_x = \frac{\partial T_{xx}}{\partial x} + \frac{\partial T_{xy}}{\partial y} + \frac{\partial T_{zx}}{\partial z}, \tag{2.76}
$$

$$
\rho \frac{\partial v_y}{\partial t} + \zeta v_y = \frac{\partial T_{xy}}{\partial x} + \frac{\partial T_{yy}}{\partial y} + \frac{\partial T_{yz}}{\partial z}, \tag{2.77}
$$

$$
\rho \frac{\partial v_z}{\partial t} + \zeta v_z = \frac{\partial T_{zx}}{\partial x} + \frac{\partial T_{yz}}{\partial y} + \frac{\partial T_{zz}}{\partial z}, \tag{2.78}
$$

$$
\begin{aligned}
\frac{\partial T_{xx}}{\partial t} &= c_{11} \frac{\partial v_x}{\partial x} + c_{12} \frac{\partial v_y}{\partial y} + c_{13} \frac{\partial v_z}{\partial z} \\
&\quad + \xi_{11} \frac{\partial^2 v_x}{\partial x \partial t} + \xi_{12} \frac{\partial^2 v_y}{\partial y \partial t} + \xi_{13} \frac{\partial^2 v_z}{\partial z \partial t},
\end{aligned} \tag{2.79}
$$

$$
\frac{\partial T_{yy}}{\partial t} = c_{12} \frac{\partial v_x}{\partial x} + c_{22} \frac{\partial v_y}{\partial y} + c_{23} \frac{\partial v_z}{\partial z}
$$

$$+\xi_{12}\frac{\partial^2 v_x}{\partial x \partial t}+\xi_{22}\frac{\partial^2 v_y}{\partial y \partial t}+\xi_{23}\frac{\partial^2 v_z}{\partial z \partial t}, \tag{2.80}$$

$$\frac{\partial T_{zz}}{\partial t}=c_{13}\frac{\partial v_x}{\partial x}+c_{23}\frac{\partial v_y}{\partial y}+c_{33}\frac{\partial v_z}{\partial z}$$

$$+\xi_{13}\frac{\partial^2 v_x}{\partial x \partial t}+\xi_{23}\frac{\partial^2 v_y}{\partial y \partial t}+\xi_{33}\frac{\partial^2 v_z}{\partial z \partial t}, \tag{2.81}$$

$$\frac{\partial T_{yz}}{\partial t}=c_{44}\frac{\partial v_z}{\partial y}+c_{44}\frac{\partial v_y}{\partial z}+\xi_{44}\frac{\partial^2 v_z}{\partial y \partial t}+\xi_{44}\frac{\partial^2 v_y}{\partial z \partial t}, \tag{2.82}$$

$$\frac{\partial T_{zx}}{\partial t}=c_{55}\frac{\partial v_x}{\partial z}+c_{55}\frac{\partial v_z}{\partial x}+\xi_{55}\frac{\partial^2 v_x}{\partial z \partial t}+\xi_{55}\frac{\partial^2 v_z}{\partial x \partial t}, \tag{2.83}$$

$$\frac{\partial T_{xy}}{\partial t}=c_{66}\frac{\partial v_y}{\partial x}+c_{66}\frac{\partial v_x}{\partial y}+\xi_{66}\frac{\partial^2 v_y}{\partial x \partial t}+\xi_{66}\frac{\partial^2 v_x}{\partial y \partial t}, \tag{2.84}$$

where v_x is the x-directional velocity, v_y is the y-directional velocity, and v_z is the z-directional velocity. T_{xx} is the x-directional normal stress, T_{yy} is the y-directional normal stress, and T_{zz} is the z-directional normal stress. T_{yz} is the shear stress defined in the yz-plane, T_{zx} is the shear stress defined in the zx-plane, and T_{xy} is the shear stress defined in the xy-plane.

2.3.1.2 Averaging of Material Parameters

Figure 2.20 shows the arrangement of reference points for the stress and velocity on a nonuniform staggered-grid system [24–27]. Δx, Δy, and Δz are the spatial intervals between the reference points of shear stress and velocity for the x, y, and z directions. Point locations of normal stress are expressed as i, j, and k, while point locations of shear stress and velocity should shift according to the staggered-grid system. For example, the point locations of the xy-shear stress are expressed as $i+0.5$, $j+0.5$, and k, whereas those of the x-velocity are expressed as $i+0.5$, j, and k. Δt is the time interval for the calculation, and the elapsed-time counter is given by superscript n for normal and shear stresses and $n+0.5$ for velocities.

The target region of a vibroacoustic problem can be filled with a heterogeneous orthotropic material, which is governed by Eqs. (2.71) and (2.72). Herein, all the material parameters are defined at the reference points of normal stress, and the mean values are used as reference points for shear stress and velocity [28–32]. The weighted arithmetic averages are employed at the reference points of velocity [33], for example,

$$\overline{\rho}\,(i+0.5,j,k)=\frac{\rho\,(i,j,k)\,\Delta x\,(i)+\rho\,(i+1,j,k)\,\Delta x\,(i+1)}{\Delta x\,(i)+\Delta x\,(i+1)}. \tag{2.85}$$

At the reference points of shear stress, the weighted harmonic averages are employed [33], for example,

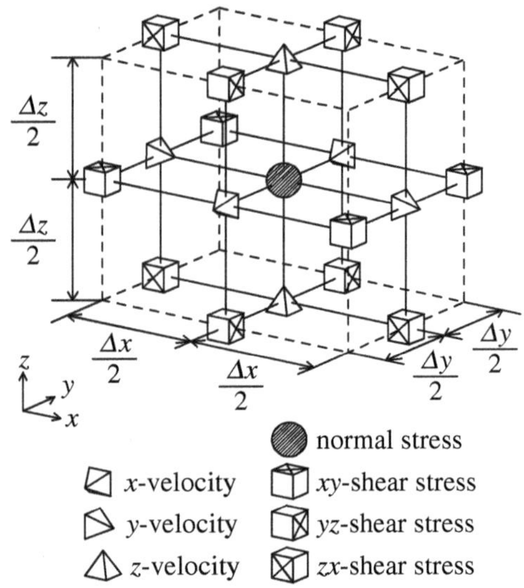

Fig. 2.20 Yee cell which has spatial intervals of Δx, Δy, and Δz

$$\overline{G_{xy}}\,(i+0.5,\,j+0.5,\,k)$$
$$=\frac{\{\Delta x\,(i)+\Delta x\,(i+1)\}\{\Delta y\,(j)+\Delta y\,(j+1)\}}{\frac{\Delta x(i)\Delta y(j)}{G_{xy}(i,j,k)}+\frac{\Delta x(i+1)\Delta y(j)}{G_{xy}(i+1,j,k)}+\frac{\Delta x(i)\Delta y(j+1)}{G_{xy}(i,j+1,k)}+\frac{\Delta x(i+1)\Delta y(j+1)}{G_{xy}(i+1,j+1,k)}}. \quad (2.86)$$

Additionally, the damping constants are given by their arithmetic averages at the velocity points and by their harmonic averages at the shear-stress points.

2.3.1.3 Boundary Conditions

A general method should accommodate a variety of boundary conditions such as a fixed boundary, free boundary, and perfectly absorptive boundary. For absorptive boundary, PML would be the most reliable option [34]. Implementation of the PML to this method is introduced in reference [19, 20]. Hence, this section focuses on two boundaries: a fixed boundary and a free boundary [33].

A fixed boundary is considered to be the boundary with the rigid body, which has infinite density and infinite shear modulus. Therefore, the averaged density as shown in Eq. (2.85) becomes infinite and then the velocity normal to the fixed surface becomes zero. This condition can be directly satisfied for the reference points of velocity located on the boundary. However, because the reference points of the velocity components parallel to the boundary surface are not defined just on the boundary, virtual reference points are assumed outside the boundary, and their values are determined so that the mean velocities on the boundary are zero. For example, if the plane $i+0.5=I+0.5$ is assumed to be a fixed boundary, then

$$v_x\,(I+0.5,\,j,\,k)=0, \quad (2.87)$$

$$v_y(I, j \pm 0.5, k) + v_y(I + 1, j \pm 0.5, k) = 0, \tag{2.88}$$

$$v_z(I, j, k \pm 0.5) + v_y(I + 1, j, k \pm 0.5) = 0. \tag{2.89}$$

Additionally, in Eq. (2.86), the related term to the rigid body in the denominator becomes zero. For example, the averaged shear modulus on the fixed boundary with a uniform mesh becomes $4G/3$ when one of the adjacent four media is the rigid body and the others have the same shear modulus G.

On the other hand, a free boundary is considered to be the boundary with vacuum, where the density and shear modulus are zero. The averaged shear modulus as shown in Eq. (2.86) becomes zero and then the shear stress defined in the free surface becomes zero. Although the normal stress to the boundary surface must be zero as a matter of course, the reference points of normal stress, which should be zero, are not arranged on the boundary. Therefore, virtual reference points are assumed outside the boundary, and their values are determined such that the mean normal stresses on the boundary are zero. For example, if the plane $i + 0.5 = I + 0.5$ is assumed to be a free boundary, then

$$T_{xy}(I + 0.5, j \pm 0.5, k) = T_{zx}(I + 0.5, j, k \pm 0.5) = 0, \tag{2.90}$$

$$T_{xx}(I, j, k) + T_{xx}(I + 1, j, k) = 0. \tag{2.91}$$

Additionally, the related term to vacuum in the numerator in Eq. (2.85) becomes zero. For example, the averaged density on the free boundary with an uniform mesh becomes $\rho/2$ when one of the adjacent two media is vacuum and the other has a density ρ.

2.3.1.4 Discretization

For example, Eq. (2.76) is discretized with a central difference as

$$\begin{aligned}
&\overline{\rho}(i + 0.5, j, k)\frac{v_x^{n+0.5}(i + 0.5, j, k) - v_x^{n-0.5}(i + 0.5, j, k)}{\Delta t}\\
&\quad + \overline{\zeta}(i + 0.5, j, k)\frac{v_x^{n+0.5}(i + 0.5, j, k) + v_x^{n-0.5}(i + 0.5, j, k)}{2}\\
&\approx \frac{T_{xx}^n(i + 1, j, k) - T_{xx}^n(i, j, k)}{\{\Delta x(i + 1) + \Delta x(i)\}/2}\\
&\quad + \frac{T_{xy}^n(i, j + 0.5, k) - T_{xy}^n(i, j - 0.5, k)}{\Delta y(j)}\\
&\quad + \frac{T_{zx}^n(i, j, k + 0.5) - T_{zx}^n(i, j, k - 0.5)}{\Delta z(k)}.
\end{aligned} \tag{2.92}$$

Eq. (2.84) is discretized with a backward difference for time derivation of the viscosity term and central difference for other terms as

$$
\begin{aligned}
&\frac{T_{xy}^{n+1}(i+0.5, j+0.5, k) - T_{xy}^{n}(i+0.5, j+0.5, k)}{\Delta t} \\
&\approx \left\{ \overline{c_{66}}(i+0.5, j+0.5, k) + \frac{\overline{\xi_{66}}(i+0.5, j+0.5, k)}{\Delta t} \right\} \\
&\times \frac{v_y^{n+0.5}(i+1, j+0.5, k) - v_y^{n+0.5}(i, j+0.5, k)}{\{\Delta x(i+1) + \Delta x(i)\}/2} \\
&- \frac{\overline{\xi_{66}}(i+0.5, j+0.5, k)}{\Delta t} \\
&\times \frac{v_y^{n-0.5}(i+1, j+0.5, k) - v_y^{n-0.5}(i, j+0.5, k)}{\{\Delta x(i+1) + \Delta x(i)\}/2} \\
&+ \left\{ \overline{c_{66}}(i+0.5, j+0.5, k) + \frac{\overline{\xi_{66}}(i+0.5, j+0.5, k)}{\Delta t} \right\} \\
&\times \frac{v_x^{n+0.5}(i+0.5, j+1, k) - v_x^{n+0.5}(i+0.5, j, k)}{\{\Delta y(j+1) + \Delta y(j)\}/2} \\
&- \frac{\overline{\xi_{66}}(i+0.5, j+0.5, k)}{\Delta t} \\
&\times \frac{v_x^{n-0.5}(i+0.5, j+1, k) - v_x^{n-0.5}(i+0.5, j, k)}{\{\Delta y(j+1) + \Delta y(j)\}/2}.
\end{aligned} \tag{2.93}
$$

Transforming these discretized equations, updating formulas for $v_x^{n+0.5}(i+0.5, j, k)$ and $T_{xy}^{n+1}(i+0.5, j+0.5, k)$ can be obtained. Updating formulas for other variables can be obtained by a similar procedure.

2.3.1.5 Stability Condition

This section focuses on stability conditions considering orthotropic media. An arbitrary wave can be expressed as a superposition of plane waves. Hence, the stability conditions for a plane wave of an arbitrary propagation angle are derived here [35]. The velocities and stress for a plane wave can be expressed as

$$v_x^{n+0.5}(i+0.5, j, k) = v_{x0}^{n+0.5} e^{\mathrm{i}\{k_x(i+0.5)\Delta x + k_y j\Delta y + k_z k\Delta z\}}, \tag{2.94}$$

$$T_{xx}^{n}(i, j, k) = T_{xx0}^{n} e^{\mathrm{i}(k_x i\Delta x + k_y j\Delta y + k_z k\Delta z)}, \tag{2.95}$$

$$T_{xy}^{n}(i+0.5, j+0.5, k) = T_{xy0}^{n} e^{\mathrm{i}\{k_x(i+0.5)\Delta x + k_y(j+0.5)\Delta y + k_z k\Delta z\}}, \tag{2.96}$$

where i is an imaginary unit, and k_x, k_y, and k_z are x, y, and z-directional wave numbers, respectively. Substituting Eqs. (2.94)–(2.96) into Eqs. (2.76)–(2.84) yields a homogeneous state-difference equation, which is expressed as

$$\boldsymbol{x}^{n+1} = \boldsymbol{A}\boldsymbol{x}^{n}, \tag{2.97}$$

where

$$\left[\boldsymbol{x}^{n}\right]^{*} = \left[\, v_{x0}^{n-0.5}\ v_{y0}^{n-0.5}\ v_{z0}^{n-0.5}\ T_{xx0}^{n}\ T_{yy0}^{n}\ T_{zz0}^{n}\ T_{yz0}^{n}\ T_{zx0}^{n}\ T_{xy0}^{n} \right], \tag{2.98}$$

$$\boldsymbol{A} = \begin{bmatrix}
r_x & 0 & 0 & R_x h_x & 0 & 0 & 0 & R_x h_z & R_x h_y \\
0 & r_y & 0 & 0 & R_y h_y & 0 & R_y h_z & 0 & R_y h_x \\
0 & 0 & r_z & 0 & 0 & R_z h_z & R_z h_y & R_z h_x & 0 \\
b_{11x}h_x & b_{12y}h_y & b_{13z}h_z & 1-B_{11x}h_x^2 & -B_{12y}h_y^2 & -B_{13z}h_z^2 & -(B_{12y}+B_{13z})h_y h_z & -(B_{13z}+B_{11x})h_z h_x & -(B_{11x}+B_{12y})h_x h_y \\
b_{12x}h_x & b_{22x}h_y & b_{23x}h_z & -B_{12x}h_x^2 & 1-B_{22x}h_y^2 & -B_{23x}h_z^2 & -(B_{22y}+B_{23z})h_y h_z & -(B_{23z}+B_{12x})h_z h_x & -(B_{12x}+B_{22y})h_x h_y \\
b_{13x}h_x & b_{23x}h_y & b_{33x}h_z & -B_{13x}h_x^2 & -B_{23x}h_y^2 & 1-B_{33x}h_z^2 & -(B_{23y}+B_{33z})h_y h_z & -(B_{33z}+B_{13x})h_z h_x & -(B_{13x}+B_{23y})h_x h_y \\
0 & b_{44x}h_z & b_{44x}h_y & 0 & -B_{44x}h_y h_z & -B_{44x}h_y h_z & 1-B_{44z}h_y^2-B_{44y}h_z^2 & -B_{44z}h_x h_y & -B_{44y}h_z h_x \\
b_{55x}h_z & 0 & b_{55x}h_x & -B_{55x}h_z h_x & 0 & -B_{55x}h_z h_x & -B_{55z}h_x h_y & 1-B_{55x}h_z^2-B_{55z}h_x^2 & -B_{55x}h_y h_z \\
b_{66x}h_y & b_{66x}h_x & 0 & -B_{66x}h_x h_y & -B_{66x}h_x h_y & 0 & -B_{66y}h_z h_x & -B_{66x}h_y h_z & 1-B_{66y}h_x^2-B_{66x}h_y^2
\end{bmatrix}, \tag{2.99}$$

$$\theta = \frac{\Delta t}{2\rho}\zeta, \tag{2.100}$$

$$r = \frac{1-\theta}{1+\theta}, \tag{2.101}$$

$$R = \frac{\mathrm{i}\Delta t}{\rho\,(1+\theta)}, \tag{2.102}$$

$$b_\alpha = \mathrm{i}\Delta t\left\{\frac{c_\alpha\,(1-\theta)-2\xi_\alpha\theta}{1+\theta}\right\}, \tag{2.103}$$

$$B_\alpha = \frac{c_\alpha\Delta t^2-\xi_\alpha\Delta t}{\rho\,(1+\theta)}, \tag{2.104}$$

$$h_d = \frac{2\sin\left(k_d \frac{\Delta d_{\min}}{2}\right)}{\Delta d_{\min}}, \tag{2.105}$$

where the asterisk denotes the transpose, $\alpha = 11, 22, 33, 12, 13, 23, 44, 55$, and 66, $d = x, y, z$, and subscript min means the minimum value. To obtain stable solutions from Eq. (2.97), all eigenvalues must be equal to or less than one for an arbitrary propagation angle, i.e., for the arbitrary coupling of k_x, k_y, and k_z. Therefore, assigning h_x, h_y, and h_z to their maximum values of $2/\Delta x_{\min}$, $2/\Delta y_{\min}$, and $2/\Delta z_{\min}$, respectively, should yield Δt such that all eigenvalues of matrix $\boldsymbol{A}$ are one or less.

2.3.1.6 Initial Condition and Excitation

Herein all initial values of velocities and stresses are set to zero and an input excitation is assumed to be a point force $F(t)$ at the cell indicated by i, j, and k. In this case, $F(n\Delta t)/\Delta S\,(i, j, k)$ should be added to the normal stress in the excitation direction, which belongs to the spatial difference term in the updating formula of velocity. $\Delta S\,(i, j, k)$ is the unit area normal to the excitation direction, for example, $\Delta S\,(i, j, k) = \Delta y\,(j)\,\Delta z\,(k)$ for x-directional excitation.

2.3.2 Plate Modeling

2.3.2.1 Basic Equations

Governing equations for the bending wave and the quasi-longitudinal wave on a plate model are described as follows. Equation (2.106) describes the bending wave propagation on the plate existing in the $x - y$ plane, and Eqs. (2.107) and (2.108) describe the in-plane wave for the x and y direction.

$$D\left(\frac{\partial^2}{\partial x^2} + \frac{\partial^2}{\partial y^2}\right)^2 w + \xi D \frac{\partial}{\partial t}\left(\frac{\partial^2}{\partial x^2} + \frac{\partial^2}{\partial y^2}\right)^2 w + \rho h \mu \frac{\partial w}{\partial t} + \rho h \frac{\partial^2 w}{\partial t^2} = q, \tag{2.106}$$

$$\frac{E}{1-\gamma^2}\left(\frac{\partial^2 u}{\partial x^2} + \gamma \frac{\partial^2 v}{\partial x \partial y}\right) + G\left(\frac{\partial^2 v}{\partial x \partial y} + \frac{\partial^2 u}{\partial y^2}\right) - \rho \frac{\partial^2 u}{\partial t^2} = 0, \tag{2.107}$$

$$\frac{E}{1-\gamma^2}\left(\frac{\partial^2 v}{\partial y^2} + \gamma \frac{\partial^2 u}{\partial x \partial y}\right) + G\left(\frac{\partial^2 u}{\partial x \partial y} + \frac{\partial^2 v}{\partial x^2}\right) - \rho \frac{\partial^2 v}{\partial t^2} = 0. \tag{2.108}$$

Here, w is the displacement of the out-of-plane bending vibration, u and v are those of the in-plane vibration in the x and y directions, ξ and μ are coefficients for modeling the damping characteristics of the material, q is an external force, D is the flexural

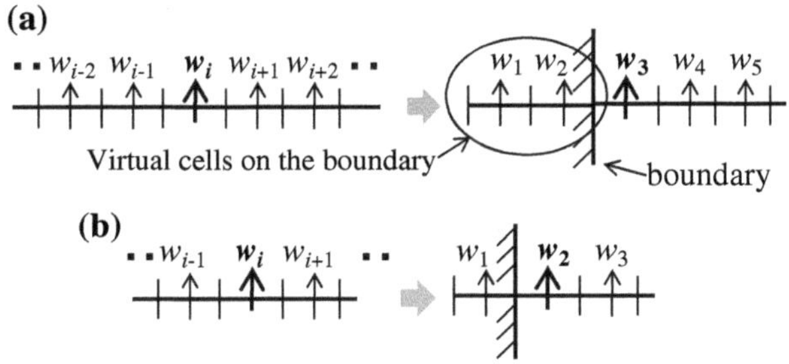

Fig. 2.21 Schematic figure for the finite-difference approximation nearby the boundary part in case of fourth-/second-order differentials. **a** 4th-order differential, **b** 2nd-order differential

rigidity ($D = Eh^3/12(1-\gamma^2)$), and E, ρ, h, γ and G are the Young's modulus, density, thickness of the plate, the Poisson's ratio, and the elastic shear modulus, respectively.

2.3.2.2 Discretization

The basic equations of Eq. (2.106)/Eq. (2.107) and (2.108) for the bending/in-plane wave has fourth-order/second-order differential system. Finite-difference approximation of the fourth-order/second-order differential for a function $w(x)$ is described as

$$\frac{\partial^4 w_i}{\partial x^4} = \frac{w_{i+2} - 4w_{i+1} + 6w_i - 4w_{i-1} + w_{i-2}}{\Delta x^4} + O\left(\Delta x^2\right), \quad (2.109)$$

$$\frac{\partial^2 w_i}{\partial x^2} = \frac{w_{i+1} - 2w_i + w_{i-1}}{\Delta x^2} + O\left(\Delta x^2\right), \quad (2.110)$$

where i is the discrete grid number in space. Equation (2.109) indicates that a parameter w_i is calculated using the neighboring five parameters including itself, as show in Fig. 2.21. As for Eq. (2.110) w_i is calculated using the neighboring three parameters. When we calculate the parameter w_3 in Fig. 2.21a or w_2 in Fig. 2.21b defined at the boundary part, two or one virtual cells must be considered for each situation. A boundary condition of the plate can be modeled by setting appropriate values for the virtual parameters. For example, the fixed edge condition for bending motion of a plate can be simulated by setting 0 for both virtual parameters on the boundary.

In case a vibration model is composed of two plate elements which are rigidly connected with each other, the vibration transmission through each element can be simulated by considering the relationship between the virtual parameters defined at the boundary belonging to each plate. Then, the relationship of the parameters is defined based on the continuity conditions concerning the bending and the in-plane wave motion of each plate. Detailed procedure is described in the next section. Spatial and time differential terms in the governing equations described above are approximated by finite difference, and time development of the bending/in-plane wave is calculated by an implicit method. The procedure of the approximation for the basic equations are shown herein. The space derivative of the parameter is approximated by central difference method, and time derivative of that is approximated by following one-sided differencing approximation:

$$\frac{\partial^2 w^{n+1}}{\partial x^2} = \frac{2w^{n+1} - 5w^n + 4w^{n-1} - w^{n-2}}{\Delta t^2} + O\left(\Delta t^2\right), \tag{2.111}$$

where n indicates the time step. As a result of the approximation, a discretized equation is obtained. Transforming the equations, updating formula for the out-of-plane displacement is described as follows. In this equation, the parameters at the time step of $n+1$ in the left side are unknown, and those at the time steps of n, $n-1$ and $n-2$ on the right side are already known parameters.

$$\begin{aligned}(2+\mu\Delta t) + \left(1+\frac{\xi}{\Delta t}\right)\Delta t^2 \frac{D}{\rho h} A^{n+1} &= \frac{q^n_{i,j}}{\rho h}\Delta t^2 + (5+\mu\Delta t)\, w^n_{i,j} \\ &\quad -4w^{n-1}_{i,j} + w^{n-2}_{i,j} + \frac{\xi D}{\rho h}\Delta t A^n,\end{aligned} \tag{2.112}$$

where

$$\begin{aligned}A^n = \Bigg(& \frac{w^n_{i+2,j} - 4w^n_{i+1,j} + 6w^n_{i,j} - 4w^n_{i-1,j} + w^n_{i-2,j}}{\Delta x^4} \\ & + \frac{w^n_{i,j+2} - 4w^n_{i,j+1} + 6w^n_{i,j} - 4w^n_{i,j-1} + w^n_{i,j-2}}{\Delta y^4} \\ & + 2\,\frac{\left(w^n_{i+1,j+1} - 2w^n_{i+1,j} + w^n_{i+1,j-1}\right) - 2\left(w^n_{i,j+1} - 2w^n_{i,j} + w^n_{i,j-1}\right)}{\Delta x^2 \Delta y^2} \\ & + \frac{\left(w^n_{i-1,j+1} - 2w^n_{i-1,j} + w^n_{i-1,j-1}\right)}{\Delta x^2 \Delta y^2}\Bigg).\end{aligned} \tag{2.113}$$

Following the same procedure, the discretized equations of the quasi-longitudinal wave for the x and y directions derived from Eqs. (2.107) and (2.108) are described as

$$
\begin{aligned}
&-2u_{i,j}^{n+1}+\frac{E\Delta t^2}{\left(1-\gamma^2\right)\rho}\left(\frac{u_{i+1,j}^{n+1}-2u_{i,j}^{n+1}+u_{i-1,j}^{n+1}}{\Delta x^2}\right.\\
&\qquad\left.+\gamma\frac{\left(v_{i+1,j+1}^{n+1}-v_{i+1,j-1}^{n+1}\right)-\left(v_{i-1,j+1}^{n+1}-v_{i-1,j-1}^{n+1}\right)}{2\Delta x\cdot 2\Delta y}\right)\\
&\qquad+\frac{G\Delta t^2}{\rho}\left(\frac{u_{i,j+1}^{n+1}-2u_{i,j}^{n+1}+u_{i,j-1}^{n+1}}{\Delta y^2}\right.\\
&\qquad\left.+\frac{\left(v_{i+1,j+1}^{n+1}-v_{i+1,j-1}^{n+1}\right)-\left(v_{i-1,j+1}^{n+1}-v_{i-1,j-1}^{n+1}\right)}{2\Delta x\cdot 2\Delta y}\right)\\
&\qquad=-5u_{i,j}^{n}+4u_{i,j}^{n-1}-u_{i,j}^{n-2},
\end{aligned}
\tag{2.114}
$$

$$
\begin{aligned}
&-2v_{i,j}^{n+1}+\frac{E\Delta t^2}{\left(1-\gamma^2\right)\rho}\left(\frac{v_{i,j+1}^{n+1}-2v_{i,j}^{n+1}+v_{i,j-1}^{n+1}}{\Delta x^2}\right.\\
&\left.+\gamma\frac{\left(u_{i+1,j+1}^{n+1}-u_{i+1,j-1}^{n+1}\right)-\left(u_{i-1,j+1}^{n+1}-u_{i-1,j-1}^{n+1}\right)}{2\Delta x\cdot 2\Delta y}\right)\\
&+\frac{G\Delta t^2}{\rho}\left(\frac{v_{i+1,j}^{n+1}-2v_{i,j}^{n+1}+v_{i-1,j}^{n+1}}{\Delta x^2}\right.\\
&\left.+\frac{\left(u_{i+1,j+1}^{n+1}-u_{i+1,j-1}^{n+1}\right)-\left(v_{i-1,j+1}^{n+1}-v_{i-1,j-1}^{n+1}\right)}{2\Delta x\cdot 2\Delta y}\right)\\
&=-5v_{i,j}^{n}+4v_{i,j}^{n-1}-v_{i,j}^{n-2},
\end{aligned}
\tag{2.115}
$$

where i and j indicatc thc discrete grid number in the x and y direction. To protect symmetry, discretization for the first-order differentiation in Eqs. (2.114) and (2.115) are performed based on the following equation:

$$
\frac{\partial w_i}{\partial x}=\frac{w_{i+1}-w_{i-1}}{2\Delta x}+O\left(4\Delta x^2\right). \tag{2.116}
$$

Finally, the obtained discrete equations and the continuity conditions for the junctions between elements are solved as simultaneous equations. Details of the procedure are described in the following section.

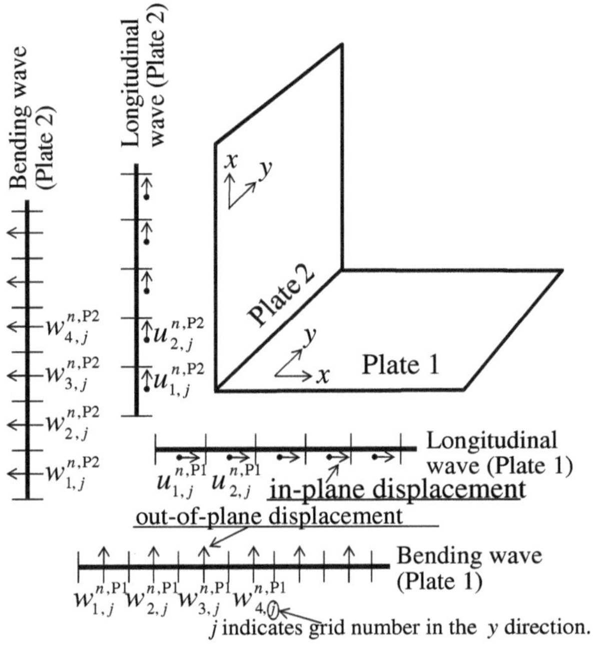

Fig. 2.22 Schematic figure and one-dimensional discrete form of a plate model

2.3.2.3 Composite Model with Multiple Plate Elements

In order to simulate a vibration model in which multiple plates are rigidly connected to each other, continuity conditions for vibration propagation at the joint part should be considered. In this section, the connecting method is described.

In Fig. 2.22, an axonometric view of the target plate model and a schematic figure, in which the three-dimensional model is illustrated in a one-dimensional discrete form, are shown. In the figure of the one-dimensional discrete form, $w_{1,j}^{n,P1}$, $w_{2,j}^{n,P1}$, $w_{3,j}^{n,P1}$ and $w_{4,j}^{n,P1}$ indicate the out-of-plane displacements in the time step of n, caused by bending deformation of Plate 1, and they are defined nearby the junction part. $w_{1,j}^{n,P2}$, $w_{2,j}^{n,P2}$, $w_{3,j}^{n,P2}$ and $w_{4,j}^{n,P2}$ indicate the out-of-plane displacements of Plate 2. $u_{1,j}^{n,P1}$, $u_{2,j}^{n,P1}$ and $u_{1,j}^{n,P2}$and $u_{2,j}^{n,P2}$ indicate the in-plane displacements caused by in-plane deformation of Plate 1 and Plate 2, respectively. The parameter j attached in the subscript of each parameter indicates the grid number in the $y-$direction. In the simulation, continuity conditions described below are considered. First, the physical parameters of the displacement and the rotation angle at the boundary of Plate 1 are made to be equal to those of Plate 2. Second, the bending moments acting at the boundary of both Plate 1 and 2 are balanced. Lastly, the shear force acting at the boundary of Plate 1 and the in-plane force acting at the boundary of Plate 2 are also balanced. Totally, four conditions are considered. Here, the rotational angle

θ_{b}, bending moment M_x, shear force F_x, and in-plane force T_x are described by the following equations.

$$\theta_{\mathrm{b}} = \frac{\partial w}{\partial x}, \tag{2.117}$$

$$M_x = -D\left(\frac{\partial^2 w}{\partial x^2} + \gamma\frac{\partial^2 w}{\partial y^2}\right), \tag{2.118}$$

$$F_x = -D\left(\frac{\partial^3 w}{\partial x^3} + (2-\gamma)\frac{\partial^3 w}{\partial x \partial y^2}\right), \tag{2.119}$$

$$T_x = \frac{Eh}{1-\gamma^2}\left(\frac{\partial u}{\partial x} + \gamma\frac{\partial v}{\partial y}\right). \tag{2.120}$$

These equations are transformed to discrete forms, and are rewritten by considering the continuity conditions.

1. Displacement:

$$\frac{3}{2}w_{2,j}^{n,\mathrm{P1}} - \frac{1}{2}w_{1,j}^{n,\mathrm{P1}} = \frac{u_{1,j}^{n,\mathrm{P2}} + u_{2,j}^{n,\mathrm{P2}}}{2}, \tag{2.121}$$

$$\frac{3}{2}w_{2,j}^{n,\mathrm{P2}} - \frac{1}{2}w_{1,j}^{n,\mathrm{P2}} = -\frac{u_{1,j}^{n,\mathrm{P1}} + u_{2,j}^{n,\mathrm{P1}}}{2}. \tag{2.122}$$

2. Rotational angle:

$$\frac{w_{2,j}^{n,\mathrm{P2}} - w_{1,j}^{n,\mathrm{P2}}}{\Delta x_{\mathrm{P2}}} = \frac{w_{2,j}^{n,\mathrm{P1}} - w_{1,j}^{n,\mathrm{P1}}}{\Delta x_{\mathrm{P1}}}. \tag{2.123}$$

3. Bending moment:

$$\frac{w_{3,j}^{n,\mathrm{P2}} - 2w_{2,j}^{n,\mathrm{P2}} + w_{1,j}^{n,\mathrm{P2}}}{\Delta x_{\mathrm{P2}}^2} + \gamma\frac{w_{2,j+1}^{n,\mathrm{P2}} - 2w_{2,j}^{n,\mathrm{P2}} + w_{2,j-1}^{n,\mathrm{P2}}}{\Delta y_{\mathrm{P2}}^2}$$
$$+\frac{w_{3,j}^{n,\mathrm{P1}} - 2w_{2,j}^{n,\mathrm{P1}} + w_{1,j}^{n,\mathrm{P1}}}{\Delta x_{\mathrm{P1}}^2} + \gamma\frac{w_{2,j+1}^{n,\mathrm{P1}} - 2w_{2,j}^{n,\mathrm{P1}} + w_{2,j-1}^{n,\mathrm{P1}}}{\Delta y_{\mathrm{P1}}^2} = 0. \tag{2.124}$$

4. Shear force and in-plane force:

$$-D\left(\frac{w_{4,j}^{n,\mathrm{P2}}-3w_{3,j}^{n,\mathrm{P2}}+3w_{2,j}^{n,\mathrm{P2}}-w_{1,j}^{n,\mathrm{P2}}}{\Delta x_{\mathrm{P2}}^{3}}\right.$$
$$\left.+(2-\gamma)\frac{\left(w_{3,j+1}^{n,\mathrm{P2}}-2w_{3,j}^{n,\mathrm{P2}}+w_{3,j-1}^{n,\mathrm{P2}}\right)-\left(w_{2,j+1}^{n,\mathrm{P2}}-2w_{2,j}^{n,\mathrm{P2}}+w_{2,j-1}^{n,\mathrm{P2}}\right)}{\Delta x_{\mathrm{P2}}\Delta y_{\mathrm{P2}}^{2}}\right)$$
$$-\frac{Eh}{1-\gamma^2}\left(\frac{u_{2,j}^{n,\mathrm{P1}}-u_{1,j}^{n,\mathrm{P1}}}{\Delta x_{\mathrm{P1}}}+\gamma\frac{u_{2,j+1}^{n,\mathrm{P1}}-u_{2,j}^{n,\mathrm{P1}}}{\Delta y_{\mathrm{P1}}}\right)=0, \tag{2.125}$$

$$-D\left(\frac{w_{4,j}^{n,\mathrm{P1}}-3w_{3,j}^{n,\mathrm{P1}}+3w_{2,j}^{n,\mathrm{P1}}-w_{1,j}^{n,\mathrm{P1}}}{\Delta x_{\mathrm{P1}}^{3}}\right.$$
$$\left.+(2-\gamma)\frac{\left(w_{3,j+1}^{n,\mathrm{P1}}-2w_{3,j}^{n,\mathrm{P1}}+w_{3,j-1}^{n,\mathrm{P1}}\right)-\left(w_{2,j+1}^{n,\mathrm{P1}}-2w_{2,j}^{n,\mathrm{P1}}+w_{2,j-1}^{n,\mathrm{P1}}\right)}{\Delta x_{\mathrm{P1}}\Delta y_{\mathrm{P1}}^{2}}\right)$$
$$+\frac{Eh}{1-\gamma^2}\left(\frac{u_{2,j}^{n,\mathrm{P2}}-u_{1,j}^{n,\mathrm{P2}}}{\Delta x_{\mathrm{P2}}}+\gamma\frac{u_{2,j+1}^{n,\mathrm{P2}}-u_{2,j}^{n,\mathrm{P2}}}{\Delta y_{\mathrm{P2}}}\right)=0. \tag{2.126}$$

Here, Pm shown in each superscript indicates Plate m, the spatial intervals Δx_{P1} and Δy_{P1} indicate the spatial intervals of Plate 1, and Δx_{P2} and Δy_{P2} indicate those of Plate 2. In these equations, each parameter indicates the in-plane or out-of-plane displacement shown in Fig. 2.22. Solutions of the six unknown parameters ($w_{1,j}^{n,P1}$,$w_{2,j}^{n,P1}$,$w_{1,j}^{n,P2}$,$w_{2,j}^{n,P2}$,$u_{1,j}^{n,P1}$ and $u_{1,j}^{n,P2}$) at the junction are solved by considering the six equations above. Then, these continuity equations and the discrete forms of Eqs. (2.112), (2.114), and (2.115) are solved together as simultaneous equations. These equations are solved in every time step, and time development of vibration is simulated. To solve the simultaneous equations, PARDISO [36], built in the Intel Math Kernel library, was used.

2.3.2.4 Vibroacoustic Coupling Method

To couple the vibration field on the plate and the sound field as shown in Fig. 2.23a, the following continuity conditions should be considered. First, the external pressure to the plate described in Eq. (2.112) is given as the difference of the sound pressure on both sides of the plate; $p_{i,j,k}^{n}$ and $p_{i-1,j,k}^{n}$, as shown in the Step 1 of Fig. 2.23b.

$$q_{j,k}^{n}=p_{i,j,k}^{n}-p_{i-1,j,k}^{n}. \tag{2.127}$$

In the next step, the bending motion of the plate is calculated based on the external pressure, and the displacement of the plate is obtained. Then, the velocity is calculated using the displacement and is given to the particle velocity in the sound field.

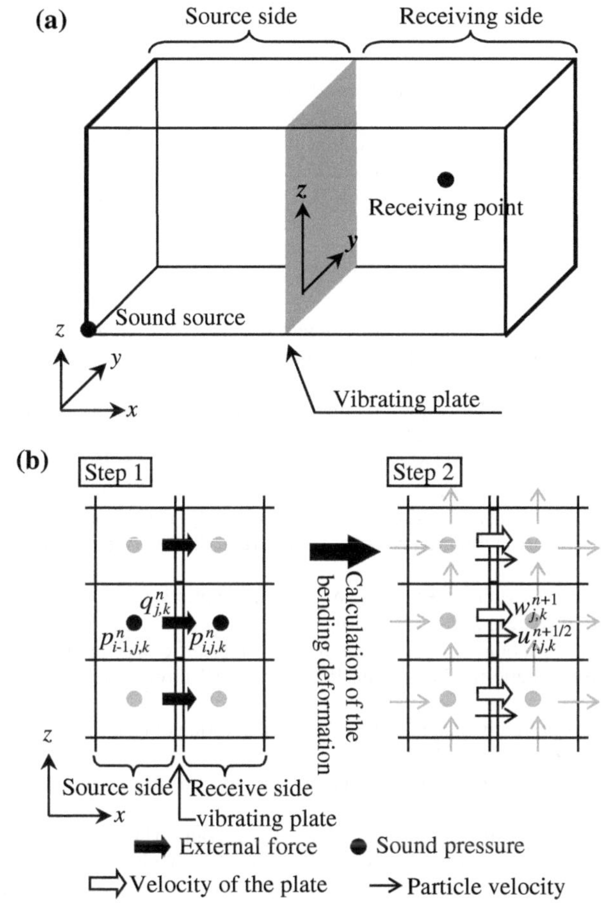

Fig. 2.23 Coupling method between the vibration and the sound

$$u_{i,j,k}^{n+1/2} = \frac{w_{j,k}^{n+1} - w_{j,k}^{n}}{\Delta t}, \tag{2.128}$$

where $w_{j,k}^{n}$ is the displacement of the plate. In case the normal impedance $Z_{\rm n}$ on the receiving side of the plate is considered, following relationship is used instead of Eq. (2.128).

$$u_{i,j,k}^{n+1/2} = \frac{w_{j,k}^{n+1} - w_{j,k}^{n}}{\Delta t} - \frac{p_{i,j,k}^{n}}{Z_{\rm n}}. \tag{2.129}$$

To update the sound pressure and particle velocity in the sound field, high-order scheme using eight reference points is applied. Detailed method is described in the Ref. [37].

References

1. K.S. Yee, Numerical solution of initial boundary value problems involving Maxwell's equations in isotropic media. IEEE Trans. Antennas Propag. **17**, 585–589 (1966)
2. T. Yokota, S. Sakamoto, H. Tachibana, Visualization of sound propagation and scattering in rooms. Acoust. Sci. Tech. **23**(1), 40–46 (2002)
3. S. Sakamoto, H. Nagatomo, A. Ushiyama, H. Tachibana, Calculation of impulse responses and acoustic parameters in a hall by the finite-difference time-domain method. Acoust. Sci. Tech. **29**(4), 256–265 (2008)
4. D.M. Sullivan, A frequency-dependent FDTD method using Z transforms. IEEE Trans. Antennas Propag. **40**, 1223–1230 (1992)
5. J. Escolano, F. Jacobsen, J. Lopez, An efficient realization of frequency dependent boundary conditions in an acoustic finite-difference time-domain model. J. Sound Vib. **316**, 234–247 (2008)
6. H. Suzuki, A. Omoto, K. Fujiwara, Treatment of boundary conditions by finite difference time domain method. Acoust. Sci. Tech. **28**(1), 16–26 (2007)
7. T. Deveze, L. Beaulieu, W. Tabbara, A forth order scheme for the FDTD algorithm applied to Maxwell's equations. IEEE Trans. Antennas Propag. **42**, 859–862 (1994)
8. M.F. Hadi, M. Piket-May, S.L. Manry Jr., A modified FDTD (2,4) scheme for modeling electrically large structures with high phase accuracy. IEEE Trans. Antennas Propag. **45**, 254–264 (1997)
9. J. Von Neumann, R. Richtmyer, A method for the numerical calculation of hydrodynamic shock. J. Appl. Phys. **21**, 232–237 (1950)
10. G. Mur, Absorbing boundary conditions for the finite difference approximation of the time domain electromagnetic-field equation. IEEE Trans. Electromagnetic Compat. **EMC-23**(4), 377–382 (1981)
11. J.-P. Berenger, A perfectly matched layer for the absorption of electromagnetic waves. J. Comput. Phys. **114**, 185–200 (1994)
12. S.K. Lele, Compact finite difference scheme with spectral-like resolution. J. Comput. Phys. **103**, 16–42 (1992)
13. R.D. Ruth, A canonical integration technique. IEEE. Trans. Mag. **37**, 3251–3254 (2001)
14. B.A. Auld, *Acoustic Fields and Waves in Solids*, vol. 1, 2nd edn. (Krieger, Malabar, 1990)
15. L. Cremer, M. Heckl, B.A.T. Petersson, *Structure-Borne Sound*, 3rd edn. (Springer, Berlin, 2005)
16. M. Sato, Fundamental investigation of shear and surface elastic waves in soft biological tissues by numerical simulation. Jpn. J. Appl. Phys. 1 **34**(5B), 2808–2811 (1995)
17. F. Iijima, T. Tsuchiya, N. Endoh, Analysis of characteristics of underwater sound propagation in the ocean by a finite difference time domain method. Jpn. J. Appl. Phys. 1 **39**(5B), 3200–3204 (2000)
18. H. Suzuki, A. Omoto, K. Fujiwara, Treatment of boundary conditions by finite difference time domain method. Acoust. Sci. Tech. **28**, 16–26 (2007)
19. M. Toyoda, D. Takahashi, Prediction for architectural structure-borne sound by the finite-difference time-domain method. Acoust. Sci. Tech. **30**, 265–276 (2009)
20. M. Toyoda, H. Miyazaki, Y. Shiba, A. Tanaka, D. Takahashi, Finite-difference time-domain method for heterogeneous orthotropic media with damping. Acoust. Sci. Tech. **33**, 77–85 (2012)
21. L. Rayleigh, *The Theory of Sound*, vol. 2. (Cambridge University Press, Cambridge, 1877)
22. R.B. Bird, W.E. Stewart, E.N. Lightfoot. *Transport Phenomena*, 2nd edn. (Wiley, New York, 2007)
23. L.N. Liebermann, The second viscosity of liquids. Phys. Rev. **75**, 1415–1422 (1949)
24. H. Jiang, H. Arai, Analysis of computation error in antenna's simulation by using non-uniform mesh fdtd. IEICE Trans. Commun. **E83-B**, 1544–1553 (2000)
25. S. Xiao, R. Vahldieck, An improved 2D-FDTD algorithm for hybrid mode analysis of quasi-planar transmission lines, in *Proceeding of IEEE MTT-S International*, (1993). pp. 421–424

26. H. Tsuru, R. Iwatsu, Accurate numerical prediction of acoustic wave propagation. Int. J. Adapt. Control Signal Process. **24**, 128–141 (2010)
27. T. Asakura, S. Sakamoto, Finite-difference time-domain analysis on leak transmission characteristics of narrow gaps. Acoust. Sci. Tech. **32**, 182–193 (2011)
28. H.W. Chang, C.J. Randall, Finite-difference timedomain modeling of elastic wave propagation in the cylindrical coordinate system, in *Proceeding of IEEE Ultrasonics Symposium*, (1988). pp. 397–402
29. C.J. Randall, D.J. Scheibner, P.T. Wu, Multipole borehole acoustic waveforms: synthetic logs with beds and borehole washouts. Geophysics **56**, 1757–1769 (1991)
30. R.W. Graves, Simulating seismic wave propagation in 3D elastic media using staggered-grid finite differences. Bull. Seismol. Soc. Am. **86**, 1091–1106 (1996)
31. Q.H. Liu, E. Schoen, F. Daube, C. Randall, H.L. Liu, P. Lee, A three-dimensional finite difference simulation of sonic logging. J. Acoust. Soc. Am. **100**, 72–79 (1996)
32. R. Vossen, J.O.A. Robertsson, C.H. Chapman, Finite-difference modeling of wave propagation in a fluid-solid configuration. Geophysics **67**, 618–624 (2002)
33. M. Toyoda, D. Takahashi, Y. Kawai, Averaged material parameters and boundary conditions for the vibroacoustic finite-difference time-domain method with a nonuniform mesh. Acoust. Sci. Tech. **33**, 273–276 (2012)
34. W.C. Chew, Q.H. Liu, Perfectly matched layers for elastodynamics: a new absorbing boundary condition. J. Comput. Acoust. **4**, 341–359 (1996)
35. T. Uno, *Finite Difference Time Domain Method for Electromagnetic Field and Antennas* (Corona Publishing Co. Ltd, San Antonio, 1998). (in Japanese)
36. O. Schenk, K. Gartner, Solving unsymmetric sparse systems of linear equations with pardiso. J. Future Gen. Comp. Sys. **20**, 475–487 (2004)
37. S. Sakamoto, Phase-error analysis of high-order finite difference time domain scheme and its influence on calculation results of impulse response in closed sound field. Acoust. Sci. Tech. **28**(4), 295–309 (2007)

Chapter 3
Finite Element Method

Toru Otsuru, Takeshi Okuzono, Noriko Okamoto and Yusuke Naka

Abstract Finite element method (FEM) is widely used in various engineering fields to solve problems with too many complexities to be dealt with by certain conventional approaches. In 1943, Courant proposed the theoretical basis of the method, and, in 1956, Turner et al. published both fundamental theory and application of FEM, namely "Stiffness deflection analysis", to solve structural problems [1, 2]. A decade and several years latter, FEM has been come to be applied to solve various acoustic problems [3–5]. Compared with other numerical techniques, FEM is advantageous in its broad range of applicability. However, FEM requires discretization of the domain, which results in huge amount of degrees-of-freedom especially when a three-dimensional domain is analyzed. Nevertheless, the matrices constructed in a standard FEM procedure have rather simple mathematical structures with sparseness. The simplicity makes their computation more efficient especially when they are processed on a parallel/vector processors. Generally speaking, the iterative methods are suitable for solving such a sparse matrix equation efficiently with far less memory space on a computer. In this chapter, fundamentals, improvements, and applications of FEM on acoustic problems are explored.

T. Otsuru (✉)
Department of Architecture and Mechatronics, Architecture Course, Faculty of Engineering, Oita University, 700 Dannoharu, Oita-shi, Oita 870–1192, Japan
e-mail: otsuru@oita-u.ac.jp

T. Okuzono
Department of Architecture, Graduate School of Engineering, Kobe University, 1-1 Rokkodai-Machi, Nada-ku, Kobe 657–8501, Japan
e-mail: okuzono@port.kobe-u.ac.jp

N. Okamoto
Department of Architecture, Ariake National College of Technology, 150 Higashihagio-Machi, Omuta, Fukuoka 836–8585, Japan
e-mail: okamoto@ariake-nct.ac.jp

Y. Naka
Institute of Aeronautical Technology, Japan Aerospace Exploration Agency, 6-13-1 Osawa, Mitaka-shi, Tokyo 181-0015, Japan
e-mail: naka.yusuke@jaxa.jp

T. Sakuma et al. (eds.), *Computational Simulation in Architectural and Environmental Acoustics*, DOI: 10.1007/978-4-431-54454-8_3,

Keywords Finite element method (FEM) · Acoustic finite elements · Discretization error · Large-scale analysis · Iterative method · Domain truncation · Dirichlet-to-Neumann (DtN) map

3.1 Fundamentals

The fundamental idea of FEM is to approximate a potential in a system with complicated shapes and/or boundary conditions by dividing the system into discretized elements that are easier to deal with on a digital computer. Then, a potential at an arbitrary point in an element is interpolated by using the potentials on discretized points (nodes) multiplied by a interpolation function. Since the function is also utilized to model the shape of the element's boundary, it is frequently called "shape function". In the following subsection, an outline of FEM formulation of three-dimensional sound field is explained.

3.1.1 Sound Field Formulation by FEM

In a three-dimensional sound field Ω with sound pressure distribution $p(x, y, z)$ as is illustrated in Fig. 3.1, kinetic energy T and potential energy U of sound at an angular frequency ω are in the forms of:

$$T = \frac{1}{2}\frac{1}{\rho\omega^2}\int_{\Omega}(\nabla p \cdot \nabla p)dV, \tag{3.1}$$

$$U = \frac{1}{2}\frac{1}{\rho c^2}\int_{\Omega} p^2 dV. \tag{3.2}$$

In the equations, ρ and c, respectively, denote air density and speed of sound. Furthermore, the work W done by an external force at a surface area Γ is obtained by

$$W = \int_{\Gamma} u_n p dS, \tag{3.3}$$

where u_n denotes normal displacement at Γ.

Then, total energy Π in the system becomes

$$\Pi = U - T - W, \tag{3.4}$$

and, based on the principle of minimum potential energy, i.e. $\delta\Pi = 0$, one can derive the sound field. Note that it is also possible to derive the same result as a weak form solution by applying Galerkin's method onto the wave equation.

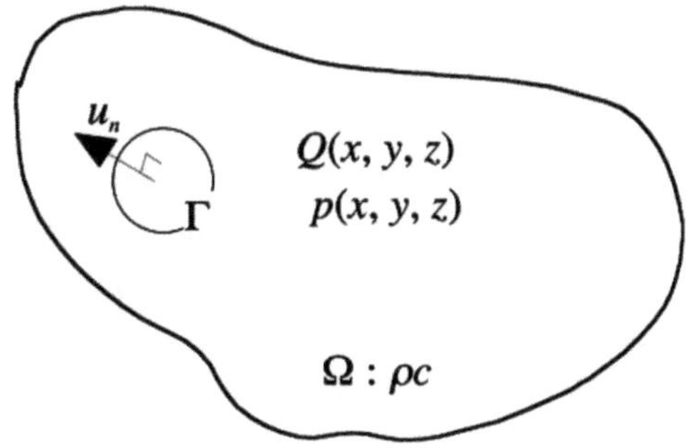

Fig. 3.1 Sound field Ω

At this point, let us discretize the system by FEM: following the standard FEM procedure, Ω is divided into a number of elements and sound pressure $p(x, y, z)$ at an arbitrary point $Q(x, y, z)$ in an element e can be approximated as

$$p(x, y, z) = \boldsymbol{N}(x, y, z)^T \boldsymbol{p}_e. \tag{3.5}$$

Here, $\boldsymbol{p}_e$ and $\boldsymbol{N}(x, y, z)$ are element nodal sound pressure vector and interpolation (shape) function vector, respectively. With the equation, we can rewrite Eqs. (3.1), (3.2), and (3.3) as

$$T = \sum_e \left(\frac{1}{2} \frac{1}{\rho\omega^2} \int_e \left\{ \left(\frac{\partial \boldsymbol{N}^T \boldsymbol{p}_e}{\partial x} \right)^2 + \left(\frac{\partial \boldsymbol{N}^T \boldsymbol{p}_e}{\partial y} \right)^2 + \left(\frac{\partial \boldsymbol{N}^T \boldsymbol{p}_e}{\partial z} \right)^2 \right\} dV \right) \tag{3.6}$$

$$= \sum_e \left(\boldsymbol{p}_e^T \frac{1}{2} \frac{1}{\rho\omega^2} \int_e \left[\boldsymbol{N}_{,x} \boldsymbol{N}_{,x}{}^T + \boldsymbol{N}_{,y} \boldsymbol{N}_{,y}{}^T + \boldsymbol{N}_{,z} \boldsymbol{N}_{,z}{}^T \right] dV \boldsymbol{p}_e \right),$$

$$U = \sum_e \left(\boldsymbol{p}_e^T \frac{1}{2} \frac{1}{\rho c^2} \int_e \left[\boldsymbol{N} \boldsymbol{N}^T \right] dV \boldsymbol{p}_e \right), \tag{3.7}$$

$$W = \sum_e \left(\boldsymbol{p}_e^T u_n \int_{\Gamma_e} \boldsymbol{N} dS \right). \tag{3.8}$$

In the equations, $\boldsymbol{N}_{,x} = \partial \boldsymbol{N}/\partial x$, $\boldsymbol{N}_{,y} = \partial \boldsymbol{N}/\partial y$, $\boldsymbol{N}_{,z} = \partial \boldsymbol{N}/\partial z$ stand. Then, the following discretized equation can be derived for each element e,

:

$$\Pi_e = \frac{1}{2\rho\omega^2} \boldsymbol{p}_e^T \boldsymbol{K}_e \boldsymbol{p}_e - \frac{1}{2\rho} \boldsymbol{p}_e^T \boldsymbol{M}_e \boldsymbol{p}_e - \boldsymbol{p}_e^T u_n \boldsymbol{W}_e. \tag{3.9}$$

Herein, element matrices, $\boldsymbol{K}_e$, $\boldsymbol{M}_e$ and $\boldsymbol{W}_e$, are respectively defined by

$$\boldsymbol{K}_e = \int_e \left[\boldsymbol{N}_{,x} \boldsymbol{N}_{,x}^T + \boldsymbol{N}_{,y} \boldsymbol{N}_{,y}^T + \boldsymbol{N}_{,z} \boldsymbol{N}_{,z}^T \right] dV, \tag{3.10}$$

$$\boldsymbol{M}_e = \frac{1}{c^2}\int_e \left[\boldsymbol{N}\boldsymbol{N}^T\right] dV, \tag{3.11}$$

$$\boldsymbol{W}_e = \int_{\Gamma_e} \boldsymbol{N} dS. \tag{3.12}$$

The total energy Π in Ω equals the summation of energies in all the elements. Therefore, by performing $\delta\Pi = 0$, one can obtain the discretized equation of motion with global matrices $\boldsymbol{K}, \boldsymbol{M}$ and $\boldsymbol{W}$ as

$$(\boldsymbol{K} - \omega^2\boldsymbol{M})\boldsymbol{p} = \rho\omega^2 u_n \boldsymbol{W}. \tag{3.13}$$

When Ω contains a dissipative wall Γ' which has normal impedance Z_n, one can derive Eq. (3.14) for the dissipative system through the same procedure as above with the help of adjoint system [6].

$$(\boldsymbol{K} + j\omega\boldsymbol{C} - \omega^2\boldsymbol{M})\mathbf{p} = \rho\omega^2 u_n \boldsymbol{W}. \tag{3.14}$$

Here, $\boldsymbol{C}$ denotes global dissipation matrix which is constructed using all the element dissipation matrices $\boldsymbol{C}_e$ in Ω. The matrix $\boldsymbol{C}_e$ for the element with Γ' is given by

$$\boldsymbol{C}_e = \frac{1}{c}\int_{\Gamma'} \frac{1}{Z_n}\left[\boldsymbol{N}\boldsymbol{N}^T\right] dS. \tag{3.15}$$

One can solve the sound field by applying the modal analysis onto Eqs. (3.13) and (3.14), whereas the direct solution is also possible and efficient when the dissipation is not very small. For practical cases on room acoustics, good agreements of both the techniques are confirmed [7].

On the other hand, the semi-discrete equation in the time-domain can be obtained with inverse Fourier transform of Eq. (3.14) as

$$\boldsymbol{M}\ddot{\boldsymbol{p}} + \boldsymbol{C}\dot{\boldsymbol{p}} + \boldsymbol{K}\boldsymbol{p} = \rho\dot{v}_n\boldsymbol{W}(=\boldsymbol{f}), \tag{3.16}$$

where $\cdot$, $\cdot\cdot$, v_n, and $\boldsymbol{f}$, respectively, signify first-order and second-order derivatives with respect to time, particle velocity, and external force vector. Applying a direct time integration method such as Newmark β method [8] to Eq. (3.16), sound pressure $\boldsymbol{p}$ at all nodes in time domain is calculable. In the Newmark β method, if $\boldsymbol{p}, \dot{\boldsymbol{p}}$ and $\ddot{\boldsymbol{p}}$ at time t are known, then $\boldsymbol{p}_{t+\Delta t}$ and $\dot{\boldsymbol{p}}_{t+\Delta t}$ at time $t + \Delta t$ can be approximated as

$$\boldsymbol{p}_{t+\Delta t} = \boldsymbol{p}_t + \Delta t\dot{\boldsymbol{p}}_t + (\Delta t^2)(\frac{1}{2} - \beta)\ddot{\boldsymbol{p}}_t + (\Delta t^2)\beta\ddot{\boldsymbol{p}}_{t+\Delta t}, \tag{3.17}$$

$$\dot{\boldsymbol{p}}_{t+\Delta t} = \dot{\boldsymbol{p}}_t + \Delta t(1-\gamma)\ddot{\boldsymbol{p}}_t + \Delta t\gamma\ddot{\boldsymbol{p}}_{t+\Delta t}, \tag{3.18}$$

where Δt is time interval between t and $t + \Delta t$, and γ, β are parameters related to the accuracy and stability of the method, in which γ is commonly set to 1/2 to maintain the second-order accuracy. Substitution of Eqs. (3.17) and (3.18) into the semi-discrete equation at $t + \Delta t$ yields

$$\left[\boldsymbol{M} + \frac{\Delta t}{2} \boldsymbol{C} + \beta(\Delta t^2) \boldsymbol{K} \right] \ddot{\boldsymbol{p}}_{t+\Delta t} = \boldsymbol{f}_{t+\Delta t} - \boldsymbol{CP} - \boldsymbol{KQ}, \tag{3.19}$$

with

$$\boldsymbol{P} = \dot{\boldsymbol{p}}_t + \frac{\Delta t}{2} \ddot{\boldsymbol{p}}_t, \boldsymbol{Q} = \boldsymbol{p}_t + \Delta t \dot{\boldsymbol{p}}_t + \left(\frac{1}{2} - \beta \right) (\Delta t^2) \ddot{\boldsymbol{p}}_t. \tag{3.20}$$

The linear system of equations of Eq. (3.19) can be solved using a direct method or an iterative method, to compute $\ddot{\boldsymbol{p}}_{t+\Delta t}$ at time $t + \Delta t$. Then, $\boldsymbol{p}_{t+\Delta t}$ and $\dot{\boldsymbol{p}}_{t+\Delta t}$ at time $t + \Delta t$ are calculable by substituting the $\ddot{\boldsymbol{p}}_{t+\Delta t}$ into Eqs. (3.17) and (3.18), respectively.

Several Newmark methods exist with different values of parameter β [9]. The following three special cases of the Newmark methods are well known: constant average acceleration method with $\beta = 1/4$, linear acceleration method with $\beta = 1/6$, and Fox-Goodwin method with $\beta = 1/12$. Here, if $\beta \geq 1/4$, then the Newmark method is unconditionally stable. For linear acceleration method and Fox-Goodwin method, the stability condition is given as

$$\Delta t \leq \frac{1}{\omega_{\max} \sqrt{1/4 - \beta}}. \tag{3.21}$$

Here, $\omega_{\max}$ is the maximum natural frequency of system, which is obtainable by solving generalized eigenvalue problem $(\boldsymbol{K}_e - \omega^2 \boldsymbol{M}_e)\boldsymbol{p}_e = 0$ for all elements.

3.1.2 3-D Acoustic Elements

When a three-dimensional sound field is solved by FEM, both tetrahedron and hexahedron elements are rather simple to construct finite element meshes. Hereafter, the formulation procedures of hexahedron 8-node (Fig. 3.2) and 27-node (Fig. 3.3) elements are briefly explained.

In general, Lagrange polynomials given by Eq. (3.22) are frequently used for constructing a finite element. When the shape of an element is interpolated by using the same function as is used for the approximation of the inner potential dealt with, the element is called "isoparametric element."

$$L_i(\xi) = \prod_{j'=1,(j' \neq i')}^{k_\xi} \frac{\xi - \xi_{j'}}{\xi_{i'} - \xi_{j'}}. \tag{3.22}$$

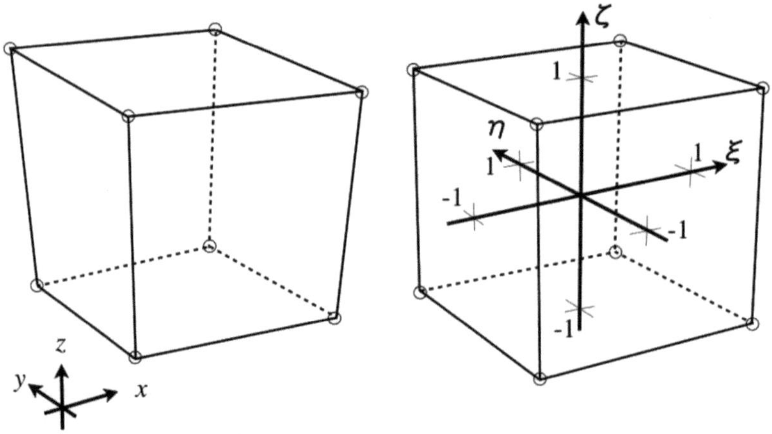

Fig. 3.2 Hexahedron 8-node element; (*left*) In the global coordinate system eight nodes are located at $P_i(x_i, y_i, z_i)$ on corners; (*right*) In the local coordinate system nodes are located at $P_i(\xi_i, \eta_i, \zeta_i)$ on corners, where $\xi_i, \eta_i, \zeta_i = 1$ or -1

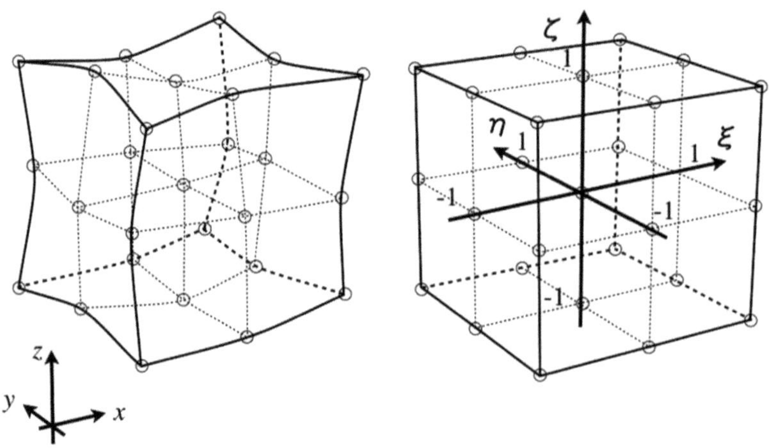

Fig. 3.3 Hexahedron 27-node element; (*left*) In the global coordinate system, (*right*) In the local coordinate system. Twenty-seven nodes are located at $P_i(\xi_i, \eta_i, \zeta_i)$, where $\xi_i, \eta_i, \zeta_i = 1$ or -1 (*edge* or *corner*), = 0 (otherwise)

Equation (3.22) stands only for a local coordinate ξ, and i denotes the global node number in ξ-direction. The corresponding local node number and its local coordinate are, respectively, written as i' and $\xi_{i'}$. The number of nodes used for the interpolation in ξ-direction is expressed as k_ξ. Likewise, one can obtain equations $L_i(\eta)$ and $L_i(\zeta)$, respectively, for η- and ζ-directions.

Thus, a shape function for a local coordinate system (ξ, η, ζ) can be defined by

$$N_i(\xi, \eta, \zeta) = L_i(\xi) L_i(\eta) L_i(\zeta). \tag{3.23}$$

One can formulate an isoparametric hexahedron 8-node element by setting $k_\xi = k_\eta = k_\zeta = 2$, or an isoparametric 27-node element by setting $k_\xi = k_\eta = k_\zeta = 3$. Therefore, a shape function vector $\boldsymbol{N}$ becomes

$$\boldsymbol{N} = \{N_1, N_2, \cdots, N_i, \cdots, N_{k_\xi k_\eta k_\zeta}\}^T. \tag{3.24}$$

Note that $\boldsymbol{N}$ acts as a "weight" function of nodal potentials in the following equation:

$$p(\xi, \eta, \zeta) = \boldsymbol{N}^T \boldsymbol{p}_e = \sum_{i=1}^{k_\xi k_\eta k_\zeta} N_i(\xi, \eta, \zeta) p_i. \tag{3.25}$$

Once $\boldsymbol{N}$ is formulated, element matrices $\boldsymbol{K}$, $\boldsymbol{M}$ and $\boldsymbol{W}$ can be calculated by Eqs. (3.10)–(3.12). In a practical calculation of the matrices, one needs to transform the coordinate system from local (ξ, η, ζ) to global (x, y, z) using Jacobian matrix, and to perform numerical integrations, e.g. Gauss-Legendre quadrature, especially when an element with some complicated shape is dealt with. The mathematics for the procedure can be found in standard textbooks on FEM [10].

The elements constructed above employs Lagrange polynomials as shape functions and are widely used in many FEM applications. On the other hand, it is also well known that the resulting accuracy of a FEM application depends strongly on the type of shape function used in the formulation of the elements. To compare the difference, another type of acoustic element with different kind of shape function is introduced below. The element is formulated by Otsuru and Tomiku [11] using natural cubic spline of degree 3 with continuity C^2 defined by the following equations:

$$(\text{if} \quad \xi_i = \pm 1) \quad S_i(\xi) = \begin{cases} 0.25\xi^3 + 0.75\xi^2 + 0.5\xi_i\xi & : \xi \in [-1, 0] \\ -0.25\xi^3 + 0.75\xi^2 + 0.5\xi_i\xi & : \xi \in [\ 0, 1] \end{cases} \tag{3.26}$$

$$(\text{if} \quad \xi_i = \ 0) \quad S_i(\xi) = \begin{cases} -0.5\xi^3 - 1.5\xi^2 + 1 & : \xi \in [-1, 0] \\ 0.5\xi^3 - 1.5\xi^2 + 1 & : \xi \in [\ 0, 1] \end{cases} \tag{3.27}$$

The shape function N_i obtained by Eq. (3.23) can be rewritten as

$$N_i(\xi, \eta, \zeta) = S_i(\xi) S_i(\eta) S_i(\zeta), \tag{3.28}$$

and related element matrices can also be obtained through the above-mentioned procedure. Figure 3.4 shows a comparison of the forms of the two shape functions and the difference might not be very obvious.

To examine the resulting differences more clearly, especially when the shape functions are practically applied, the relative errors ε in eigenfrequency computation of a one-dimensional sound field are computed to compare each other using three types of elements with different kinds of shape functions: hexahedron 8- and 27-node elements with Lagrange polynomial functions (Lin8, Lag27) and hexahedron 27-node element with spline polynomial function (Spl27).

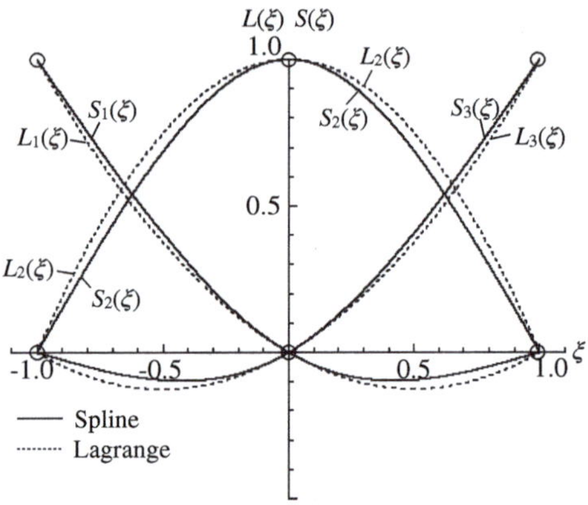

Fig. 3.4 Comparison of shape functions $L(\xi)$ and $S(\xi)$. $k_\xi = 3$; $\xi_1 = -1, \xi_2 = 0, \xi_3 = 1$

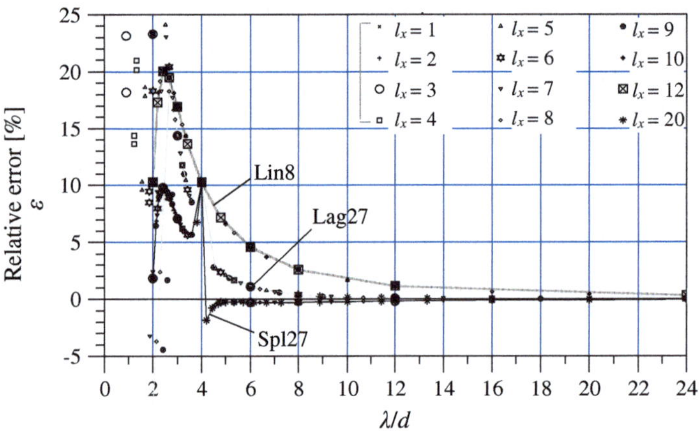

Fig. 3.5 Comparison of relative errors (ε) in eigenfrequency computation between element types, and between spatial resolutions. Hexahedron 8- and 27 node elements with Lagrange polynomial functions (Lin8, Lag27) and hexahedron 27-node element with spline polynomial function (Spl27) are compared. Spatial resolution is defined by wavelength (λ)/nodal-distance(d). The parameter l_x denotes the number of elements in x-direction ($l_x = 1 \sim 20$) [11]

Figure 3.5 shows that the relative error at the spatial resolution of "$\lambda/d \leq 4$" becomes large and unstable regardless of element types. The relative error decreases as the increase of the number of element's node from 8 to 27, and the values of Spl27 are smaller and stabler than those of Lin27 in the region of "$\lambda/d > 4$".

In practical application stages of FEM, one needs to choose the solver technique as well as the element type considering the balance of the efficiency and resulting accuracy so that the results satisfy one's primary purposes. Although spatial resolution λ/d gives a rough but practical estimation, note that the resulting accuracy clearly depends on the element type employed.

3.2 Efficient Computation with Iterative Solvers

The iterative methods, which are a technique to obtain a solution by converging the value that results from iterative calculations made with an arbitrary initial value given at the beginning, have been used recently in various fields to solve large systems of linear equations that result from the large-scale analysis model with complex geometrical structures because the method might accomplish rapid computation and reduce the required memory size. Numerical sound fields analyses using FEM have the same aspects, and therefore it can also be expected by applying the method. This section presents numerical aspects of the linear system of equations in finite-element sound field analysis in both frequency and time domains and actual convergence behaviors of applied iterative methods. For the details of iterative methods, see Refs. [12–14].

3.2.1 Linear System of Equations and Solvers

3.2.1.1 Numerical Aspects of The Linear System of Equations

As presented in the previous section, the discretized matrix equation in frequency domain for sound field with a sound source of an angular frequency ω is formulated as

$$(\boldsymbol{K} + j\omega \boldsymbol{C} - \omega^2 \boldsymbol{M})\boldsymbol{p} = j\omega\rho v_0 \boldsymbol{W}(=\boldsymbol{f}). \tag{3.29}$$

On the other hand, the following linear system of equations needs to be solved at each time step for time domain analysis:

$$\left[\boldsymbol{M} + \frac{\Delta t}{2}\boldsymbol{C} + \beta(\Delta t^2)\boldsymbol{K}\right]\ddot{\boldsymbol{p}}_{t+\Delta t} = \boldsymbol{f}_{t+\Delta t} - \boldsymbol{C}\boldsymbol{P} - \boldsymbol{K}\boldsymbol{Q}. \tag{3.30}$$

For simplicity, the above-presented equations represent as

$$\boldsymbol{A}\boldsymbol{x} = \boldsymbol{b}. \tag{3.31}$$

For either formulation, the coefficient matrix $\boldsymbol{A}$ has sparse structure with many zero elements as shown in Fig. 3.6. The maximum number of nonzero elements of a row in a matrix is dependent on finite-elements used for spatial discretization. In the discretization using hexahedral 27-node isoparametric elements, the number becomes 125, whereas 27 nonzero elements appear for 8-node isoparametric elements. Further, $\boldsymbol{A}$ is a complex symmetric matrix, i.e. a non-Hermitian matrix $\boldsymbol{A} \neq \boldsymbol{A}^H$, because stiffness matrix $\boldsymbol{K}$ and mass matrix $\boldsymbol{M}$ are real symmetric matrices and dissipation matrix $\boldsymbol{C}$ is a complex matrix. Here, if the lumped dissipation matrix is used, the complex number appears in only diagonal part of $\boldsymbol{A}$. However,

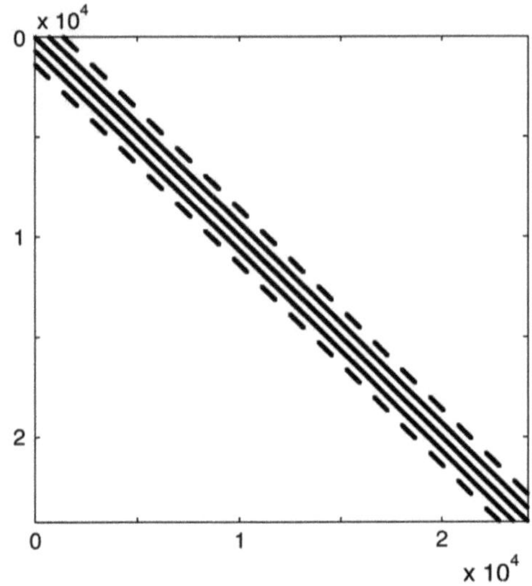

Fig. 3.6 An example of structure of coefficient matrix $\boldsymbol{A}$ with degrees of freedom of 25,641

normalized acoustic impedance in real number is given for impedance boundary condition in time domain analysis, $\boldsymbol{A}$ becomes real symmetric matrix with dissipation matrix having elements of real number. This feature is different from frequency domain analysis, in this case, numeric operation with only real number is possible. In room acoustics computation, external force is typically given at a node or some nodes assuming the omnidirectional point source. Consequently, most of the elements in external force vector $\boldsymbol{f}$ has zero elements.

3.2.1.2 Solution of The Linear System of Equations

The solution technique of the linear system of equations is classified into two types, namely direct methods and iterative methods. The direct methods, which are based on Gauss elimination or LU/Cholesky decomposition, yield unique solution and are robust. However, the methods produce nonzero elements, called fill-in, in a matrix after factorizations. Due to the property, application of direct methods to large-scale problem is generally not suitable from the perspective of memory requirement.

On the other hand, the iterative methods require merely the retention of nonzero elements of the coefficient matrix of calculation according to the algorithm. For the sparse matrix, it can greatly reduce the required memory compared to direct methods. There are many sparse storage formats such as CRS (Compressed Row Storage) or CCS (Compressed Column Storage) formats [12]. Further advantage of iterative methods is easiness of parallelization, and this feature is also suitable for large-scale analysis. However, a problem is the uncertainty [12] by which the convergence differs for every type of iterative methods used and for every object of analysis. Because of that uncertainty, it is necessary to know how each iterative method converges.

There are two types of iterative methods, namely stationary iterative methods and Krylov subspace methods. Stationary iterative methods represented by SOR (Successive Over Relaxation) method have a long history. They are easy to understand and to implement, but generally not effective [12]. On the other hand, Krylov subspace methods are a relatively recently developed technique [12], and approximate solution of linear system of equations are searched in Krylov subspace ($K_n(\boldsymbol{A};\boldsymbol{r}_0) = \mathrm{Span}\{\boldsymbol{r}_0, \boldsymbol{A}\boldsymbol{r}_0, \ldots, \boldsymbol{A}_{n-1}\boldsymbol{r}_0\}$). Here, $\boldsymbol{r}_0$ represents initial residual $\boldsymbol{b} - \boldsymbol{A}\boldsymbol{x}_0$ defined with initial approximate solution $\boldsymbol{x}_0$.

The CG (Conjugate Gradient) method is the best known Krylov subspace methods having superior convergence property for sparse matrix, and theoretically converges within n iterations where n is dimension number. However, CG method cannot be applied in a straightforward way to linear system of equations with non-Hermitian coefficient matrix that arise from finite-element sound field analysis because it is only applicable for linear systems with Hermitian coefficient matrix.

There exist the extended methods of the CG method, namely iterative methods based on Arnoldi process or based on Lanczos process, which are applicable to linear system with non-Hermitian coefficient matrix, and these are still area of active research. The iterative methods based on the Arnoldi process such as GMRes (Generalized Minimal Residual) method [15] have considerable disadvantages in computational costs due to long recurrence formula, i.e., memory and operations required per iteration increase linearly with the iteration count. The Lanczos process is viewed as a simplification for the case when the Anoldi process is applied to Hermitian matrix. In iterative methods based on Lanczos process, Bi-Conjugate Gradient (BiCG) method [16] is the most popular. Since these types of methods do not increase memory and operations per iteration with the number of iterations, it is favorable from the perspective of computational costs.

3.2.1.3 Preconditioning

In the use of the iterative methods, a preconditioning is generally adopted to the linear system $\boldsymbol{A}\boldsymbol{x} = \boldsymbol{b}$ to improve the convergence of the iterative method. Preconditioning is a process to transforming the $\boldsymbol{A}\boldsymbol{x} = \boldsymbol{b}$ into another more suitable linear system to solve, but has same solution. Using a preconditioning matrix $\boldsymbol{M}$, the $\boldsymbol{A}\boldsymbol{x} = \boldsymbol{b}$ is transformed into

$$\boldsymbol{M}^{-1}\boldsymbol{A}\boldsymbol{x} = \boldsymbol{M}^{-1}\boldsymbol{b}. \tag{3.32}$$

In designing of preconditioning matrix, it is said that $\boldsymbol{M}$ must be a good approximation to $\boldsymbol{A}$ in some measure. The simplest preconditioning is a diagonal scaling with $\boldsymbol{M} = \mathrm{diag}(\boldsymbol{A})$. Other sophisticated preconditionings are incomplete LU/Cholesky factorization with $\boldsymbol{M}$ that approximates $\boldsymbol{A}$ or sparse approximate inverse with $\boldsymbol{M}$ that approximates $\boldsymbol{A}^{-1}$ [12, 13]. For sound field analysis using FEM, absolute diagonal scaling with $\boldsymbol{M} = \mathrm{diag}(|\boldsymbol{A}|)$ has been proposed [17]. This preconditioning does not require extra memory in construction of preconditioning matrix.

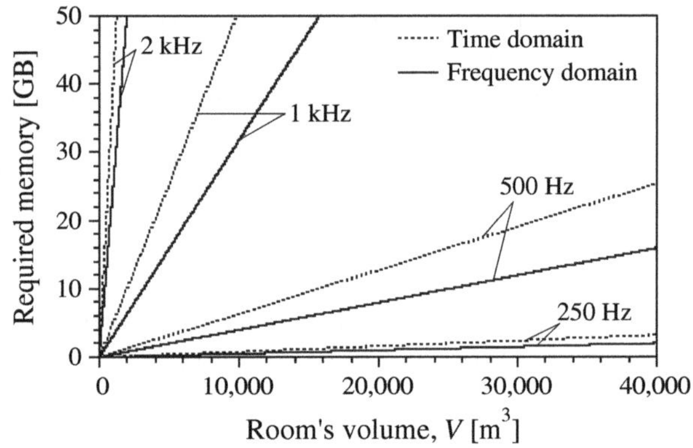

Fig. 3.7 Required memory for FE analysis in time and frequency domains

3.2.2 Estimation of Required Memory

Estimations of required memories in both frequency and time domains analyses using the iterative methods based on the Lanzcos process are presented, in which we assume symmetricity of matrices, use of hexahedral 27-node finite elements, and use of lumped dissipation matrix based on locally reactive model. For either analysis, the required memory spaces *RM* can be estimated roughly using [18, 19]

$$RM \approx a \cdot DOF\,[\text{bytes}], \tag{3.33}$$

where a and *DOF*, respectively, represent constant and degrees of freedom. For frequency and time domains analyses, $a = 1,000$ and $a = 1,600$, respectively. Further, assuming the analysis of cubic cavity, the *RM* is represented with upper limit frequency $f_{\max}$ Hz, room-volume V m^3, and wavelength per nodal distance λ/d representing spatial resolution of FE mesh as

$$RM \approx a \cdot V \cdot \left(\frac{\lambda/d \cdot f_{\max}}{c}\right)^3 \,[\text{bytes}]. \tag{3.34}$$

Figure 3.7 presents the relationship between V and *RM* for both domain analyses as a function of $f_{\max}$ when the FE mesh with $\lambda/d = 5.0$ is used. From the figure, the *RM* increase significantly with increasing $f_{\max}$, rather than V. As an example, if one predicts sound field in a room with volume of 10,000 m^3 up to 500 Hz, memory requirements for frequency and time domain analyses become 4 and 6.4 GB, respectively.

3.2.3 Convergence of Iterative Methods

3.2.3.1 Frequency Domain Analysis

Numerical examples of sound field analysis in two existing reverberation rooms and a small concert hall are presented to demonstrate the convergence characteristics of the iterative methods applied to frequency domain analysis [18]. The intended reverberation rooms are a rectangular-shaped reverberation room with the volume of 220 m^3 (RR) and a heptahedral irregular-shaped reverberation room with the volume of the 168 m^3 (IR). Also, the small concert hall is listed in the benchmark platform on computational methods for architectural/environmental acoustics as a benchmark problem B1-1F [20], and has complicated geometry with a volume of 3,000 m^3.

In numerical analysis in the reverberation rooms, omnidirectional point source is located at corner points of each room, and sound pressures at 125, 250, and 500 Hz pure tones are analyzed. Two kinds of boundary conditions are assumed, namely Cond. 1: all boundaries are assumed to be typical concrete walls and Cond. 2 : as a absorbing material, glass-wool is installed to a part of the wall. On the other hand, sound pressure at 125 and 250 Hz pure tones from a omnidirectional point source placed at 1.5 m above the floor are computed in a small hall analysis. For all analyses, Spl27 is used to discretize the spatial domains, where the respective FE models are created to satisfy the spatial discretization requirement $\lambda/d \geq 4.5$.

Four iterative methods, namely COCG (Conjugate Orthogonal Conjugate Gradient) method [21], CGS (Conjugate Gradient Squared) method [22], BiCGStab (Bi-Conjugate Gradient Stabilized) method [23] and GPBiCG (Generalized Product type Bi-CG) method [24] are selected from the iterative methods based on Lanczos process. COCG is an effective iterative solver for linear system with complex symmetric coefficient matrix, which has iterative schemes equivalent to the BiCG, but require less memory and few operations per iteration than BiCG. CGS, BiCGStab, and GP-BiCG are BiCG-like solvers developed to improve irregular convergence of the BiCG. For the four iterative methods, absolute diagonal scaling [17] is adopted as a preconditioning method.

In BiCG-like solvers, initial solution $\boldsymbol{p}_0$ and initial shadow residual $\boldsymbol{r}_0^*$ must be provided for initial setting. Generally, initial residual $\boldsymbol{r}_0$ is given to the $\boldsymbol{r}_0^*$, whereas previous study [25] shows the use of uniform random numbers to $\boldsymbol{r}_0^*$ yields better convergence for some problems. Therefore, two initial settings ($\boldsymbol{p}_0 = \boldsymbol{0}, \boldsymbol{r}_0^* = \boldsymbol{r}_0$) and ($\boldsymbol{p}_0 = \boldsymbol{0}, \boldsymbol{r}_0^* =$ random) are used here, and these are represented as BiCG-r_0 and BiCG-random where random is pseudorandom numbers. The $\boldsymbol{p}_0$ for COCG is set to $\boldsymbol{0}$.

As the stopping criterions for iterative methods, the iteration process is terminated when relative residual 2-norm becomes less than $\varepsilon_{\text{iter}}$ or when the iterations is reached maximum number of iterations, in which $\varepsilon_{\text{iter}}$ is set to 10^{-12}, which is sufficiently small value, and maximum iteration numbers is set to *DOF*.

Table 3.1 lists numbers of sparse matrix-vector products of four iterative methods at 125–500 Hz in two absorbing conditions for RR and IR. Because these iterative

Table 3.1 The number of sparse matrix-vector products of four iterative methods

	Iterative solver-r_0^*	Number of matrix-vector products					
		RR			IR		
		125 (Hz)	250 (Hz)	500 (Hz)	125 (Hz)	250 (Hz)	500 (Hz)
Cond. 1	COCG	579	4,108	32,215	701	5,126	38,061
	CGS-r_0	1,242	9,014	70,112	1,448	19,862	236,398
	CGS-random	1,274	9,350	70,520	1,496	13,674	78,612
	BiCGStab-r_0	max	max	179 554	4424	max	max
	BiCGStab-random	max	max	170,272	3,628	max	191,282
	GPBiCG-r_0	div.	div.	div.	div.	div.	div.
	GPBiCG-random	div.	div.	div.	div.	div.	div.
Cond. 2	COCG	566	4,736	34,991	691	5,476	37,663
	CGS-r_0	1,108	11,800	div.	1,460	13,318	65,160
	CGS-random	1,152	8,816	50,878	1,488	11,886	45,118
	BiCGStab-r_0	max	20,038	91,530	4,486	34,048	77,224
	BiCGStab-random	3,416	20,130	82,134	4,184	32,524	70,916
	GPBiCG-r_0	div.	div.	div.	div.	div.	div.
	GPBiCG-random	div.	div.	div.	div.	div.	div.

DOF (RR): 1,755 (125 Hz), 12,325 (250 Hz), 92,169 (500 Hz)
DOF (IR): 2,475 (125 Hz), 17,661 (250 Hz), 133,209 (500 Hz)

methods have different amounts of operations per iteration, the number of matrix-vector products, which comprise a large percent of operations, is counted instead of number of iterations. Also, "max." and "div." in the list, respectively, represent termination of operations by maximum iteration numbers and divergence. From the table, the number of sparse matrix-vector products for convergence is significantly different among the iterative methods. Only two iterative methods COCG and CGS-*random* have robust convergence characteristics irrespective of room geometries, absorbing conditions, and frequencies. Also, the use of random numbers to $\boldsymbol{r}_0^*$ yields stable and faster convergence in CGS from which importance of proper initial setting is confirmed.

Further, convergence characteristics of the two methods in small concert hall analysis are presented to show the effectiveness in large-scale analysis with complicated geometry. Numerical results show that COCG and CGS-random, respectively, converge with a number of matrix-vector products of 18,945 and 27,076 at 125 Hz. For 250 Hz, the numbers are 23,440 in COCG and 29,872 in CGS-random. COCG converges with less number of matrix-vector products irrespective of frequencies from which it can be said that COCG is an effective solver for frequency domain analysis. Detailed discussion on convergence property of COCG applied to sound field analysis is described in Ref. [26].

3.2.3.2 Time Domain Analysis

To demonstrate the convergence characteristics of the iterative methods applied to time domain analysis, numerical examples of sound field analysis in a small

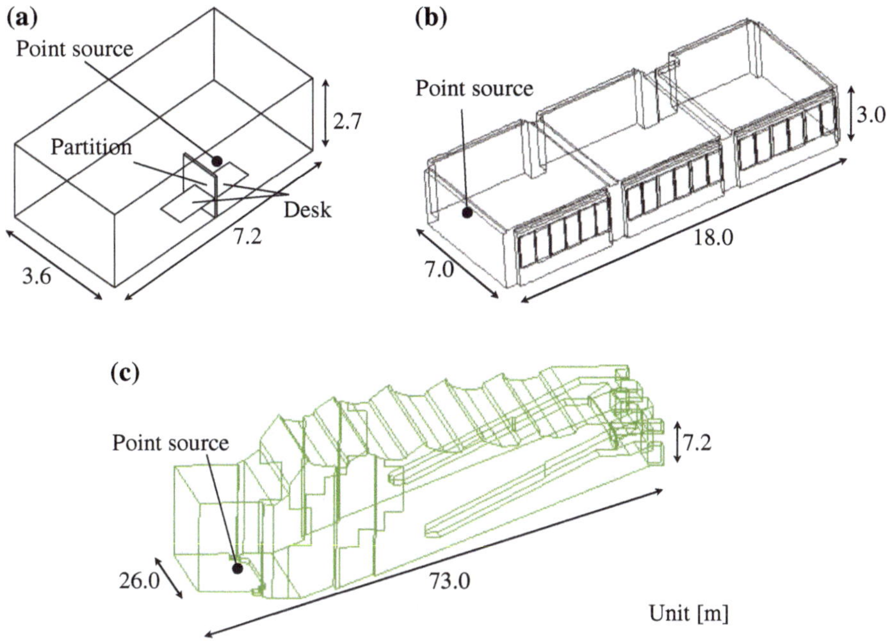

Fig. 3.8 Schematic drawing of **a** small office, **b** lecture room, and **c** large-scale multi-purpose hall to be analyzed

office ($V = 70\,\mathrm{m}^3$), a lecture room ($V = 360\,\mathrm{m}^3$) and a large-scale multi-purpose hall ($V = 37,000\,\mathrm{m}^3$) illustrated in Fig. 3.8 are presented [27, 28], in which sound pressures inside the rooms for band-limited source signals emitted from an omni-directional point source placed as shown in Fig. 3.8 are computed. First, convergence characteristics of five iterative methods applied to sound field analysis in a small office are compared where COCG, Conjugate A-Orthogonal Conjugate Residual (COCR) method [29], Quasi-Minimal Residual for complex symmetric matrix (QMR) method [30] for linear system with complex symmetric coefficient matrix, CGS and BiCGStab are adopted. Then, the convergence characteristics in sound field analysis of a lecture room and a large-scale multi-purpose hall, which can be considered as more practical application, are presented focusing on only COCG. Finally, effectiveness of IC factorization preconditioning is demonstrated.

In the all numerical examples, Spl27 is used for spatial discretization where the respective FE mesh is created to satisfy the spatial discretization requirement $\lambda/d \geq 4.8$. As for the boundary conditions of each analysis, the small office and the multi-purpose hall analyses use normalized acoustic impedance ratio z_n corresponding to absorption coefficient from literatures for each boundary, whereas the lecture room analysis uses z_n measured *in-situ*. A direct time integration method, namely constant average acceleration method is adopted for time discretization. Table 3.2 lists *DOF* of analysis, time interval Δt, analyzed time length T and total number of time steps

Table 3.2 Setting of FE analysis of sound field in a small office, a lecture room and a multi-purpose hall

Room	Frequency (Hz)	*DOF*	Δt (ms)	T (ms)	N_{step}
Small office	250	15,207	0.023	1,000	44,100
	500	94,845	0.023	1,000	44,100
	1,000	593,519	0.023	500	22,050
Lecture room	250	50,851	0.023	1,000	44,100
	500	275,619	0.023	1,000	44,100
Multi-purpose hall	1,000	4,282,841	0.023	2,000	88,200
	125	2,630,435	0.015	2,000	131,072

Table 3.3 Comparison of performance among five iterative methods applied to sound field analysis in a small office

Iterative solver	Average number of matrix-vector products		
	250 (Hz)	500 (Hz)	1,000 (Hz)
COCG	14.7	13.2	14.4
COCR	14.3	12.9	13.8
QMR	14.3	12.9	13.9
CGS	16.6	15.3	16.0
BiCGStab	18.3	15.3	17.3

N_{step} for respective analyses. The iterations of iterative methods are terminated when the relative residual 2-norm at each time step becomes less than 10^{-6}. The initial solution for starting iterative process is set to $\mathbf{0}$ at first time step, and solution before 1 time step is given at subsequent time steps.

Table 3.3 shows a comparison of mean number of sparse matrix-vector products per iteration among five iterative methods on the analysis of sound field in the small office. Here, absolute diagonal scaling is used as a preconditioning. The results show that all iterative methods converge rapidly with the number of matrix-vector products less than 20 in spite of the use of simple preconditioning. It can be said that application of iterative methods to time domain analysis is significantly effective because of the convergence with sufficiently small iteration numbers compared to *DOF* of problem. Further, COCG, COCR, and QMR for complex symmetric matrix are more effective than CGS and BiCGStab with the smaller numbers of sparse matrix-vector products.

Figure 3.9a and b shows the number of sparse matrix-vector products at each time step of COCG applied to sound field analysis in the lecture room and the multi-purpose hall where the absolute diagonal scaling preconditioning is adopted. The solver also converges rapidly irrespective of sound fields with the small number of matrix-vector products from which application of the iterative method is effective for practical problems.

Further, improvement of convergence by using more sophisticated preconditioning techniques, namely IC factorization preconditionings, is presented in comparison

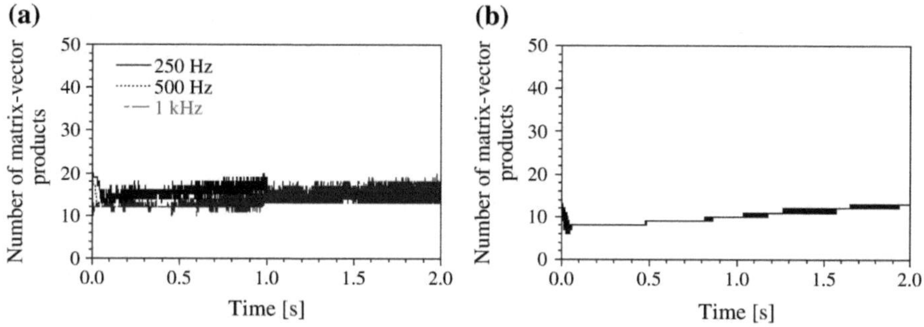

Fig. 3.9 Number of sparse matrix-vector products of COCG method applied to sound field analysis in **a** lecture room and **b** multi-purpose hall

Table 3.4 Performance of two IC factorization preconditionings applied to sound field analysis in a small office and a lecture room

Room	Average number of matrix-vector products			
	Frequency (Hz)	COCG	IC(0)-COCG	DIC-COCG
Small office	250	12.8	2.2	5.2
	500	12.8	2.2	5.2
	1,000	13.2	2.0	5.0
Lecture room	250	15.5	3.4	5.9
	500	12.5	3.4	5.0

with absolute diagonal scaled COCG when the sound fields in the small office and the lecture room are predicted. Among IC factorization preconditionings, IC(0) factorization without consideration of fill-in and simpler DIC factorization are adopted here.

Table 3.4 shows a comparison of average number of matrix-vector products per iteration among the preconditioned iterative methods. IC(0)-COCG and DIC-COCG show better performance than COCG with smaller number of matrix-vector products. Specifically, IC(0)-COCG is a powerful preconditioned iterative method for time domain analysis with a few number of matrix-vector products.

3.3 Application to Exterior Problems

In principle, the finite element method is not applicable to an exterior problem in an unbounded domain in a normal way, since it is not feasible to discretize an infinitely large domain. However, the finite element method can be extended to analyze acoustic field in an unbounded domain by appropriately modeling the infinite, open region away from the sound sources and scattering objects.

In this section, some approaches for applying the finite element method to an exterior problem are briefly reviewed, and then the Dirichlet-to-Neumann (DtN) method [31–33] is explained as an example of such methods.

3.3.1 Approaches to Exterior Problems

For applying the finite element method to an exterior problem, the infinite domain is divided into a bounded and an unbounded domain. The finite element method is applied to the bounded domain, and the unbounded domain, which is the rest of the whole infinite domain outside of the bounded domain, is modeled in another way. Several such methods have been proposed [34].

One of such methods is combining the finite element method and the boundary element method. The finite element method is applied in the bounded domain, and the boundary element is used for solving the exterior problem in the unbounded domain.

Another commonly used approach is to use the infinite elements [35–39]. The infinite elements are extension of the finite elements. The unbounded domains is discretized into the elements with semi-infinite sizes. As with the finite elements, the basis functions are associated with the infinite elements. The basis functions are determined to satisfy the radiation condition.

In some other approaches, the acoustic field in the outer, unbounded domain is modeled and expressed as the condition on the boundary of the inner bounded domain, in which the finite element analysis is applied. Such a boundary condition is called a nonreflecting boundary condition (NRBC), since it is derived to diminish the reflection back into the bounded domain. Some of the NRBCs are local conditions [40–42], and the others are nonlocal ones [43–45]. While all of the local conditions are approximate, some of the nonlocal conditions are approximate and the others are exact. The DtN boundary condition described below in this section is an example of a nonlocal, exact boundary condition.

Unwanted reflection into the bounded domain can be reduced also by placing absorbing layers right outside of the bounded domain. Such a boundary condition is called an absorbing boundary condition (ABC). The perfectly matched layer (PML) [46] is an example of such an approach. Although the PML is commonly used with the finite difference method, it can be used with the finite element method as well.

3.3.2 Dirichlet-to-Neumann Method

3.3.2.1 DtN Map

The procedure of the DtN method for analyzing the exterior problem shown in Fig. 3.10 by the finite element method is reviewed here. In the frequency domain,

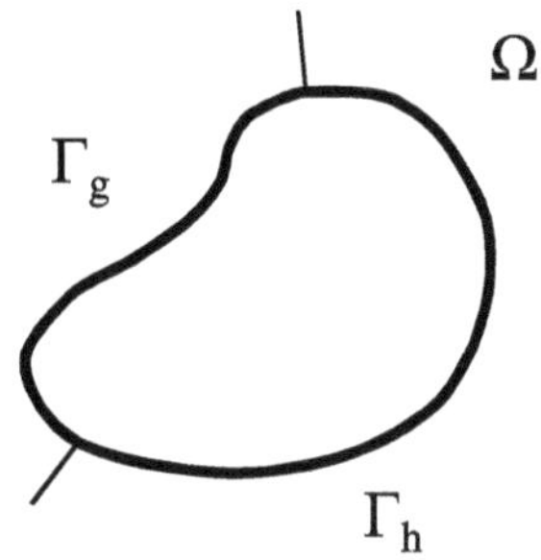

Fig. 3.10 Exterior problem

the exterior problem is described as follows:

$$\nabla^2 p + k^2 p + f = 0 \qquad \text{in } \Omega, \tag{3.35}$$

$$p = g \qquad \text{on } \Gamma_{\mathrm{g}}, \tag{3.36}$$

$$\frac{\partial p}{\partial n} = h \qquad \text{on } \Gamma_{\mathrm{h}}, \tag{3.37}$$

$$\lim_{r \to \infty} r^{(d-1)/2} \left(\frac{\partial p}{\partial r} - jkp \right) = 0. \tag{3.38}$$

Here, Ω is the entire unbounded domain. Γ_g and Γ_h are the parts of the boundary of the object on which the Dirichlet and Neumann boundary conditions are imposed, respectively.

Equation (3.38) is the Sommerfeld radiation condition, which describes that the acoustic wave propagates outward only. In this equation, r is the distance from the coordinate origin, and d is the spatial dimension (2 or 3).

For applying the DtN method to this exterior problem, the artificial boundary Γ_{dtn} is introduced to divide the entire domain Ω into a bounded domain Ω_{b} and an unbounded domain Ω_{ω} as shown in Fig. 3.11. The finite element method is used only in the bounded domain. The interior problem in Ω_{b} is described as

$$\nabla^2 p + k^2 p + f = 0 \qquad \text{in } \Omega_b, \tag{3.39}$$

$$p = g \qquad \text{on } \Gamma_{\mathrm{g}}, \tag{3.40}$$

$$\frac{\partial p}{\partial n} = h \qquad \text{on } \Gamma_{\mathrm{h}}, \tag{3.41}$$

$$\frac{\partial p}{\partial n} = M_{\mathrm{dtn}}(p) \qquad \text{on } \Gamma_{\mathrm{dtn}}. \tag{3.42}$$

The specific expression of Eq. (3.42), which is the condition on the artificial boundary Γ_{dtn}, is determined so that Eqs. (3.39)–(3.42) become equivalent to Eqs. (3.35)–(3.38). M_{dtn} in Eq. (3.42) is called the Dirichlet-to-Neumann (DtN) map, since it is a mapping from the Dirichlet datum p to the Neumann datum $\partial p / \partial n$.

Fig. 3.11 Domain truncation

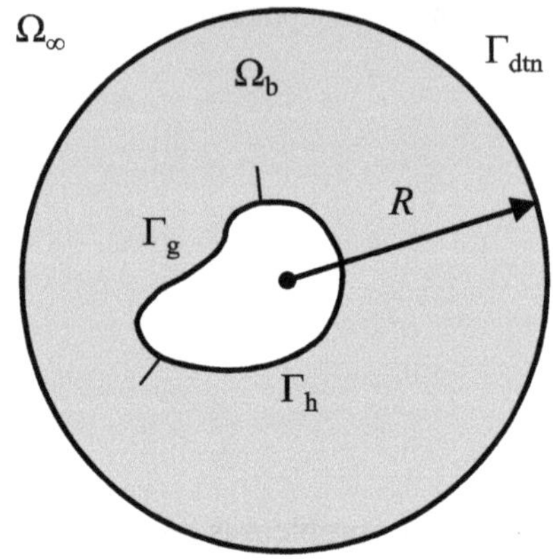

The general procedure of the DtN method is described in the following:

Step 1 Truncation of Domain
Divide the whole domain Ω into a bounded domain Ω_b and an unbounded domain Ω_∞ by introducing an artificial boundary Γ_{dtn} as shown in Fig. 3.11.

Step 2 Analysis in Unbounded Domain
Obtain the analytical solution in the unbounded domain Ω_∞.

Step 3 Derivation of DtN Map
Derive the DtN map, which is the relationship between the acoustic pressure and its normal derivative on the artificial boundary Γ_{dtn}, by using the analytical solution obtained in Step 2.

Step 4 Numerical Analysis in Bounded Domain
Solve the interior problem in the bounded domain Ω_b by a numerical method. Use the DtN map as the boundary condition on the artificial boundary Γ_{dtn}.

By following these steps, the DtN map for a two-dimensional problem is obtained below:

Step 1 Truncation of Domain
In order to derive the explicit form of the DtN map M_{dtn}, the analytical solution in the unbounded domain Ω_∞ is needed. To obtain the analytical solution, the shape of the artificial boundary Γ_{dtn} have to be simple. Circular and spherical boundaries are generally used for two- and a three-dimensional problems, respectively. Another necessary condition for deriving the analytical solution in Ω_∞ is that no sound sources or objects are located in this domain. Therefore, the artificial boundary Γ_{dtn} needs to be placed such that all of the sources and objects are involved inside it. In the problem shown in Fig. 3.11, a circular artificial boundary with the radius of R is introduced to divide the domain.

Step 2 Analysis in Unbounded Domain
In the cylindrical coordinate system (r, θ) with the origin at the center of

the circle of the artificial boundary, the exterior problem in the unbounded domain Ω_∞ with the Dirichlet boundary condition on the artificial boundary Γ_{dtn} is described as follows:

$$\nabla^2 p + k^2 p = 0 \qquad \text{in } \Omega_\infty, \tag{3.43}$$

$$p = p(R, \theta) \qquad \text{on } \Gamma_{\mathrm{dtn}}, \tag{3.44}$$

$$\lim_{r\to\infty} r^{(d-1)/2}\left(\frac{\partial p}{\partial n} - jkp\right) = 0. \tag{3.45}$$

The analytical solution of this problem is obtained as

$$p(r,\theta) = \sum_{n=-\infty}^{\infty} \frac{1}{2\pi R} \int_{\Gamma_{\mathrm{dtn}}} \frac{H_{|n|}^{(1)}(kr)}{H_{|n|}^{(1)}(kR)} e^{jn\theta} e^{-jn\theta'} p(R,\theta')\, d\Gamma', \tag{3.46}$$

where

$$d\Gamma' = R\, d\theta', \tag{3.47}$$

and $H_n^{(1)}$ is the Hankel function of the first kind.

Equation (3.46) implies that the solution in the unbounded domain Ω_∞ can be obtained by using the solution on the artificial boundary $p(R,\theta)$. The solution of the artificial boundary is obtained numerically when solving the interior problem in Ω_{b}. Thus, by using the numerical solution on Γ_{dtn}, the solution in the unbounded domain Ω_∞ can be calculated.

Step 3 Derivation of DtN Map

By taking the derivative of Eq. (3.46) with respect to r and evaluate it on the artificial boundary $r = R$, the following DtN map is obtained.

$$\frac{\partial p}{\partial n}(R,\theta) = M_{\mathrm{dtn}}(p(R,\theta))$$
$$= \sum_{n=-\infty}^{\infty} \frac{1}{2\pi R} \frac{kH_{|n|}^{(1)\prime}(kR)}{H_{|n|}^{(1)}(kR)} e^{jn\theta} \int_{\Gamma_{\mathrm{dtn}}} e^{-jn\theta'} p(R,\theta')\, d\Gamma', \tag{3.48}$$

where the prime (′) after the Hankel function indicates the derivative.

The DtN map in a three-dimensional problem can also be derived in the same procedure:

Step 1 Truncation of Domain

A sphere with the radius of R is introduced as the artificial boundary. The spherical coordinate system (r, θ, ϕ) with the origin at the center of sphere of the artificial boundary is used.

Step 2 Analysis in Unbounded Domain

The analytical solution in the unbounded domain Ω_∞ is obtained as

$$p(r,\theta,\phi) = \sum_{n=0}^{\infty}\sum_{i=-n}^{n} \frac{(2n+1)(n-|i|)!}{4\pi R^2 (n+|i|)!} \frac{h_n^{(1)}(kr)}{h_n^{(1)}(kR)} \times \int_{\Gamma_{\mathrm{dtn}}} P_n^{|i|}(\cos\theta) e^{ji\phi} P_n^{|i|}(\cos\theta') e^{-ji\phi'} p(R,\theta',\phi')\, d\Gamma', \quad (3.49)$$

where

$$d\Gamma' = R^2 \sin\theta' d\theta' d\phi', \quad (3.50)$$

and $h_n^{(1)}$ is the spherical Hankel function of the first kind, and P_n^i is the associated Legendre polynomial.

Step 3 Derivation of DtN Map

Taking the normal derivative of Eq. (3.49) leads to the following DtN map:

$$\frac{\partial p}{\partial n}(R,\theta) = M_{\mathrm{dtn}}(p) = \sum_{n=0}^{\infty}\sum_{i=-n}^{n} \frac{(2n+1)(n-|i|)!}{4\pi R^2 (n+|i|)!} \frac{kh_n^{(1)'}(kR)}{h_n^{(1)}(kR)} P_n^{|i|}(\cos\theta) e^{ji\phi} \times \int_{\Gamma_{\mathrm{dtn}}} P_n^{|i|}(\cos\theta') e^{-ji\phi'} p(R,\theta',\phi')\, d\Gamma'. \quad (3.51)$$

From Eqs. (3.48) and (3.51), the DtN map can be expressed in a general form as

$$M_{\mathrm{dtn}} = \sum_{n=\underline{N}(d)}^{\infty} \sum_{i=-I(d)}^{I(d)} z_{ni}(k,R)\, \psi_{ni}^{*}(\boldsymbol{x}) \int_{\Gamma_{\mathrm{dtn}}} \psi_{ni}(\boldsymbol{x}')\, p(\boldsymbol{x}')\, d\Gamma'. \quad (3.52)$$

In this equation, $\underline{N}$ and I depend on the spatial dimension d. $\underline{N} = -\infty$ and $I = 0$ for $d = 2$, and $\underline{N} = 0$ and $I = n$ for $d = 3$.

3.3.2.2 DtN Finite Element Method

In Step 4 of the DtN method, the finite element analysis is conducted for the interior problem in the bounded domain Ω_{b}. The DtN map obtained in Step 3 is used as the boundary condition on the artificial boundary Γ_{dtn}.

By introducing the weighting function w, the weak form of the interior problem in Ω_{b} is written as

$$a(w,p) - (w, M_{\mathrm{dtn}}(p))_{\Gamma_{\mathrm{dtn}}} = (w,f) + (w,h)_{\Gamma_{\mathrm{h}}}, \quad (3.53)$$

where

$$a(w,p) = \int_{\Omega_b} \left(\nabla w^* \cdot \nabla p - k^2 w p \right) d\Omega, \tag{3.54}$$

$$(w,p)_{\Gamma_{dtn}} = \int_{\Gamma_{dtn}} w^* p \, d\Gamma, \tag{3.55}$$

$$(w,p) = \int_{\Omega_b} w^* p \, d\Omega, \tag{3.56}$$

$$(w,p)_{\Gamma_h} = \int_{\Gamma_h} w^* p \, d\Gamma. \tag{3.57}$$

The finite element discretization of this weak form leads to the following linear system:

$$(\boldsymbol{K} - \boldsymbol{B}_{dtn}) \boldsymbol{p} = \boldsymbol{f}. \tag{3.58}$$

In this equation, $\boldsymbol{K}$ is the stiffness matrix resulting from the normal finite element discretization, and $\boldsymbol{B}_{dtn}$ is the additional matrix associated with the DtN boundary condition. The (a, b) element of $\boldsymbol{B}_{dtn}$ is obtained as

$$[\boldsymbol{B}_{dtn}]_{ab} = (N_a, M_{dtn}(N_b))_{\Gamma_{dtn}}, \tag{3.59}$$

where N_a is the basis function of the finite element method. The element is expressed in the form of Eq. (3.52) as

$$[\boldsymbol{B}_{dtn}]_{ab} = \sum_{n=\underline{N}(d)}^{\infty} \sum_{i=-I(d)}^{I(d)} z_{ni}(k,R) \int_{\Gamma_{dtn}} \psi^*_{ni}(\boldsymbol{x}) N^*_a(\boldsymbol{x}) \, d\Gamma \int_{\Gamma_{dtn}} \psi_{ni}(\boldsymbol{x}') N_b(\boldsymbol{x}') \, d\Gamma'. \tag{3.60}$$

This equation implies that the elements of $\boldsymbol{B}_{dtn}$ can be obtained by the integration in the dimension one less than the spatial dimension of the problem.

Unlike the usual finite element stiff matrix, the (a, b) element of $\boldsymbol{B}_{dtn}$ is nonzero when the nodes a and b are both on the artificial boundary, even if these nodes are not adjacent. Since the number of the nodes on the artificial boundary Γ_{dtn} is much smaller than that in the entire numerical domain Ω_b, the matrix on the left-hand side of Eq. (3.58) is still sparse and the computational penalty due to the DtN method is not significant in many cases.

Once $\boldsymbol{B}_{dtn}$ is obtained by Eq. (3.60), the acoustic field in the bounded domain Ω_b as well as on the artificial boundary Γ_{dtn} can be calculated by solving the linear system Eq. (3.58). As with the normal finite element method, iterative methods can be utilized for solving the linear system efficiently. Some algorithms for the DtN finite element method suitable for the iterative methods have been proposed [33, 47, 48].

In practical implementation, the infinite series in the DtN maps in the forms of Eqs. (3.48) and (3.51) need to be truncated after a finite number of terms. Although the DtN maps expressed by Eqs. (3.48) and (3.51) are the exact boundary conditions, the truncation of the infinite series in practical computation produces some error. With the increase in the number of terms N, the error decreases, but the computational cost grows. Therefore, the number of terms N needs to be determined by considering the trade-off between the accuracy and the computational cost. Since the error converges to a certain level with increasing N, understanding the convergence behavior helps determine the truncation criterion.

The number of terms N also affects the uniqueness of the solution [49, 50]. If the exact DtN map without the series truncation is used, the uniqueness of the solution is always assured. On the other hand, if the truncated series after N terms is used to express the DtN map, the solution is unique only when N is larger than kR. Although the necessary number of terms N and the resulting computational burden increase as the wavenumber k becomes larger, the unique solution can be obtained by simply increasing N, without changing the codes.

In this section, the basics of the DtN method are reviewed. Beyond the basics, the DtN method is being extended in several ways. For instance, the modified DtN method for reducing the truncation error [51] and the localization of the DtN map [52] are studies. Also, the application of the DtN method is not limited to an exterior problem in the frequency domain as described in this section. Not only for an infinite domain, the DtN method can be used also for a large finite domain with similar domain truncation [34]. The DtN method can be applied in the time domain as well [34].

References

1. R. Courant, Variational methods for the solution of problems of equilibrium and vibrations. Bull. Amer. Math. Soc. **49**, 1–23 (1943)
2. R.W. Clough, H.C. Martin, M.J. Turner, L. J. Topp, Stiffness and deflection analysis of complex structures. J. Aeronaut. Sci., **23**:805–823, 854 (1956)
3. G. Gladwell, A finite element method for acoustics, in *Proceedings of Fifth International Congress of Acoustics, Liege, No. Paper L33* (1965)
4. A. Craggs, The use of simple three-dimensional acoustic finite elements for determining the natural modes and frequencies of complex shaped enclosures. J. Sound Vib. **23**(3), 331–339 (1972)
5. T. Shuku, K. Ishihara, The analysis of the acoustic field in irregularly shaped rooms by the finite element method. J. Sound Vib. **29**(1), 67–75 (1973)
6. A. Hauck, M. Kaltenbacker, T. Otsuru, G. Cohen, *Different Type of Finite Elements in Computational Acoustics of Noise Propagation in Fluids-Finite and Boundary Element Methods.* (Springer, Berlin, 2008)
7. T. Okuzono, T. Kondo, T. Otsuru, N. Okamoto, Y. Kurogi, Accuracy of direct frequency response analysis in sound field of a cavity using finite element method by comparison with modal frequency response analysis, in *Proceedings in the 19th International Congress on Acoustics (Madrid), No. COM-05-003*, 2007
8. N.M. Newmark, A method of computation for structural dynamics. J. Eng. Mech. Div. **85**, 67–94 (1959)
9. T.J.R. Hughes (ed.), The finite element method linear static and dynamic finite element analysis. (Dover Publications, New York, 2000)

10. J.Z. Zhu, O.C. Zienkiewicz, R. L. Taylor. *The Finite Element Method: Its Basis and Fundamentals*, 7th edn. (Butterworth-Heinemann, Oxford (UK), 2013)
11. T. Otsuru, R. Tomiku, Basic characteristics and accuracy of acoustic element using spline function in finite element sound field analysis. Acoust. Sci. Tech. **21**(2), 87–95 (2000)
12. R. Barrett, M. Berry, T.F. Chan, J. Demmel, J. Donato, J. Dongarra, V. Eijkhout, R. Pozo, C. Romine, H. van der Vorst, *Templates for the Solution of Linear Systems: Building Blocks for Iterative Methods*. (SIAM, Philadelphia, 1994)
13. H.A. van der Vorst, *Iterative Krylov Methods for Large Linear Systems*. (Cambridge University press, Cambridge, 2003)
14. Y. Saad, *Iterative Methods for Sparse Linear Systems*, 2nd edn. (SIAM, Philadelphia, 2003)
15. Y. Saad, M.H. Schultz, GMRES: A generalized minimal residual algorithm for solving nonsymmetric linear systems. SIAM J. Sci. Stat. comput. **7**(3), 856–869 (1986)
16. R. Fletcher, Conjugate gradient methods for indefinite systems. Lec. Notes Math. **506**, 73–89 (1976)
17. N. Moethuthu, A. Okamoto, M. Shiode, S. Kakihara, Fujino, T. Otsuru, A particular preconditioning technique for symmetric complex matrix. Information **10**(2), 193–198 (2007)
18. N. Okamoto, R. Tomiku, T. Otsuru, Y. Yasuda, Numerical analysis of large-scale sound fields using iterative methods part II: application of Krylov subspace methods to finite element analysis. J. Comput. Acoust. **15**(4), 473–493 (2007)
19. T. Okuzono, T. Otsuru, N. Okamoto, R. Tomiku, Sound field analysis of rooms by time domain finite element method with an iterative method. J. Environ. Eng. **73**(628), 701–706 (2008). (in Japanese)
20. http://gacoust.hwe.oita-u.ac.jp/aij-bpca/
21. H.A. van der Vorst, J. Melissen, A Petrov-Galerkin type method for solving $Ax = b$, where A is symmetric complex. IEEE Trans. Magn. **26**(2), 706–708 (1992)
22. P. Sonneveld, A fast Lanczos-type solver for nonsymmetric linear system. SIAM J. Sci. Stat. Comput. **10**, 36–52 (1989)
23. H.A. van der Vorst, Bi-CGStab, a fast and smoothly converging variant of Bi-CG for the solution of nonsymmetric linear system. SIAM J. Sci. Stat. Comput. **13**, 631–644 (1992)
24. S.-L. Zhang, GPBi-CG: generalized product-type methods based on Bi-CG for solving nonsymmetric linear system. SIAM J. Sci. Stat. Comput. **18**, 537–551 (1997)
25. S. Fujino, K. Abe, Effective strategy for improvement of convergence of BiCG-like, methods, in *Proceedings of HPCS2002*, pp. 51–58 (2002) (in Japanese)
26. N. Okamoto, T. Otsuru, R. Tomiku, S. Fujino, Convergence of cocg method applied to finite element sound field analysis of rooms. Trans. JSCES **2005002**, 2005 (2005). (in Japanese)
27. T. Okuzono, T. Otsuru, R. Tomiku, N. Okamoto, T. Minokuchi, Speedup of time domain finite element sound field analysis of rooms, *in Proceedings of the Inter-Noise 2008,* Shanghai, No. 0877 (2008)
28. T. Okuzono, Construction of high efficient wave based numerical analysis based on time domain finite element method for predicting architectural acoustic environment. Ph.D. thesis, Oita University (2010)
29. T. Sogabe, S.L. Zhang, A COCR method for solving complex symmetric linear systems. J. Comput. Appl. Math. **199**(2), 297–303 (2007)
30. R. W. Freund, N.M. Nachtigal, An implementation of the QMR method based on coupled two-term recurrences. Technical Report 92.15 RIACS (1992)
31. K. Feng, Asymptotic radiation conditions for reduced wave equation. J. Comput. Math. **2**, 1–4 (1984)
32. J.B. Keller, D. Givoli, Exact non-reflecting boundary conditions. J. Comput. Phys. **82**, 172–192 (1989)
33. A.A. Oberai, M. Malhotra, P. Pinsky, On the implementation of the Dirichlet-to-Neumann radiation condition for iterative solution of the Helmholtz equation. Appl. Num. Math. **27**, 443–464 (1998)
34. D. Givoli, *Numerical methods for problems in infinite domains* (Elsevier, Amsterdam, 1992)
35. P. Bettess, Infinite elements. Int. J. Num. Meth. Eng. **8**, 53–64 (1977)

36. P. Bettess, O.C. Zienkiewicz, Diffraction and refraction of surface waves using finite and infinite elements. Int. J. Num. Meth. Eng. **11**, 1271–1290 (1977)
37. P. Bettess, More on infinite elements. Int. J. Num. Meth. Eng. **15**, 1613–1626 (1980)
38. O.C. Zienkiewicz, K. Bando, P. Bettess, C. Emson, T.C. Chiam, Mapped infinite elements for exterior wave problems. Int. J. Num. Meth. Eng. **21**, 1229–1251 (1985)
39. K. Gerdes, A review of infinite element methods for exterior Helmholtz problems. J. Comput. Acoust. **8**, 43–62 (2000)
40. B. Engquist, A. Majda, Absorbing boundary conditions for the numerical simulation of waves. Math. Comput. **31**, 629–651 (1977)
41. A. Bayliss, E. Turkel, Radiation boundary conditions for wave-like equations. Commun. Pure Appl. Math. **33**, 707–725 (1980)
42. K. Feng, Finite element method and natural boundary reduction, in *Proceedings of the International Congress of Mathematicians Warsaw*, pp.1439–1453,(1983)
43. B. Gustafsson, H.-O. Kreiss, Boundary conditions for time dependent problems with an artificial boundary. J. Comput. Phys. **30**, 333–351 (1979)
44. T. Hagstrom, H.B. Keller, Exact boundary conditions at an artificial boundary for partial differential equations in cylinders. SIAM J. Math. Anal. **17**, 322–341 (1986)
45. L. Ting, M.J. Miksis, Exact boundary conditions for scattering problems. J. Acoust. Soc. Am. **80**, 1825–1827 (1986)
46. J.-P. Berenger, A perfectly matched layer for the absorption of electromagnetic waves. J. Comput. Phys. **114**, 185–200 (1994)
47. M. Malhotra, P.M. Pinsky, A matrix-free interpretation of the nonlocal Dirichlet-to-Neumann radiation boundary condition. Int. J. Numer. Meth. Eng. **39**, 3705–3713 (1996)
48. M. Malhotra, P.M. Pinsky, Parallel preconditioning based on it h-hierarchical finite elements with application to acoustics. Comput. Meth. Appl. Mech. Eng. **155**, 97–117 (1998)
49. I. Harari, Computational methods for problems of acoustics with particular reference to exterior domains, Ph.D. thesis, Stanford University (1988)
50. I. Harari, T.J.R. Hughes, Analysis of continuous formulations underlying the computation of time-harmonic acoustics in exterior domains. Comput. Meth. Appl. Mech. Eng. **97**, 103–124 (1992)
51. M.J. Grote, J.B. Keller, On nonreflecting boundary conditions. J. Comput. Phys. **122**, 231–243 (1995)
52. D. Givoli, I. Patlashenko, J.B. Keller, High-order boundary conditions and finite elements for infinite domains. Comput. Meth. Appl. Mech. Eng. **143**, 13–39 (1997)

Chapter 4
Boundary Element Method

Yosuke Yasuda and Tetsuya Sakuma

Abstract This chapter gives an outline of acoustic analysis using the boundary element method (BEM). In the first section, the fundamentals of the BEM and its application to sound field analysis are explained. The second section presents two advanced techniques, the indirect approach with degenerate boundary and the domain decomposition method. The third section introduces a new application of the fast multipole method, as a fast solution technique for large-scale problems.

Keywords Boundary element method (BEM) · Boundary integral equation (BIE) · Nonuniqueness problem · Indirect method · Domain decomposition method (DDM) · Fast multipole method (FMM) · Fast multipole boundary element method (FMBEM) · Iterative method

4.1 Fundamentals

The boundary element method (BEM) is a method of solving the boundary value problem for the field with boundary discretization, which is different from the FEM and the FDTD method that employ spatial discretization. Therefore the BEM requires fewer elements and nodes, significantly reducing the number of DOFs of problems. In particular, it is powerful for exterior problems, such as radiation and scattering

Y. Yasuda (✉)
Faculty of Engineering, Kanagawa University, 3-27-1 Rokkakubashi, Kanagawa-ku, Yokohama 221-8686, Japan
e-mail: yyasuda@kanagawa-u.ac.jp

T. Sakuma
Graduate School of Frontier Sciences, The University of Tokyo, 5-1-5 Kashiwanoha, Kashiwa-shi, Chiba 277-8563, Japan
e-mail: sakuma@k.u-tokyo.ac.jp

T. Sakuma et al. (eds.), *Computational Simulation in Architectural and Environmental Acoustics*, DOI: 10.1007/978-4-431-54454-8_4,

problems, in the free field [1–5]. This section briefly describes the fundamentals of the BEM in the frequency domain and in the time domain.

4.1.1 Frequency Domain BEM

4.1.1.1 Boundary Integral Equations

We consider sound field Ω and boundary Γ with inward normal vector $\boldsymbol{n}$, as shown in Fig. 4.1. Introducing an arbitrary test function Φ to the Helmholtz equation gives a weak formulation as

$$\int_{\Omega} \Phi(\boldsymbol{r}) \left[\nabla^2 p(\boldsymbol{r}) + k^2 p(\boldsymbol{r}) \right] d\Omega = 0. \tag{4.1}$$

According to the Green's theorem, the above equation can be transformed as follows:

$$\begin{aligned}
&\int_{\Omega} \Phi(\boldsymbol{r}) \left[\nabla^2 p(\boldsymbol{r}) + k^2 p(\boldsymbol{r}) \right] d\Omega \\
&\quad = -\int_{\Gamma} \Phi(\boldsymbol{r}) \frac{\partial p(\boldsymbol{r})}{\partial n(\boldsymbol{r})} d\Gamma - \int_{\Omega} \Phi(\boldsymbol{r}) \left[\nabla \Phi(\boldsymbol{r}) \cdot \nabla p(\boldsymbol{r}) - k^2 \Phi(\boldsymbol{r}) p(\boldsymbol{r}) \right] d\Omega \\
&\quad = \int_{\Gamma} \left[p(\boldsymbol{r}) \frac{\partial \Phi(\boldsymbol{r})}{\partial n(\boldsymbol{r})} - \frac{\partial p(\boldsymbol{r})}{\partial n(\boldsymbol{r})} \Phi(\boldsymbol{r}) \right] d\Gamma + \int_{\Omega} p(\boldsymbol{r}) \left[\nabla^2 \Phi(\boldsymbol{r}) + k^2 \Phi(\boldsymbol{r}) \right] d\Omega \\
&\quad = 0,
\end{aligned} \tag{4.2}$$

where $\partial/\partial n$ denotes the inward normal derivative. Here, as the test function, we use fundamental solutions for arbitrary combinations of observation point p and source point q, that satisfies

$$\nabla^2 G(\boldsymbol{r}_{\mathrm{p}}, \boldsymbol{r}_{\mathrm{q}}) + k^2 G(\boldsymbol{r}_{\mathrm{p}}, \boldsymbol{r}_{\mathrm{q}}) = -\delta(r_{\mathrm{pq}}), \tag{4.3}$$

where $r_{\mathrm{pq}} = |\boldsymbol{r}_{\mathrm{p}} - \boldsymbol{r}_{\mathrm{q}}|$, and δ is the Dirac's delta function. In three-dimensional space, the fundamental solution is given as

$$G(\boldsymbol{r}_{\mathrm{p}}, \boldsymbol{r}_{\mathrm{q}}) = \frac{1}{4\pi} \frac{\exp(jkr_{\mathrm{pq}})}{r_{\mathrm{pq}}}, \tag{4.4}$$

while in two-dimensional space,

$$G(\boldsymbol{r}_{\mathrm{p}}, \boldsymbol{r}_{\mathrm{q}}) = \frac{j}{4} H_0^{(1)} \left(kr_{\mathrm{pq}} \right). \tag{4.5}$$

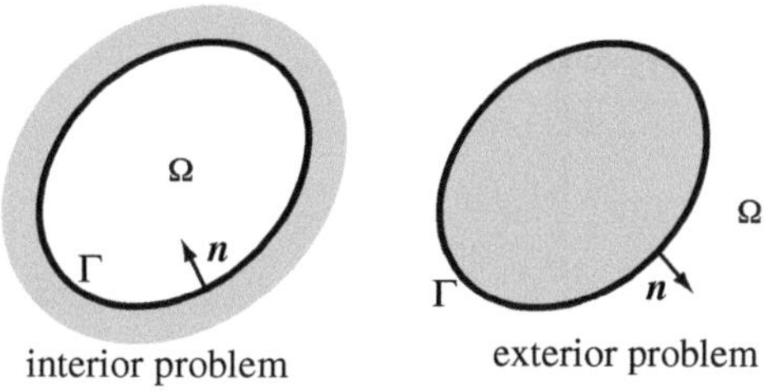

Fig. 4.1 Domain and boundary in sound field

Substituting the above solution into Eq. (4.2) gives the boundary integral equation as

$$\varepsilon(\boldsymbol{r}_{\mathrm{p}})p(\boldsymbol{r}_{\mathrm{p}}) = \int_{\Gamma} \left[p(\boldsymbol{r}_{\mathrm{q}}) \frac{\partial G(\boldsymbol{r}_{\mathrm{p}}, \boldsymbol{r}_{\mathrm{q}})}{\partial n_{\mathrm{q}}} - \frac{\partial p(\boldsymbol{r}_{\mathrm{q}})}{\partial n_{\mathrm{q}}} G(\boldsymbol{r}_{\mathrm{p}}, \boldsymbol{r}_{\mathrm{q}}) \right] d\Gamma, \tag{4.6}$$

where $\varepsilon(\boldsymbol{r}_{\mathrm{p}})$ represents the proportion of opening angle from observation point p to domain Ω, normalized by 4π in three dimensions, and by 2π in two dimensions. If the point p is located inside the domain, $\varepsilon = 1$; on a smooth boundary, $\varepsilon = 1/2$; and outside the domain, $\varepsilon = 0$. The above equation means that sound pressure at an arbitrary point is determined by sound pressure and its normal derivative (pressure gradient) on all boundaries. In the case that the point p is located on the boundary, it is known as the Kirchhoff-Helmholtz boundary integral equation.

Supposing that the point p is located on a smooth boundary, Eq. (4.6) can be differentiated with respect to the inward normal direction at the point p, as follows:

$$\frac{1}{2}\frac{\partial p(\boldsymbol{r}_{\mathrm{p}})}{\partial n_{\mathrm{p}}} = \int_{\Gamma} \left[p(\boldsymbol{r}_{\mathrm{q}}) \frac{\partial^2 G(\boldsymbol{r}_{\mathrm{p}}, \boldsymbol{r}_{\mathrm{q}})}{\partial n_{\mathrm{p}} \partial n_{\mathrm{q}}} - \frac{\partial p(\boldsymbol{r}_{\mathrm{q}})}{\partial n_{\mathrm{q}}} \frac{\partial G(\boldsymbol{r}_{\mathrm{p}}, \boldsymbol{r}_{\mathrm{q}})}{\partial n_{\mathrm{p}}} \right] d\Gamma. \tag{4.7}$$

Equation (4.6) is called the basic form, while Eq. (4.7) is called the normal derivative form or hypersingular formulation. The boundary element method is applicable to each formulation.

4.1.1.2 Sound Source and Boundary Conditions

If a sound source is located in the domain, the direct contribution to observation point p, $p_{\mathrm{d}}(\boldsymbol{r}_{\mathrm{p}})$, is added to the right side of Eq. (4.6) in the basic form, and its gradient, $\partial p_{\mathrm{d}}(\boldsymbol{r}_{\mathrm{p}})/\partial n_{\mathrm{p}}$, is added to the right side of Eq. (4.7) in the normal derivative form. For a monopole source at point s with the volume velocity Q, the direct sound pressure is given by

$$p_{\mathrm{d}}(\boldsymbol{r}_{\mathrm{p}}) = -j\omega\rho Q \frac{\exp(jkr_{\mathrm{ps}})}{4\pi r_{\mathrm{ps}}}. \tag{4.8}$$

When multiple sources exist, each contribution can be simply added.

Fig. 4.2 Three types of boundary condition

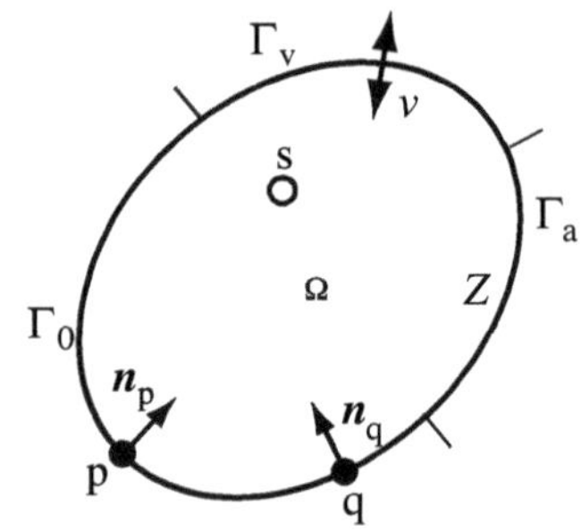

Regarding boundary Γ, as shown in Fig. 4.2, we consider three types of boundary: rigid boundary Γ_0, vibrating boundary Γ_v, and absorbing boundary Γ_a. Assuming local reaction on the boundaries, the boundary conditions are given by

$$\frac{\partial p(\boldsymbol{r}_\mathrm{q})}{\partial n_\mathrm{q}} = \begin{cases} 0 & \mathrm{q} \in \Gamma_0 \\ j\omega\rho v(\boldsymbol{r}_\mathrm{q}) & \mathrm{q} \in \Gamma_\mathrm{v} \\ -j\omega\rho Y(\boldsymbol{r}_\mathrm{q}) p(\boldsymbol{r}_\mathrm{q}) & \mathrm{q} \in \Gamma_\mathrm{a} \end{cases}, \tag{4.9}$$

where v is the normal vibration velocity (inward is positive), and Y is the normal acoustic admittance (reciprocal of normal impedance Z).

Reflecting the above boundary conditions, and adding the direct contribution of sound sources to Eq. (4.6) gives

$$\begin{aligned} \frac{1}{2}p(\boldsymbol{r}_\mathrm{p}) = p_\mathrm{d}(\boldsymbol{r}_\mathrm{p}) &+ \int_\Gamma \left[\frac{\partial G(\boldsymbol{r}_\mathrm{p}, \boldsymbol{r}_\mathrm{q})}{\partial n_\mathrm{q}} + aY(\boldsymbol{r}_\mathrm{q}) G(\boldsymbol{r}_\mathrm{p}, \boldsymbol{r}_\mathrm{q}) \right] p(\boldsymbol{r}_\mathrm{q})\, d\Gamma \\ &- a \int_\Gamma G(\boldsymbol{r}_\mathrm{p}, \boldsymbol{r}_\mathrm{q}) v(\boldsymbol{r}_\mathrm{q})\, d\Gamma, \end{aligned} \tag{4.10}$$

where $a = j\omega\rho$, and the point p is on a smooth boundary. On the other hand, Eq. (4.7) in the normal derivative form can be transformed into

$$\begin{aligned} \frac{a}{2} \left[v(\boldsymbol{r}_\mathrm{p}) - Y(\boldsymbol{r}_\mathrm{p}) p(\boldsymbol{r}_\mathrm{p}) \right] = &\frac{\partial p_\mathrm{d}(\boldsymbol{r}_\mathrm{p})}{\partial n_\mathrm{p}} \\ &+ \int_\Gamma \left[\frac{\partial^2 G(\boldsymbol{r}_\mathrm{p}, \boldsymbol{r}_\mathrm{q})}{\partial n_\mathrm{p} \partial n_\mathrm{q}} + aY(\boldsymbol{r}_\mathrm{q}) \frac{\partial G(\boldsymbol{r}_\mathrm{p}, \boldsymbol{r}_\mathrm{q})}{\partial n_\mathrm{p}} \right] p(\boldsymbol{r}_\mathrm{q})\, d\Gamma \\ &- a \int_\Gamma \frac{\partial G(\boldsymbol{r}_\mathrm{p}, \boldsymbol{r}_\mathrm{q})}{\partial n_\mathrm{p}} v(\boldsymbol{r}_\mathrm{q})\, d\Gamma, \end{aligned} \tag{4.11}$$

where $v(\boldsymbol{r}_\mathrm{p}) = 0$ if $\mathrm{p} \in \Gamma_0, \Gamma_\mathrm{a}$, and $Y(\boldsymbol{r}_\mathrm{p}) = 0$ if $\mathrm{p} \in \Gamma_0, \Gamma_\mathrm{v}$.

4.1.1.3 Discretization of Integral Equations

To numerically solve a boundary integral equation, it is discretized by dividing the boundary into elements, and by piecewise approximation of variables with nodal values. Constructing a linear system of discretized equations with the same number of nodes, and solving the system gives all nodal values on the boundary. As a discretization scheme, the collocation method is commonly used, and the Galerkin method [6, 7] is also used for a special formulation (see Sect. 4.2.1). In general, a variable on the boundary is approximately given by

$$x(\boldsymbol{r}) = \sum_{i=1}^{N} \varphi_i(\boldsymbol{r}) x_i = \boldsymbol{\varphi}^T(\boldsymbol{r})\boldsymbol{x}, \tag{4.12}$$

where N is the total number of nodes, x_i and φ_i are the nodal value and the interpolation function of the i-th node, respectively. Based on the above expression, discretization by the collocation method is explained in the following.

In the collocation method, a discretized equation is given by collocating the observation point p at each node on the boundary in the boundary integral equation. In the basic form, Eq. (4.10) is discretized with respect to the i-th node, as follows:

$$\begin{aligned}\frac{1}{2}p_i = p_{\mathrm{d},i} &+ \int_\Gamma \left\{ \frac{\partial G(\boldsymbol{r}_i, \boldsymbol{r}_\mathrm{q})}{\partial n_\mathrm{q}} + aG(\boldsymbol{r}_i, \boldsymbol{r}_\mathrm{q}) \left[\boldsymbol{\varphi}^T(\boldsymbol{r}_\mathrm{q})\boldsymbol{y}\right] \right\} \left[\boldsymbol{\varphi}^T(\boldsymbol{r}_\mathrm{q})\boldsymbol{p}\right] d\Gamma \\ &- a \int_\Gamma G(\boldsymbol{r}_i, \boldsymbol{r}_\mathrm{q}) \left[\boldsymbol{\varphi}^T(\boldsymbol{r}_\mathrm{q})\boldsymbol{v}\right] d\Gamma.\end{aligned} \tag{4.13}$$

Thus, we can compose the equations for all nodes into a linear system such that

$$\left(-\frac{1}{2}\boldsymbol{I} + \boldsymbol{H} + \boldsymbol{D}\right)\boldsymbol{p} = \boldsymbol{G}\boldsymbol{v} - \boldsymbol{p}_\mathrm{d}, \tag{4.14}$$

where $\boldsymbol{I}$ is the identity matrix, and the matrices $\boldsymbol{H}$, $\boldsymbol{G}$, and $\boldsymbol{D}$ have the following entries:

$$H_{ij} = \int_\Gamma \frac{\partial G(\boldsymbol{r}_i, \boldsymbol{r}_\mathrm{q})}{\partial n_\mathrm{q}} \varphi_j(\boldsymbol{r}_\mathrm{q})\, d\Gamma, \tag{4.15}$$

$$G_{ij} = a \int_\Gamma G(\boldsymbol{r}_i, \boldsymbol{r}_\mathrm{q}) \varphi_j(\boldsymbol{r}_\mathrm{q})\, d\Gamma, \tag{4.16}$$

$$D_{ij} = a \int_\Gamma G(\boldsymbol{r}_i, \boldsymbol{r}_\mathrm{q}) \left[\boldsymbol{\varphi}^T(\boldsymbol{r}_\mathrm{q})\boldsymbol{y}\right] \varphi_j(\boldsymbol{r}_\mathrm{q})\, d\Gamma. \tag{4.17}$$

In the case of using constant elements that approximate a constant value in each element, $\boldsymbol{D} = \boldsymbol{G}\boldsymbol{Y}$, with the diagonal admittance matrix $\boldsymbol{Y}$.

In the normal derivative form, Eq. (4.11) is discretized as

$$\frac{a}{2}(v_i - Y_i p_i) = \frac{\partial p_{\mathrm{d},i}}{\partial n_i} + \int_{\Gamma} \left\{ \frac{\partial^2 G(\boldsymbol{r}_i, \boldsymbol{r}_{\mathrm{q}})}{\partial n_i \partial n_{\mathrm{q}}} + a \frac{\partial G(\boldsymbol{r}_i, \boldsymbol{r}_{\mathrm{q}})}{\partial n_i} \left[\boldsymbol{\varphi}^T(\boldsymbol{r}_{\mathrm{q}}) \boldsymbol{y} \right] \right\} \left[\boldsymbol{\varphi}^T(\boldsymbol{r}_{\mathrm{q}}) \boldsymbol{p} \right] d\Gamma - a \int_{\Gamma} \frac{\partial G(\boldsymbol{r}_i, \boldsymbol{r}_{\mathrm{q}})}{\partial n_i} \left[\boldsymbol{\varphi}^T(\boldsymbol{r}_{\mathrm{q}}) \boldsymbol{v} \right] d\Gamma, \tag{4.18}$$

and then a linear system is composed as

$$\left(\boldsymbol{H}' + \boldsymbol{D}' + \frac{a}{2} \boldsymbol{Y} \right) \boldsymbol{p} = \left(\boldsymbol{G}' + \frac{a}{2} \boldsymbol{I} \right) \boldsymbol{v} - \boldsymbol{p}'_{\mathrm{d}}, \tag{4.19}$$

where $\boldsymbol{p}'_{\mathrm{d}}$ is the pressure gradient vector of the direct sound, and the matrices $\boldsymbol{H}'$, $\boldsymbol{G}'$, and $\boldsymbol{D}'$ have the following entries:

$$H'_{ij} = \int_{\Gamma} \frac{\partial^2 G(\boldsymbol{r}_i, \boldsymbol{r}_{\mathrm{q}})}{\partial n_i \partial n_{\mathrm{q}}} \varphi_j(\boldsymbol{r}_{\mathrm{q}}) \, d\Gamma, \tag{4.20}$$

$$G'_{ij} = a \int_{\Gamma} \frac{\partial G(\boldsymbol{r}_i, \boldsymbol{r}_{\mathrm{q}})}{\partial n_i} \varphi_j(\boldsymbol{r}_{\mathrm{q}}) \, d\Gamma, \tag{4.21}$$

$$D'_{ij} = a \int_{\Gamma} \frac{\partial G(\boldsymbol{r}_i, \boldsymbol{r}_{\mathrm{q}})}{\partial n_i} \left[\boldsymbol{\varphi}^T(\boldsymbol{r}_{\mathrm{q}}) \boldsymbol{y} \right] \varphi_j(\boldsymbol{r}_{\mathrm{q}}) \, d\Gamma. \tag{4.22}$$

In the case using constant elements, $\boldsymbol{D}' = \boldsymbol{G}'\boldsymbol{Y}$, as well as in the basic form.

Solving Eq. (4.14) or (4.19) gives nodal sound pressures on the boundary, however, the basic form is used except for a special case, because the treatment of singular integrals is complicated in the normal derivative form. For exterior problems, the nonuniqueness problem occurs at specific frequencies, so that special treatments are needed as is described later. In principle, the finite element method usually handles symmetric band matrices, whereas the boundary element method handles nonsymmetric and non-Hermitian dense matrices.

A post-processing is needed to calculate sound pressure in the domain from nodal sound pressures on the boundary. Sound pressure at a point, p, in the domain is expressed by

$$p(\boldsymbol{r}_\mathrm{p}) = p_\mathrm{d}(\boldsymbol{r}_\mathrm{p}) + \int_\Gamma \left\{ \frac{\partial G(\boldsymbol{r}_\mathrm{p}, \boldsymbol{r}_\mathrm{q})}{\partial n_\mathrm{q}} + aG(\boldsymbol{r}_\mathrm{p}, \boldsymbol{r}_\mathrm{q}) \left[\boldsymbol{\varphi}^T(\boldsymbol{r}_\mathrm{q})\boldsymbol{y} \right] \right\} \left[\boldsymbol{\varphi}^T(\boldsymbol{r}_\mathrm{q})\boldsymbol{p} \right] d\Gamma$$
$$- a \int_\Gamma G(\boldsymbol{r}_\mathrm{p}, \boldsymbol{r}_\mathrm{q}) \left[\boldsymbol{\varphi}^T(\boldsymbol{r}_\mathrm{q})\boldsymbol{v} \right] d\Gamma. \quad (4.23)$$

Let the row vectors $\boldsymbol{h}^T$, $\boldsymbol{g}^T$, and $\boldsymbol{d}^T$ have the entries replacing the i-th node with point p in Eqs. (4.15), (4.16), and (4.17), respectively, and the above equation is rewritten as

$$p(\boldsymbol{r}_\mathrm{p}) = p_\mathrm{d}(\boldsymbol{r}_\mathrm{p}) + \left[\boldsymbol{h}^T(\boldsymbol{r}_\mathrm{p}) + \boldsymbol{d}^T(\boldsymbol{r}_\mathrm{p}) \right] \boldsymbol{p} - \boldsymbol{g}^T(\boldsymbol{r}_\mathrm{p})\boldsymbol{v}. \quad (4.24)$$

Therefore, generating the row vectors for the observation point, and substituting the obtained nodal sound pressures, sound pressure in the domain can be calculated.

4.1.1.4 Boundary Elements and Numerical Integration

In the boundary element method, triangular or quadrangular elements are used in three-dimensional space. As shown in Fig. 4.3, there exists a variety of elements such as constant element, linear element, and quadratic element, and particularly, those using the same set of interpolation functions and shape functions are known as isoparametric elements. Regarding continuity of variables between elements, the elements are classified into the conforming or nonconforming type, and the constant and high-order elements that have no node on the sides are of the latter type. If the collocation method is applied to the normal derivative form, nonconforming elements are used since nodes should be located on smooth surface. In the following, a simple and an easy way to numerically treat constant elements is explained.

The constant element has one node in the center of the element, and the interpolation function, $\varphi_i = 1$ or $\varphi_i = 0$, if inside or outside the element. Accordingly, in the matrices generation for the basic form or the normal derivative form, each entry is calculated as follows:

$$H_{ij} = \int_{\Gamma_j} \frac{\partial G(\boldsymbol{r}_i, \boldsymbol{r}_\mathrm{q})}{\partial n_j} d\Gamma = -\frac{1}{4\pi} \int_{\Gamma_j} \left(1 - jkr_{i\mathrm{q}}\right) \frac{\partial r_{i\mathrm{q}}}{\partial n_j} \frac{\exp(jkr_{i\mathrm{q}})}{r_{i\mathrm{q}}^2} d\Gamma, \quad (4.25)$$

$$G_{ij} = a \int_{\Gamma_j} G(\boldsymbol{r}_i, \boldsymbol{r}_\mathrm{q})\, d\Gamma = -\frac{a}{4\pi} \int_{\Gamma_j} \frac{\exp(jkr_{i\mathrm{q}})}{r_{i\mathrm{q}}} d\Gamma, \quad (4.26)$$

or

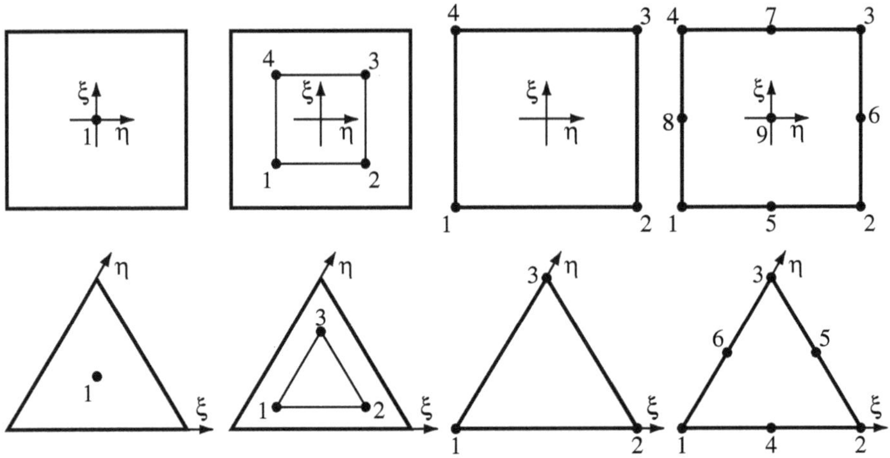

Fig. 4.3 Variation of boundary elements: constant and linear elements of nonconforming type, and linear and quadratic elements of conforming type (in *left*-to-*right*)

$$H'_{ij} = \int_{\Gamma_j} \frac{\partial^2 G(\boldsymbol{r}_i, \boldsymbol{r}_\mathrm{q})}{\partial n_i \partial n_j} \, d\Gamma$$
$$= \frac{1}{4\pi} \int_{\Gamma_j} \left[\left(3 - jkr_{i\mathrm{q}} - k^2 r_{i\mathrm{q}}^2\right) \frac{\partial r_{i\mathrm{q}}}{\partial n_i} \frac{\partial r_{i\mathrm{q}}}{\partial n_j} + (1 - jkr_{i\mathrm{q}}) \boldsymbol{n}_i \cdot \boldsymbol{n}_j \right] \cdot \frac{\exp(jkr_{i\mathrm{q}})}{r_{i\mathrm{q}}^3} \, d\Gamma, \tag{4.27}$$

$$G'_{ij} = a \int_{\Gamma_j} \frac{\partial G(\boldsymbol{r}_i, \boldsymbol{r}_\mathrm{q})}{\partial n_i} \, d\Gamma = -\frac{a}{4\pi} \int_{\Gamma_j} \left(1 - jkr_{i\mathrm{q}}\right) \frac{\partial r_{i\mathrm{q}}}{\partial n_i} \frac{\exp(jkr_{i\mathrm{q}})}{r_{i\mathrm{q}}^2} \, d\Gamma, \tag{4.28}$$

where Γ_j denotes the element for the j-th node, and $\partial r/\partial n = \boldsymbol{n} \cdot \boldsymbol{r}/r$. For the above calculations, numerical integration inside each element is needed, and the Gaussian quadrature [8] is commonly used except for the singular case of $i = j$. For example, after the transformation of three-dimensional global coordinates into two-dimensional local coordinates, the integral inside a quadrangular element is obtained by

$$\int_{\Gamma_j} f(\boldsymbol{r}) \, d\Gamma = \int_{-1}^{1} \int_{-1}^{1} f(\xi, \eta) \, |\boldsymbol{J}(\xi, \eta)| \, d\xi d\eta$$
$$\approx \sum_{m=1}^{K} \sum_{n=1}^{K} w_m w_n f(\xi_m, \eta_n) \, |\boldsymbol{J}(\xi_m, \eta_n)| \,, \tag{4.29}$$

where w_m and w_n are the weights, ξ_m and η_n are the Gauss nodes, K is the number of nodes, and the Jacobian determinant is

$$|\boldsymbol{J}(\xi,\eta)| = \left| \frac{d\boldsymbol{r}}{d\xi} \times \frac{d\boldsymbol{r}}{d\eta} \right|$$
$$= \sqrt{\left(\frac{dx}{d\xi}\frac{dy}{d\eta} - \frac{dx}{d\eta}\frac{dy}{d\xi}\right)^2 + \left(\frac{dy}{d\xi}\frac{dz}{d\eta} - \frac{dy}{d\eta}\frac{dz}{d\xi}\right)^2 + \left(\frac{dz}{d\xi}\frac{dx}{d\eta} - \frac{dz}{d\eta}\frac{dx}{d\xi}\right)^2}. \quad (4.30)$$

For triangular elements, Hammer's quadrature formula [9] is applicable in a similar way.

In the case of $i = j$, corresponding to diagonal entries, the Gaussian quadrature is not applicable due to the singularity of the integral kernels. With constant elements, $H_{ii} = 0$ and $G'_{ii} = 0$, whereas a special treatment is needed for G_{ii} and H'_{ii}. In polar coordinates with the origin at the node, letting R be the distance from the origin to a point on the sides, Eqs. (4.26) and (4.27) are converted to the contour integrals around the element [10], such that

$$G_{ii} = \frac{\rho c}{4\pi}\left\{\oint \exp[jkR(\theta)]\, d\theta - 2\pi\right\}, \quad (4.31)$$

$$H'_{ii} = \frac{1}{4\pi}\left\{2\pi jk - \oint \frac{\exp[jkR(\theta)]}{R(\theta)}\, d\theta\right\}. \quad (4.32)$$

The above integrals include no singularity, and can be calculated by piecewise approximation. Moreover, even if $i \neq j$, the use of the Gaussian quadrature yields low accuracy for two nearby nodes. For the nearly singular integrals, an advanced scheme called the projection and angular and radial transformation (PART) method [11] is available.

4.1.1.5 Nonuniqueness Problem

It is known that the boundary integral equation theoretically involves a nonuniqueness problem that a solution is not determined for exterior problems at fictitious eigenfrequencies. In the basic form, it occurs at the interior eigenfrequencies of the Neumann problems, while at those of the Dirichlet problems in the normal derivative form. Consequently, in the application of the boundary element method, computational accuracy becomes considerably worse at the frequencies. To avoid the problem, the Burton–Miller formulation [12] and the combined Helmholtz integral equation formulation (CHIEF) [13] are commonly used, and an alternative formulation based on the indirect method is proposed (see Sect. 4.2.1).

In the Burton–Miller method, two integral equations in the basic form and in the normal derivative form are linearly combined, and the following system is solved:

$$\left[\left(-\frac{1}{2}\boldsymbol{I}+\boldsymbol{H}+\boldsymbol{D}\right)+\alpha\left(\boldsymbol{H}'+\boldsymbol{D}'+\frac{a}{2}\boldsymbol{Y}\right)\right]\boldsymbol{p}=\left[\boldsymbol{G}+\alpha\left(\boldsymbol{G}'+\frac{a}{2}\boldsymbol{I}\right)\right]\boldsymbol{v}-\left(\boldsymbol{p}_{\mathrm{d}}+\alpha\boldsymbol{p}'_{\mathrm{d}}\right), \tag{4.33}$$

where α is the coupling factor with an imaginary part, and $\alpha = 1/jk$ is commonly used [14]. This method is theoretically simple, however, computational cost for matrices generation becomes greater.

In the CHIEF, multiple points are located inside a body in an exterior problem, and additional integral equations for the points are given with $\varepsilon = 0$ in Eq. (4.6). Then, an overdetermined system is constructed as

$$\begin{bmatrix} -\frac{1}{2}\boldsymbol{I}+\boldsymbol{H}+\boldsymbol{D} \\ \breve{\boldsymbol{H}}+\breve{\boldsymbol{D}} \end{bmatrix}\boldsymbol{p}=\begin{bmatrix} \boldsymbol{G} \\ \breve{\boldsymbol{G}} \end{bmatrix}\boldsymbol{v}-\begin{Bmatrix} \boldsymbol{p}_{\mathrm{d}} \\ \breve{\boldsymbol{p}}_{\mathrm{d}} \end{Bmatrix}, \tag{4.34}$$

where $\breve{\boldsymbol{H}}, \breve{\boldsymbol{G}}, \breve{\boldsymbol{D}}$, and $\breve{\boldsymbol{p}}_{\mathrm{d}}$ are the matrices and the vector for additional interior points. Representing the global matrix in the left side by $\boldsymbol{A}$, and applying the least squares method gives

$$\left(\boldsymbol{A}^H\boldsymbol{A}\right)\boldsymbol{p}=\boldsymbol{A}^H\left\{\begin{bmatrix} \boldsymbol{G} \\ \breve{\boldsymbol{G}} \end{bmatrix}\boldsymbol{v}-\begin{Bmatrix} \boldsymbol{p}_{\mathrm{d}} \\ \breve{\boldsymbol{p}}_{\mathrm{d}} \end{Bmatrix}\right\}, \tag{4.35}$$

where $(\ )^H$ denotes the Hermitian matrix. The above system can be solved with smaller computational cost than the Burton–Miller method, however, some strategies are required to the location and the number of additional points.

4.1.2 Time Domain BEM

4.1.2.1 Time Domain Boundary Integral Equation

In a similar way to the Helmholtz equation in the frequency domain, a test function Φ is introduced to the wave equation in the time domain, and applying the Green's theorem gives

$$\int_{\Gamma}\left[p(\boldsymbol{r},t)\frac{\partial\Phi(\boldsymbol{r},t)}{\partial n(\boldsymbol{r})}-\frac{\partial p(\boldsymbol{r},t)}{\partial n(\boldsymbol{r})}\Phi(\boldsymbol{r},t)\right]d\Gamma$$
$$+\int_{\Omega}p(\boldsymbol{r},t)\left[\nabla^2\Phi(\boldsymbol{r},t)-\frac{1}{c^2}\frac{\partial^2}{\partial t^2}\Phi(\boldsymbol{r},t)\right]d\Omega=0. \tag{4.36}$$

For the test function, we use fundamental solutions of the wave equation $\hat{G}$, that satisfy

$$\nabla^2 \tilde{G}(\boldsymbol{r}_\mathrm{p}, t|\boldsymbol{r}_\mathrm{q}, t_0) - \frac{1}{c^2}\frac{\partial^2}{\partial t^2}\tilde{G}(\boldsymbol{r}_\mathrm{p}, t|\boldsymbol{r}_\mathrm{q}, t_0) = -\delta(r_\mathrm{pq})\delta(t - t_0), \tag{4.37}$$

where t and t_0 are time at the observation point p and at the source point q. In three-dimensional space, the fundamental solution is given as

$$\tilde{G}(\boldsymbol{r}_\mathrm{p}, t|\boldsymbol{r}_\mathrm{q}, t_0) = \frac{1}{4\pi r_\mathrm{pq}}\delta\left(t - t_0 - \frac{r_\mathrm{pq}}{c}\right). \tag{4.38}$$

Substituting it into Eq. (4.36), and integrating the equation with respect to t_0 give

$$\begin{aligned}\varepsilon(\boldsymbol{r}_\mathrm{p})p(\boldsymbol{r}_\mathrm{p}, t) &= \int_0^t \int_\Gamma \left[p(\boldsymbol{r}_\mathrm{q}, t_0)\frac{\partial \tilde{G}(\boldsymbol{r}_\mathrm{p}, t|\boldsymbol{r}_\mathrm{q}, t_0)}{\partial n_\mathrm{q}} - \frac{\partial p(\boldsymbol{r}_\mathrm{q}, t_0)}{\partial n_\mathrm{q}}\tilde{G}(\boldsymbol{r}_\mathrm{p}, t|\boldsymbol{r}_\mathrm{q}, t_0)\right] d\Gamma\, dt_0 \\ &= -\frac{1}{4\pi}\int_\Gamma \left[\frac{\partial}{\partial n_\mathrm{q}} p\left(\boldsymbol{r}_\mathrm{q}, t - \frac{r_\mathrm{pq}}{c}\right)\frac{1}{r_\mathrm{pq}} + p\left(\boldsymbol{r}_\mathrm{q}, t - \frac{r_\mathrm{pq}}{c}\right)\frac{1}{r_\mathrm{pq}^2}\frac{\partial r_\mathrm{pq}}{\partial n_\mathrm{q}}\right. \\ &\quad \left. + \dot{p}\left(\boldsymbol{r}_\mathrm{q}, t - \frac{r_\mathrm{pq}}{c}\right)\frac{1}{c r_\mathrm{pq}}\frac{\partial r_\mathrm{pq}}{\partial n_\mathrm{q}}\right] d\Gamma, \end{aligned} \tag{4.39}$$

where $\dot{(\,)}$ denotes the time differential $\partial/\partial t$.

If a sound source is located in the domain, the direct contribution to observation point p, $p_\mathrm{d}(\boldsymbol{r}_\mathrm{p}, t)$, is added to the right side of Eq. (4.39). For a monopole source at point s with the volume velocity $Q(t)$, the direct sound pressure is given by

$$p_\mathrm{d}(\boldsymbol{r}_\mathrm{p}, t) = -\rho\dot{Q}\left(t - \frac{r_\mathrm{ps}}{c}\right)\frac{1}{4\pi r_\mathrm{ps}}. \tag{4.40}$$

Assuming local reaction on the three types of boundary, rigid boundary Γ_0, vibrating boundary Γ_v, and absorbing boundary Γ_a, the boundary conditions in the time domain are given by

$$\frac{\partial p(\boldsymbol{r}_\mathrm{q}, t)}{\partial n_\mathrm{q}} = \begin{cases} 0 & \mathrm{q} \in \Gamma_0 \\ -\rho\dot{v}(\boldsymbol{r}_\mathrm{q}, t) & \mathrm{q} \in \Gamma_\mathrm{v} \\ \rho\int_0^t y(\boldsymbol{r}_\mathrm{q}, \tau)p(\boldsymbol{r}_\mathrm{q}, t - \tau)\, d\tau & \mathrm{q} \in \Gamma_\mathrm{a} \end{cases}, \tag{4.41}$$

where $y(\tau)$ is the impulse admittance [15], that is, the inverse Fourier transform of the normal acoustic admittance $Y(f)$. Considering the frequency characteristics of admittance on the absorbing boundary, the convolution integral with the impulse admittance is involved as expressed above. If a frequency-independent real admittance is assumed, the boundary condition is simply given by $\partial p(\boldsymbol{r}_\mathrm{q}, t)/\partial n_\mathrm{q} = \rho Y(\boldsymbol{r}_\mathrm{q})\dot{p}(\boldsymbol{r}_\mathrm{q}, t)$. Reflecting the above boundary conditions, and adding the direct contribution of sound sources to Eq. (4.39), the following equation is given for the observation point p on a smooth boundary:

$$\frac{1}{2}p(\boldsymbol{r}_\mathrm{p},t) = p_\mathrm{d}(\boldsymbol{r}_\mathrm{p},t) - \frac{1}{4\pi}\int_\Gamma \frac{\partial r_\mathrm{pq}}{\partial n_\mathrm{q}}\left[\frac{1}{r_\mathrm{pq}^2}p\left(\boldsymbol{r}_\mathrm{q},t-\frac{r_\mathrm{pq}}{c}\right)+\frac{1}{cr_\mathrm{pq}}\dot{p}\left(\boldsymbol{r}_\mathrm{q},t-\frac{r_\mathrm{pq}}{c}\right)\right]d\Gamma - \frac{\rho}{4\pi}\int_\Gamma \frac{1}{r_\mathrm{pq}}\left[\int_0^{t-\frac{r_\mathrm{pq}}{c}} y\left(\boldsymbol{r}_\mathrm{q},\tau\right)\dot{p}\left(\boldsymbol{r}_\mathrm{q},t-\tau-\frac{r_\mathrm{pq}}{c}\right)d\tau - \dot{v}\left(\boldsymbol{r}_\mathrm{q},t-\frac{r_\mathrm{pq}}{c}\right)\right]d\Gamma. \tag{4.42}$$

4.1.2.2 Discretization of Time Domain Integral Equation

In the spatial domain, an interpolation function of each node φ_i is used as explained in Sect. 4.1.1.3, while in the time domain, the time interval $[0, T]$ is discretized with equal time steps Δt, and a continuous interpolation functions Ψ_k is introduced for each time step $k\Delta t$ $(k = 0, \ldots, M)$. Consequently, a variable on the boundary is approximately given by

$$x(\boldsymbol{r},t) = \sum_{i=1}^{N}\varphi(\boldsymbol{r})x_i(t) = \sum_{i=1}^{N}\varphi(\boldsymbol{r})\left[\sum_{k=0}^{M}\Psi_k(t)x_i^k\right] = \sum_{k=0}^{M}\Psi_k(t)\left[\boldsymbol{\varphi}^T(\boldsymbol{r})\boldsymbol{x}^k\right]. \tag{4.43}$$

Here, the time domain interpolation functions hold $\Psi_k(t) = \Psi(t - t_k)$ with a mother function Ψ, and the use of a high-order piecewise polynomial for the mother function improves numerical stability [16, 17].

Applying the collocation method in the spatial domain, Eq. (4.42) can be discretized at the time steps $t_m = m\Delta t$, as follows:

$$\frac{1}{2}p_i^m = p_{\mathrm{d},i}^m - \frac{1}{4\pi}\sum_{k=0}^{m}\int_\Gamma \frac{\partial r_\mathrm{iq}}{\partial n_\mathrm{q}}\left[\frac{1}{r_\mathrm{iq}^2}\Psi_k\left(t_m-\frac{r_\mathrm{iq}}{c}\right)+\frac{1}{cr_\mathrm{iq}}\dot{\Psi}_k\left(t_m-\frac{r_\mathrm{iq}}{c}\right)\right]\cdot\left[\boldsymbol{\varphi}^T(\boldsymbol{r}_\mathrm{q})\boldsymbol{p}^k\right]d\Gamma - \frac{\rho}{4\pi}\sum_{k=0}^{m}\int_\Gamma \frac{1}{r_\mathrm{iq}}\left\{\dot{\Psi}_k\left(t_m-\frac{r_\mathrm{iq}}{c}\right)\sum_{l=0}^{k}\Delta t\left[\boldsymbol{\varphi}^T(\boldsymbol{r}_\mathrm{q})\boldsymbol{y}^l\right]\left[\boldsymbol{\varphi}^T(\boldsymbol{r}_\mathrm{q})\boldsymbol{p}^{k-l}\right] - \dot{\Psi}_k\left(t_m-\frac{r_\mathrm{iq}}{c}\right)\left[\boldsymbol{\varphi}^T(\boldsymbol{r}_\mathrm{q})\boldsymbol{v}^k\right]\right\}d\Gamma, \tag{4.44}$$

where $\boldsymbol{y}^l$ is the vector of impulse admittance at the time step $\tau_l = l\Delta t$. In the last term on the right side, the summation for the absorbing boundary is transformed into

$$S_\mathrm{a} = \sum_{k=0}^{m}\int_\Gamma \frac{1}{r_\mathrm{iq}}\left\{\sum_{l=0}^{m-k}\Delta t\dot{\Psi}_{k+l}\left(t_m-\frac{r_\mathrm{iq}}{c}\right)\left[\boldsymbol{\varphi}^T(\boldsymbol{r}_\mathrm{q})\boldsymbol{y}^l\right]\right\}\left[\boldsymbol{\varphi}^T(\boldsymbol{r}_\mathrm{q})\boldsymbol{p}^k\right]d\Gamma, \tag{4.45}$$

and if a frequency-independent real admittance is assumed, it is simplified as

$$S_{\mathrm{a}}=\sum_{k=0}^{m}\int_{\Gamma}\frac{1}{r_{i\mathrm{q}}}\dot{\Psi}_{k}\left(t_{m}-\frac{r_{i\mathrm{q}}}{c}\right)\left[\boldsymbol{\varphi}^{T}(\boldsymbol{r}_{\mathrm{q}})\boldsymbol{y}\right]\left[\boldsymbol{\varphi}^{T}(\boldsymbol{r}_{\mathrm{q}})\boldsymbol{p}^{k}\right]d\Gamma. \tag{4.46}$$

Due to the time shift rule for Ψ_k, the integrands in Eq. (4.44) are only dependent on the relative time t_{m-k}, thus a recursive linear system is constructed as follows:

$$\left(\frac{1}{2}\boldsymbol{I}+\boldsymbol{A}^{0}+\boldsymbol{C}^{0}\right)\boldsymbol{p}^{m}=\boldsymbol{p}_{\mathrm{d}}^{m}-\sum_{k=1}^{m}\left(\boldsymbol{A}^{k}+\boldsymbol{C}^{k}\right)\boldsymbol{p}^{m-k}+\sum_{k=0}^{m}\boldsymbol{B}^{k}\boldsymbol{v}^{m-k}, \tag{4.47}$$

where $\boldsymbol{I}$ is the identity matrix, and the matrices $\boldsymbol{A}^k$, $\boldsymbol{B}^k$, and $\boldsymbol{C}^k$ have the following entries:

$$A_{ij}^{k}=\frac{1}{4\pi}\int_{\Gamma}\frac{\partial r_{i\mathrm{q}}}{\partial n_{\mathrm{q}}}\left[\frac{1}{r_{i\mathrm{q}}^{2}}\Psi\left(t_{k}-\frac{r_{i\mathrm{q}}}{c}\right)+\frac{1}{c r_{i\mathrm{q}}}\dot{\Psi}\left(t_{k}-\frac{r_{i\mathrm{q}}}{c}\right)\right]\varphi_{j}(\boldsymbol{r}_{\mathrm{q}})\,d\Gamma, \tag{4.48}$$

$$B_{ij}^{k}=\frac{\rho}{4\pi}\int_{\Gamma}\frac{1}{r_{i\mathrm{q}}}\dot{\Psi}\left(t_{k}-\frac{r_{i\mathrm{q}}}{c}\right)\varphi_{j}(\boldsymbol{r}_{\mathrm{q}})\,d\Gamma, \tag{4.49}$$

$$C_{ij}^{k}=\frac{\rho}{4\pi}\int_{\Gamma}\frac{1}{r_{i\mathrm{q}}}\left\{\sum_{l=0}^{k}\Delta t\dot{\Psi}\left(t_{k-l}-\frac{r_{i\mathrm{q}}}{c}\right)\left[\boldsymbol{\varphi}^{T}(\boldsymbol{r}_{\mathrm{q}})\boldsymbol{y}^{l}\right]\right\}\varphi_{j}(\boldsymbol{r}_{\mathrm{q}})\,d\Gamma. \tag{4.50}$$

In boundary integrals, the Gaussian quadrature can be applied to nondiagonal entries, while regarding diagonal entries, $A_{ii}^k = 0$, and a special treatment is needed for B_{ii}^k and C_{ii}^k due to the singularity. In a similar way to the treatment in the frequency domain, considering polar coordinates with the origin at the node of a constant element, and letting R be the distance from the origin to a point on the sides convert Eq. (4.49) to thc following contour integral around the element:

$$B_{ii}^{k}=\frac{\rho}{4\pi}\oint\left[\Psi\left(t_{k}-\frac{R(\theta)}{c}\right)-\Psi(t_{k})\right]d\theta \tag{4.51}$$

In the case with constant elements, $\boldsymbol{C}^k=\sum_{l=0}^{k}\Delta t\boldsymbol{B}^{k-l}\boldsymbol{Y}^l$ where $\boldsymbol{Y}^l$ is the diagonal matrix having the entries of the impulse admittance vector $\boldsymbol{y}^l$.

Numerically solving Eq. (4.47) at each time step, nodal sound pressures are recursively obtained. This method is known as the marching-on-in-time (MOT), and stability of recursive calculations can be improved by using a high-order quadrature rule with a relatively large time step. For example, good results were reported on the conditions using the three-order polynomial for Ψ, setting $c\Delta t \approx \Delta x$ for the element size Δx, and $\Delta t > 1/10f_{\max}$ for the upper limit frequency $f_{\max}$ [17, 18].

Furthermore, other time domain methods were proposed for the normal derivative form and the Burton–Miller formulation, and applied to scattering problems and so on [15–19].

4.2 Indirect Method and Domain Decomposition Method

This section introduces two advanced techniques in the frequency domain BEM. In the first half, an indirect method using degenerate boundary for thin bodies is presented with its applications and advantages in computation [3–5, 10, 20]. In the latter half, a domain decomposition method that treats sound field as a coupled system with subdivided domains is presented, giving its effective applications to such problems connecting closed spaces, hemi-free field, and so on [21–26].

4.2.1 Indirect BEM

4.2.1.1 Degenerate Boundary

If the conventional BEM is applied to thin bodies, computational accuracy is considerably spoiled due to close proximity of elements on the two sides of bodies. To overcome this difficulty, degenerate boundary is introduced assuming an extreme condition that a body has zero thickness, as shown in Fig. 4.4. Arbitrarily defining the front and back sides for degenerate boundary Γ_{d}, the two kinds of boundary integral equation for the observation point p on a smooth boundary, Eqs. (4.6) and (4.7), are transformed into

$$\frac{1}{2}\hat{p}(\boldsymbol{r}_{\mathrm{p}}) = \int_{\Gamma_{\mathrm{d}}} \left[\tilde{p}(\boldsymbol{r}_{\mathrm{q}}) \frac{\partial G(\boldsymbol{r}_{\mathrm{p}}, \boldsymbol{r}_{\mathrm{q}})}{\partial n_{\mathrm{q}}^{+}} - \frac{\partial \tilde{p}(\boldsymbol{r}_{\mathrm{q}})}{\partial n_{\mathrm{q}}^{+}} G(\boldsymbol{r}_{\mathrm{p}}, \boldsymbol{r}_{\mathrm{q}}) \right] d\Gamma, \tag{4.52}$$

$$\frac{1}{2}\frac{\partial \hat{p}(\boldsymbol{r}_{\mathrm{p}})}{\partial n_{\mathrm{p}}^{+}} = \int_{\Gamma_{\mathrm{d}}} \left[\tilde{p}(\boldsymbol{r}_{\mathrm{q}}) \frac{\partial^2 G(\boldsymbol{r}_{\mathrm{p}}, \boldsymbol{r}_{\mathrm{q}})}{\partial n_{\mathrm{p}}^{+} \partial n_{\mathrm{q}}^{+}} - \frac{\partial \tilde{p}(\boldsymbol{r}_{\mathrm{q}})}{\partial n_{\mathrm{q}}^{+}} \frac{\partial G(\boldsymbol{r}_{\mathrm{p}}, \boldsymbol{r}_{\mathrm{q}})}{\partial n_{\mathrm{p}}^{+}} \right] d\Gamma, \tag{4.53}$$

where $\hat{p} = p^{+} + p^{-}$, $\tilde{p} = p^{+} - p^{-}$, $p^{\pm}$ are the sound pressures on the front and back sides, and $\partial/\partial n^{+}$ denotes the inward normal derivative for the front side.

Giving three types of boundary condition on each side, the above two equations are expressed as

$$\frac{1}{2}\hat{p}(\boldsymbol{r}_{\mathrm{p}}) = p_{\mathrm{d}}(\boldsymbol{r}_{\mathrm{p}}) + \int_{\Gamma_{\mathrm{d}}} \frac{\partial G(\boldsymbol{r}_{\mathrm{p}}, \boldsymbol{r}_{\mathrm{q}})}{\partial n_{\mathrm{q}}^{+}} \tilde{p}(\boldsymbol{r}_{\mathrm{q}})\, d\Gamma$$

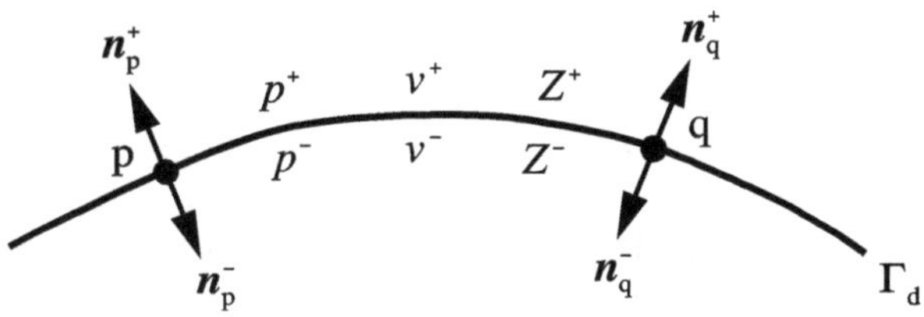

Fig. 4.4 Degenerate boundary

$$+a\int_{\Gamma_d} G(\boldsymbol{r}_p,\boldsymbol{r}_q)\left[\hat{Y}(\boldsymbol{r}_q)\hat{p}(\boldsymbol{r}_q)+\tilde{Y}(\boldsymbol{r}_q)\tilde{p}(\boldsymbol{r}_q)\right]d\Gamma$$
$$-a\int_{\Gamma_d} G(\boldsymbol{r}_p,\boldsymbol{r}_q)\hat{v}(\boldsymbol{r}_q)\,d\Gamma, \tag{4.54}$$

$$\frac{a}{2}\left[\tilde{v}(\boldsymbol{r}_p)-\tilde{Y}(\boldsymbol{r}_p)\hat{p}(\boldsymbol{r}_p)-\hat{Y}(\boldsymbol{r}_p)\tilde{p}(\boldsymbol{r}_p)\right]=\frac{\partial p_d(\boldsymbol{r}_p)}{\partial n_p^+}+\int_{\Gamma_d}\frac{\partial^2 G(\boldsymbol{r}_p,\boldsymbol{r}_q)}{\partial n_p^+\partial n_q^+}\tilde{p}(\boldsymbol{r}_q)\,d\Gamma$$
$$+a\int_{\Gamma_d}\frac{\partial G(\boldsymbol{r}_p,\boldsymbol{r}_q)}{\partial n_p^+}\left[\hat{Y}(\boldsymbol{r}_q)\hat{p}(\boldsymbol{r}_q)+\tilde{Y}(\boldsymbol{r}_q)\tilde{p}(\boldsymbol{r}_q)\right]d\Gamma$$
$$-a\int_{\Gamma_d}\frac{\partial G(\boldsymbol{r}_p,\boldsymbol{r}_q)}{\partial n_p^+}\hat{v}(\boldsymbol{r}_q)\,d\Gamma, \tag{4.55}$$

where $\hat{v}=v^+ + v^-$, $\tilde{v}=v^+ - v^-$, $\hat{Y}=(Y^+ + Y^-)/2$, $\tilde{Y}=(Y^+ - Y^-)/2$, $v^\pm$ and $Y^\pm$ are the normal vibration velocities and the normal acoustic admittances on the two sides. In the above forms, the sum and the difference of sound pressure are employed as variables on degenerate boundary, and the application of the BEM is commonly called as the indirect BEM.

4.2.1.2 Indirect Formulations

Dividing degenerate boundary into two-sided elements, and applying the collocation method to Eqs. (4.55) and (4.56) give

$$\left(-\frac{1}{2}\boldsymbol{I}+\hat{\boldsymbol{D}}\right)\hat{\boldsymbol{p}}+\left(\boldsymbol{H}+\tilde{\boldsymbol{D}}\right)\tilde{\boldsymbol{p}}=\boldsymbol{G}\hat{\boldsymbol{v}}-\boldsymbol{p}_d, \tag{4.56}$$

$$\left(\hat{\boldsymbol{D}}'+\frac{a}{2}\tilde{\boldsymbol{Y}}\right)\hat{\boldsymbol{p}}+\left(\boldsymbol{H}'+\tilde{\boldsymbol{D}}'+\frac{a}{2}\hat{\boldsymbol{Y}}\right)\tilde{\boldsymbol{p}}=\boldsymbol{G}'\hat{\boldsymbol{v}}+\frac{a}{2}\tilde{\boldsymbol{v}}-\boldsymbol{p}'_d, \tag{4.57}$$

where the matrices $\boldsymbol{H}$, $\boldsymbol{G}$, $\boldsymbol{H}'$, and $\boldsymbol{G}'$ correspond to Eqs. (4.15), (4.16), (4.20), and (4.21), $\hat{\boldsymbol{D}}$, $\tilde{\boldsymbol{D}}$, $\hat{\boldsymbol{D}}'$, and $\tilde{\boldsymbol{D}}'$ are given by substituting $\hat{Y}$ and $\tilde{Y}$ in Eqs. (4.17) and (4.22), respectively. Each of the above systems has equations of the same number of nodes, and the double number of unknowns for and. Therefore we cannot solve only either system in the basic form or in the normal derivative form. An approach to solve the problem is to couple the two kinds of system, which gives a greater system called the dual form [3, 10] as

$$\begin{bmatrix} -\frac{1}{2}\boldsymbol{I}+\hat{\boldsymbol{D}} & \boldsymbol{H}+\tilde{\boldsymbol{D}} \\ \hat{\boldsymbol{D}}'+\frac{a}{2}\tilde{\boldsymbol{Y}} & \boldsymbol{H}'+\tilde{\boldsymbol{D}}'+\frac{a}{2}\hat{\boldsymbol{Y}} \end{bmatrix} \begin{Bmatrix} \hat{\boldsymbol{p}} \\ \tilde{\boldsymbol{p}} \end{Bmatrix} = \begin{bmatrix} \boldsymbol{G} & \boldsymbol{0} \\ \boldsymbol{G}' & \frac{a}{2}\boldsymbol{I} \end{bmatrix} \begin{Bmatrix} \hat{\boldsymbol{v}} \\ \tilde{\boldsymbol{v}} \end{Bmatrix} - \begin{Bmatrix} \boldsymbol{p}_{\mathrm{d}} \\ \boldsymbol{p}'_{\mathrm{d}} \end{Bmatrix}. \tag{4.58}$$

This formulation requires the number of DOFs equal to the double number of nodes, however, it can be surely solved. In post-processing, sound pressure on each side of degenerate boundary is immediately obtained by $p^{\pm} = (\hat{p} \pm \tilde{p})/2$, while sound pressure in the domain needs to be calculated in a similar way as Eq. (4.24), by

$$p(\boldsymbol{r}_{\mathrm{p}}) = p_{\mathrm{d}}(\boldsymbol{r}_{\mathrm{p}}) + \left[\boldsymbol{h}^T(\boldsymbol{r}_{\mathrm{p}}) + \tilde{\boldsymbol{d}}^T(\boldsymbol{r}_{\mathrm{p}})\right]\tilde{\boldsymbol{p}} + \hat{\boldsymbol{d}}^T(\boldsymbol{r}_{\mathrm{p}})\hat{\boldsymbol{p}} - \boldsymbol{g}^T(\boldsymbol{r}_{\mathrm{p}})\hat{\boldsymbol{v}}, \tag{4.59}$$

where the row vectors $\hat{\boldsymbol{d}}^T$ and $\tilde{\boldsymbol{d}}^T$ are given by replacing the i-th node with point p, and substituting $\hat{Y}$ and $\tilde{Y}$ in Eq. (4.17), respectively.

As an ordinary case for thin plates, we consider the boundary condition where the two sides are rigid or vibrating together. In this case, the sum of vibration velocity on the two sides $\hat{v} = 0$, and the difference $\tilde{v} = 2v^+$, thus Eq. (4.57) is simplified in the following without $\hat{\boldsymbol{p}}$,

$$\boldsymbol{H}'\tilde{p} = a\boldsymbol{v}^+ - \boldsymbol{p}'_{\mathrm{d}}. \tag{4.60}$$

In a different way from the dual form, the above normal derivative form can be solved itself, which gives the difference of sound pressure on the degenerate boundary. In post-processing, sound pressure in the domain can be calculated by

$$p(\boldsymbol{r}_{\mathrm{p}}) = p_{\mathrm{d}}(\boldsymbol{r}_{\mathrm{p}}) + \boldsymbol{h}^T(\boldsymbol{r}_{\mathrm{p}})\tilde{\boldsymbol{p}} - 2\boldsymbol{g}^T(\boldsymbol{r}_{\mathrm{p}})\boldsymbol{v}^+, \tag{4.61}$$

and also, sound pressure on each side of degenerate boundary needs to be calculated by

$$p^{\pm}(\boldsymbol{r}_{\mathrm{p}}) = p_{\mathrm{d}}(\boldsymbol{r}_{\mathrm{p}}) + \boldsymbol{h}^T(\boldsymbol{r}_{\mathrm{p}})\tilde{\boldsymbol{p}} - 2\boldsymbol{g}^T(\boldsymbol{r}_{\mathrm{p}})\boldsymbol{v}^+ \pm \frac{1}{2}\tilde{p}(\boldsymbol{r}_{\mathrm{p}}). \tag{4.62}$$

As an alternative approach to the above case, the Galerkin method can be applied in discretization of the boundary integral equation [6, 7]. Reflecting the boundary condition to Eq. (4.55) gives the following normal derivative form:

$$av^+(\boldsymbol{r}_{\mathrm{p}}) = \frac{\partial p_{\mathrm{d}}(\boldsymbol{r}_{\mathrm{p}})}{\partial n_{\mathrm{p}}^+} + \int_{\Gamma_{\mathrm{d}}} \frac{\partial^2 G(\boldsymbol{r}_{\mathrm{p}}, \boldsymbol{r}_{\mathrm{q}})}{\partial n_{\mathrm{p}}^+ \partial n_{\mathrm{q}}^+}\tilde{p}(\boldsymbol{r}_{\mathrm{q}})\, d\Gamma. \tag{4.63}$$

Introducing the interpolation function of each node φ_i as a test function, the Galerkin method gives

$$a\int_{\Gamma_{\rm d}} \varphi_i(\boldsymbol{r}_{\rm p})\left[\boldsymbol{\varphi}^T(\boldsymbol{r}_{\rm p})\boldsymbol{v}^+\right] d\Gamma = \int_{\Gamma_{\rm d}} \varphi_i(\boldsymbol{r}_{\rm p})\left[\boldsymbol{\varphi}^T(\boldsymbol{r}_{\rm p})\boldsymbol{p'}_{\rm d}\right] d\Gamma$$
$$+\int_{\Gamma_{\rm d}} \varphi_i(\boldsymbol{r}_{\rm p})\left\{\int_{\Gamma_{\rm d}} \frac{\partial^2 G(\boldsymbol{r}_{\rm p},\boldsymbol{r}_{\rm q})}{\partial n_{\rm p}^+\partial n_{\rm q}^+}\left[\boldsymbol{\varphi}^T(\boldsymbol{r}_{\rm q})\tilde{\boldsymbol{p}}\right] d\Gamma_{\rm q}\right\} d\Gamma_{\rm p}. \tag{4.64}$$

Thus, we can compose the equations for all nodes into a linear system such that

$$\bar{\boldsymbol{H}}'\tilde{\boldsymbol{p}} = \boldsymbol{E}\left(a\boldsymbol{v}^+ - \boldsymbol{p'}_{\rm d}\right), \tag{4.65}$$

where the matrices $\bar{\boldsymbol{H}}'$ and $\boldsymbol{E}$ have the following entries:

$$\bar{H}'_{ij} = \int_{\Gamma}\int_{\Gamma} \frac{\partial^2 G(\boldsymbol{r}_{\rm p},\boldsymbol{r}_{\rm q})}{\partial n_{\rm p}\partial n_{\rm q}}\varphi_i(\boldsymbol{r}_{\rm p})\varphi_j(\boldsymbol{r}_{\rm q})\, d\Gamma_{\rm p}\, d\Gamma_{\rm q}, \tag{4.66}$$

$$E_{ij} = \int_{\Gamma} \varphi_i(\boldsymbol{r}_{\rm p})\varphi_j(\boldsymbol{r}_{\rm p})\, d\Gamma. \tag{4.67}$$

The above matrices are symmetric matrices, accordingly the Galerkin method brings advantages over the collocation method in memory requirements and solution of linear system, although computational time for matrix generation increases due to the double integrals. It is noted that the Galerkin method is applicable not only to the indirect formulation but also to the conventional formulation [7].

4.2.1.3 Application of the Dual Form to Exterior Problems

In the application of the BEM to exterior problems, computational accuracy becomes considerably worse at fictitious eigenfrequencies of the corresponding interior problems (see Sect. 4.1.1.5). As common measures to the difficulty, the Burton–Miller formulation and the combined Helmholtz integral equation formulation (CHIEF) are well known, while an alternative approach using the indirect method in the dual form has been proposed. As shown in Fig. 4.5, degenerate boundary is employed supposing an imaginary space just inside the body, and assuming that the acoustic impedance on the interior side, $Z = \rho c$ as absorbing boundary. This approach is also able to ensure computational accuracy [27], although a greater amount of memory storage is required with doubling the number of unknowns. In addition, if an iterative

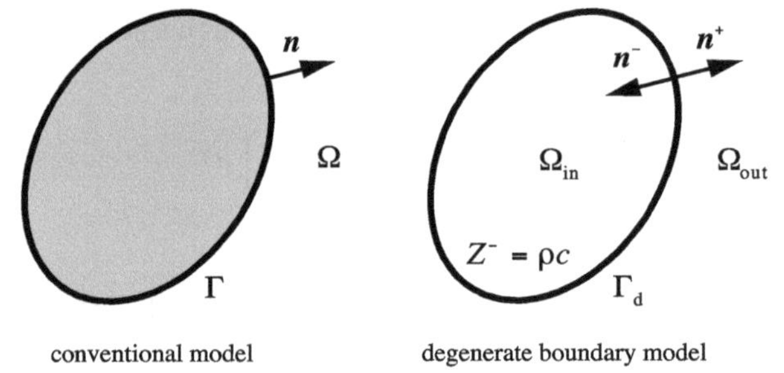

Fig. 4.5 Modeling of an exterior problem using degenerate boundary

solver is applied to the linear system, convergency becomes fairly good, resulting in an advantage in computational time [28].

4.2.2 Domain Decomposition Method

The FEM, FDM, and FDTD method can easily deal with fields composed of two or more media and temperature-gradient fields, because these methods discretize the analysis domain itself. On the other hand, the BEM can basically deal only with uniform-medium fields, because this method discretizes the boundary of the domain. In order to analyze such fields using the BEM, the domain decomposition method is applied. In this method, the whole domain is decomposed into subdomains, and the boundary integral equations for the subdomains are solved by using the continuity conditions on the interface boundaries. In the field of acoustics, the domain decomposition method is often applied not only to coupled fields composed of two or more media, such as air fields with porous-type absorbing materials [21, 25], but also to single-medium fields, such as coupled interior/exterior fields [22], fields including exterior half spaces [26] (see Sects. 7.3.3 and 7.5), and fields with thin bodies [23, 24].

4.2.2.1 Formulation

Figure 4.6 shows a sound field decomposed into two subdomains, Ω_I and Ω_II. Here subdomain Ω_I is bounded by two kinds of boundary: the boundary belonging only to the subdomain Ω_I, denoted by $\Gamma_{\mathrm{I,I}}$, and the interface boundary between subdomains Ω_I and Ω_II, denoted by $\Gamma_{\mathrm{I,II}}$. The same kind of denotation is used for subdomain Ω_II, the sound pressure p and the particle velocity v on boundaries. Numbers N_I and N_II denote those of boundary nodes for subdomains Ω_I and Ω_II, respectively, and N_a $(= N_\mathrm{I} + N_\mathrm{II})$ denotes the number of unknowns for the whole domain.

Discretizing the boundary integral equation for each subdomain gives the following two linear systems, which are other expressions of Eq. (4.14); matrix $(-\frac{1}{2}\boldsymbol{I} +$

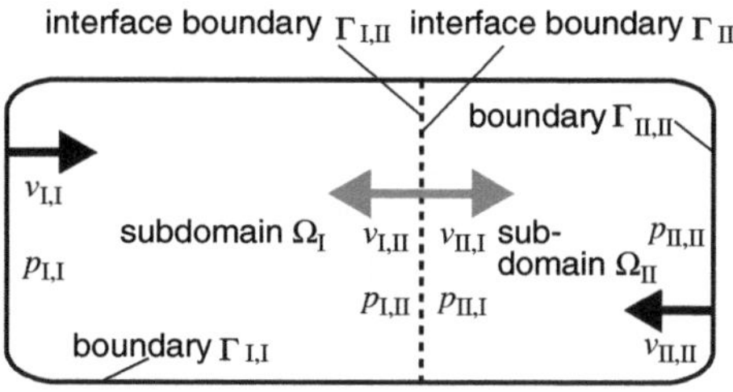

Fig. 4.6 Dedomposition of sound field

$\boldsymbol{H}+\boldsymbol{D}$) is denoted by $\boldsymbol{A}$, and matrices and vectors are divided according to kinds of boundary.

$$\begin{bmatrix} \boldsymbol{A}_{I1} & \boldsymbol{A}_{I2} \end{bmatrix} \begin{Bmatrix} \boldsymbol{p}_{I,I} \\ \boldsymbol{p}_{I,II} \end{Bmatrix} = \begin{bmatrix} \boldsymbol{G}_{I1} & \boldsymbol{G}_{I2} \end{bmatrix} \begin{Bmatrix} \boldsymbol{v}_{I,I} \\ \boldsymbol{v}_{I,II} \end{Bmatrix} - \boldsymbol{p}_{d,I}, \tag{4.68}$$

$$\begin{bmatrix} \boldsymbol{A}_{II1} & \boldsymbol{A}_{II2} \end{bmatrix} \begin{Bmatrix} \boldsymbol{p}_{II,II} \\ \boldsymbol{p}_{II,I} \end{Bmatrix} = \begin{bmatrix} \boldsymbol{G}_{II1} & \boldsymbol{G}_{II2} \end{bmatrix} \begin{Bmatrix} \boldsymbol{v}_{II,II} \\ \boldsymbol{v}_{II,I} \end{Bmatrix} - \boldsymbol{p}_{d,II}, \tag{4.69}$$

where $\boldsymbol{p}_{d,I}$ and $\boldsymbol{p}_{d,II}$ denote the direct sound vectors in subdomains Ω_I and Ω_{II}, respectively. Unknown vectors are $\boldsymbol{p}_{I,I}$, $\boldsymbol{p}_{I,II}$, and $\boldsymbol{v}_{I,II}$ in subdomain Ω_I, and $\boldsymbol{p}_{II,II}$, $\boldsymbol{p}_{II,I}$, and $\boldsymbol{v}_{II,I}$ in subdomain Ω_{II}.

The following equations of continuity for sound pressure and particle velocity are satisfied on the interface boundary:

$$\boldsymbol{p}_{I,II} = \boldsymbol{p}_{II,I}, \tag{4.70}$$

$$\boldsymbol{v}_{I,II} = -\boldsymbol{v}_{II,I}. \tag{4.71}$$

Applying Eqs. (4.70) and (4.71) to Eqs. (4.68) and (4.69) leads to a single linear system, including all unknown vectors in the whole domain, as:

$$\begin{bmatrix} \boldsymbol{A}_{I1} & \boldsymbol{0} & \boldsymbol{A}_{I2} & -\boldsymbol{G}_{I2} \\ \boldsymbol{0} & \boldsymbol{A}_{II1} & \boldsymbol{A}_{II2} & \boldsymbol{G}_{II2} \end{bmatrix} \begin{Bmatrix} \boldsymbol{p}_{I,I} \\ \boldsymbol{p}_{II,II} \\ \boldsymbol{p}_{I,II} \\ \boldsymbol{v}_{I,II} \end{Bmatrix} = \begin{bmatrix} \boldsymbol{G}_{I1} & \boldsymbol{0} \\ \boldsymbol{0} & \boldsymbol{G}_{II1} \end{bmatrix} \begin{Bmatrix} \boldsymbol{v}_{I,I} \\ \boldsymbol{v}_{II,II} \end{Bmatrix} - \begin{Bmatrix} \boldsymbol{p}_{d,I} \\ \boldsymbol{p}_{d,II} \end{Bmatrix}. \tag{4.72}$$

Solving Eq. (4.72) gives all values of unknowns on all boundaries including the interface boundary. Sound pressure at an observation point in a subdomain is calculated by substitution of the given values on the boundaries to the linear system obtained from the integral equation for the corresponding subdomain.

There are some other approaches to solve this type of problem. One approach is based on elimination of unknown vectors on the interface boundary from Eqs. (4.68) and (4.69). Another approach is using the iterative domain coupling technique [29], where separate computation for each subdomain and successive renewal of values on the interface boundaries between subdomains are performed until the whole domain

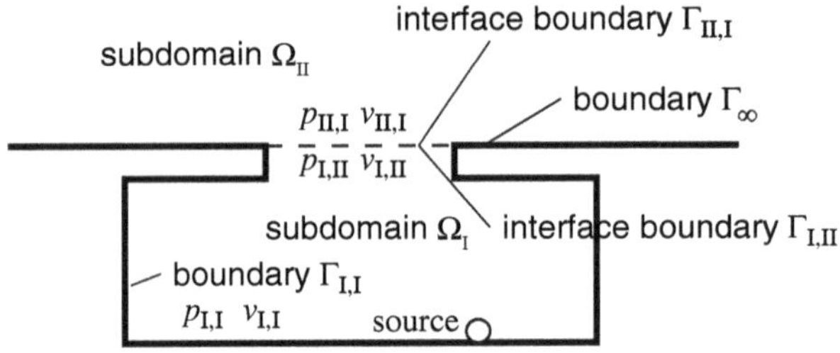

Fig. 4.7 Half space and depressed space under infinite rigid plane

converges. It has been reported that the latter one did not show good convergence for the acoustic BEM in the frequency domain [26].

4.2.2.2 Analysis of Sound Field Including Half Space

The BEM is a powerful method for half-space problems. When the analysis field is composed not only of a half space but also of a depressed space beyond an infinitely large plane, the domain decomposition method is often applied. Some of the application examples are analysis of sound propagation from a depressed road [26] (see Sect. 7.3), and analysis of flanking transmission sound through open windows in a building (see Sect. 7.5).

Figure 4.7 shows a sound field composed of a depressed space under an infinite rigid plane (subdomain Ω_{I}) and a half space above (subdomain Ω_{II}). A linear system Eq. (4.68) is finally obtained for subdomain Ω_{I} by a general domain decomposition method. As for subdomain Ω_{II}, the following fundamental solution for a half space, G_{inf}, is adopted instead of G in the integral equation Eq. (4.6):

$$G_{\mathrm{inf}}(\boldsymbol{r}_{\mathrm{p}}, \boldsymbol{r}_{\mathrm{q}}) = G(\boldsymbol{r}_{\mathrm{p}}, \boldsymbol{r}_{\mathrm{q}}) + G(\boldsymbol{r}_{\mathrm{p}'}, \boldsymbol{r}_{\mathrm{q}}), \tag{4.73}$$

where point p′ is the image of point p with respect to the infinite rigid plane. Equation $\partial G_{\mathrm{inf}}/\partial n = 0$ is satisfied when points p and q are on the plane including boundary Γ_∞. Since $\partial p(\boldsymbol{r}_{\mathrm{q}})/\partial n_{\mathrm{q}} = 0$ when $\mathrm{q} \in \Gamma_\infty$ and since $r_{\mathrm{pq}} = r_{\mathrm{p'q}}$ when $\mathrm{q} \in \Gamma_\infty, \Gamma_{\mathrm{II,I}}$, Eq. (4.6) can be simplified as the following, where the integration area is limited to boundary $\Gamma_{\mathrm{II,I}}$:

$$\begin{aligned} p(\boldsymbol{r}_{\mathrm{p}}) &= -\int_{\Gamma_{\mathrm{II,I}}} \frac{\partial p(\boldsymbol{r}_{\mathrm{q}})}{\partial n_{\mathrm{q}}} G_{\mathrm{inf}}(\boldsymbol{r}_{\mathrm{p}}, \boldsymbol{r}_{\mathrm{q}})\, d\Gamma \\ &= -2a \int_{\Gamma_{\mathrm{II,I}}} v_{\mathrm{II,I}}(\boldsymbol{r}_{\mathrm{q}}) G(\boldsymbol{r}_{\mathrm{p}}, \boldsymbol{r}_{\mathrm{q}})\, d\Gamma. \end{aligned} \tag{4.74}$$

Equation (4.74) is satisfied when point p is located in subdomain Ω_{II}, on boundary $\Gamma_{\mathrm{II,I}}$ or Γ_∞. The left hand side of Eq. (4.74) is not multiplied by $1/2$ even when point p is on $\Gamma_{\mathrm{II,I}}$ or Γ_∞, because point p and its image p′ coincide with each other.

The following linear system is obtained by discretizing Eq. (4.74):

$$\boldsymbol{p}_{\mathrm{II,I}} = -2\boldsymbol{G}_{\mathrm{II}}\boldsymbol{v}_{\mathrm{II,I}} = \boldsymbol{Z}_{\mathrm{II}}\boldsymbol{v}_{\mathrm{II,I}}, \tag{4.75}$$

where

$$G_{\mathrm{II},ij} = a \int_{\Gamma_{\mathrm{II,I}}} G(\boldsymbol{r}_i, \boldsymbol{r}_{\mathrm{q}})\varphi_j(\boldsymbol{r}_{\mathrm{q}})\, d\Gamma. \tag{4.76}$$

Matrix $\mathbf{Z}_{\mathrm{II}}$ is called an impedance matrix.

Applying Eqs. (4.70) and (4.71) to Eqs. (4.68) and (4.75) leads to a single linear system, including all unknown vectors in the whole domain, as:

$$\begin{bmatrix} \boldsymbol{A}_{\mathrm{I1}} & \boldsymbol{A}_{\mathrm{I2}} & -\boldsymbol{G}_{\mathrm{I2}} \\ \boldsymbol{0} & \boldsymbol{I} & \boldsymbol{Z}_{\mathrm{II}} \end{bmatrix} \begin{Bmatrix} \boldsymbol{p}_{\mathrm{I,I}} \\ \boldsymbol{p}_{\mathrm{I,II}} \\ \boldsymbol{v}_{\mathrm{I,II}} \end{Bmatrix} = \begin{bmatrix} \boldsymbol{G}_{\mathrm{I1}} \\ \boldsymbol{0} \end{bmatrix} \boldsymbol{v}_{\mathrm{I,I}} - \begin{Bmatrix} \boldsymbol{p}_{\mathrm{d,I}} \\ \boldsymbol{0} \end{Bmatrix}. \tag{4.77}$$

Solving Eq. (4.77) gives values of unknown vectors $\boldsymbol{p}_{\mathrm{I,II}}$ and $\boldsymbol{v}_{\mathrm{I,II}}$ on the interface boundary and $\boldsymbol{p}_{\mathrm{I,I}}$ on boundary $\Gamma_{\mathrm{I,I}}$. Sound pressure at an observation point in a subdomain is calculated by substitution of the given values on the boundaries to the linear system obtained from the integral equation for the corresponding subdomain.

4.2.2.3 Coupling Analysis with Sound Field in Porous-Type Absorbing Material

A sound field in a porous-type absorbing material satisfies the Helmholtz equation with the corresponding effective sound speed and effective medium density. Hence, not only an analysis of a sound field in a porous-type material but also a coupling analysis with such fields using the domain decomposition method can be easily realized using the BEM [21, 25]. The values of the effective sound speed and effective medium density in a porous-type material are generally complex numbers, which are obtained by Delany and Bazley's formula [30], Miki's formula [31], or the Rayleigh model [32] (see Sect. 8.1).

4.2.2.4 Analysis of Sound Field with Thin Bodies

When the standard BEM is applied to sound fields with thin bodies such as mufflers and ducts, its computational accuracy may be considerably worse. This is because boundary elements on both sides of the thin body are much closer to each other compared to element size, while the kernel function of the boundary element integration is singular or hypersingular. This problem can be avoided by applying the domain decomposition method, where subdomains are fields of both sides of the thin body [23, 24]. As mentioned in Sect. 4.2.1, an alternative approach for avoiding this problem is application of the indirect method in the dual form.

4.3 Application of Fast Multipole Method

In the BEM, a linear system with a dense matrix is required to be solved. This causes large amount of operation count and memory requirement, making it difficult to apply the BEM to large-scale problems. On the other hand, the fast multipole method (FMM) proposed by Rokhlin [33] has been widely applied as a fast solution for large-scale potential problems in a variety of fields [34–37], and its application to the BEM has been also progressed [38–54]. The BEM accelerated by the FMM is called the fast multipole BEM (FMBEM). Many studies have been conducted on the FMBEM for the Helmholtz equation (i.e., in the frequency domain) in the field of acoustics [42–55].

The points of the FMBEM are: (i) application of an iterative method to the linear system obtained from the boundary integral equation and (ii) efficient calculation of matrix-vector products, which is the largest computational load in the iterative process, using the FMM without producing the system matrix.

Here we present the outline of the FMBEM for the Helmholtz equation with some related topics. For more details of the FMM for the Helmholtz equation, refer to a state-of-the-art book [56].

4.3.1 Application of Iterative Method to BEM

The FMBEM requires an iterative method when solving a linear system obtained from the boundary integral equation, such as Eqs. (4.14), (4.19), and (4.33). Iterative methods are techniques that use successive approximations to obtain more accurate solutions to a linear system at each step, and Krylov subspace methods [57] are the main type of the methods. In a Krylov method, iterative calculation of matrix-vector products for the system matrix occupies large amount of total operations. The operation count for calculation of a matrix-vector product is $O(N^2)$, where N is the number of unknowns, because the system matrix for the BEM is dense. Hence, if the number of iterations is sufficiently smaller than N, application of an iterative method simply reduces computation time compared to a classical direct method such as Gaussian elimination, which requires an $O(N^3)$ operation count. However, the $O(N^2)$ memory requirement for storing the system matrix is not reduced.

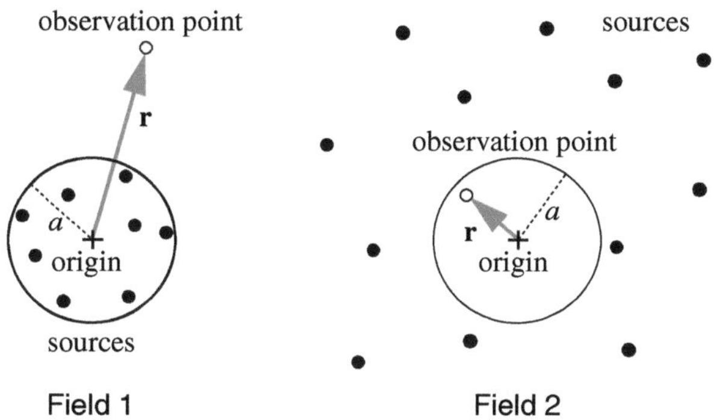

Fig. 4.8 Two kinds of steady-state sound field

4.3.2 Fast Multipole Method

4.3.2.1 Multipole Expansions and Local Expansions

Figure 4.8 shows two kinds of steady-state sound field. When all of the source points are nearer to the origin than an observation point, $\boldsymbol{r}$ (Field 1), the potential U at $\boldsymbol{r}$ can be represented as a function of $\boldsymbol{r}$ by multipole expansions at the origin as follows:

$$U(\boldsymbol{r}) = \sum_{n=0}^{\infty} \sum_{m=-n}^{n} M_n^m h_n^{(1)}(kr) Y_n^m(\theta, \varphi), \tag{4.78}$$

where $\boldsymbol{r} = (r, \theta, \varphi)$, $h_n^{(1)}$ are the spherical Hankel functions of the first kind, M_n^m are the multipole expansion coefficients at the origin, and Y_n^m are the spherical harmonics. On the other hand, when observation point $\boldsymbol{r}$ is nearer to the origin than all source points (Field 2), the potential U at $\boldsymbol{r}$ can be represented as another function of $\boldsymbol{r}$ by local expansions at the origin as follows:

$$U(\boldsymbol{r}) = \sum_{n=0}^{\infty} \sum_{m=-n}^{n} L_n^m j_n(kr) Y_n^m(\theta, \varphi), \tag{4.79}$$

where j_n are the spherical Bessel functions, and L_n^m are the local expansion coefficients at the origin. These two kinds of sound field can be expanded at other points. The translation of the multipole/local expansion coefficients at one point, $\boldsymbol{r}_1$, to those at another point, $\boldsymbol{r}_2$, can be generally expressed in a matrix-vector form as follows:

$$\tilde{\boldsymbol{C}} = \boldsymbol{W}(\boldsymbol{t})\boldsymbol{C}, \tag{4.80}$$

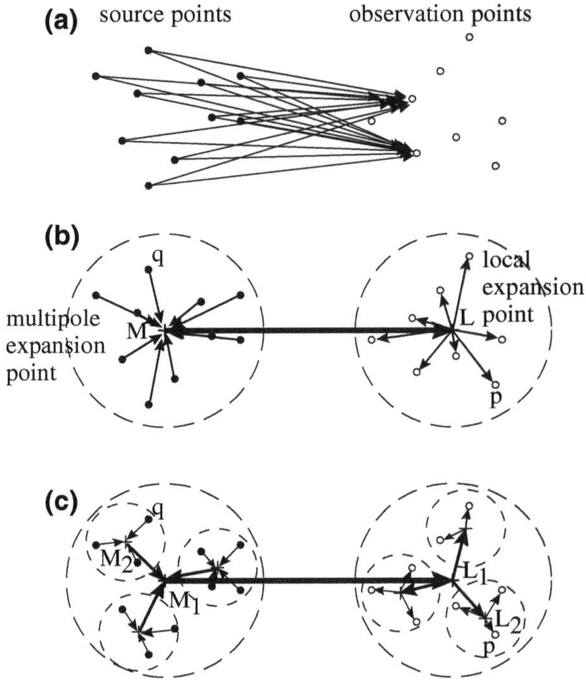

Fig. 4.9 Evaluation of potentials at observation points by: **a** computing interactions among all points, **b** FMM, and **c** MLFMM

where $\boldsymbol{C}$ and $\tilde{\boldsymbol{C}}$ are the vectors of multipole/local expansion coefficients at expansion points $\boldsymbol{r}_1$ and $\boldsymbol{r}_2$, respectively, $\boldsymbol{W}$ is the dense matrix for the translation of expansion coefficients, and $\boldsymbol{t} = \boldsymbol{r}_2 - \boldsymbol{r}_1$.

Calculation of matrix-vector products in a iterative method for the BEM is physically a summation of the potentials from all of the boundary elements. In the FMBEM, boundary elements and nodes are regarded as sources and observation points, respectively, and contributions from far-field elements are efficiently evaluated using multipole and local expansions.

4.3.2.2 Multilevel Expansions

Figure 4.9 shows three cases of potential evaluation at observation points. Instead of computing interactions among all points (Fig. 4.9a), contributions from far-field elements are evaluated in the FMM as follows (Fig. 4.9b):

- q2M: contributions from source points lying close together are represented by a multipole expansion at a multipole expansion point, M, where $r_{\mathrm{Mq}} < r_{\mathrm{pM}}$. Here the multipole expansion coefficients are calculated.

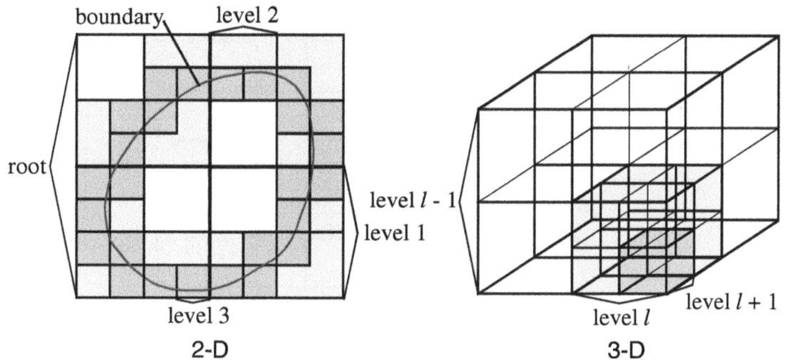

Fig. 4.10 Hierarchical cell structures

- M2L: contribution from the expansion point M is represented by a local expansion at a local expansion point, L, where $r_{\mathrm{pL}} < r_{\mathrm{LM}}$. Here the multipole expansion coefficients are translated to the local expansion ones.
- L2p: The potential at each observation point is computed by the local expansion.

As mentioned above, multipole and local expansions can be re-expanded at other points with translation of expansion coefficients. More rapid evaluation of potentials can be done by multilevel applications of multipole and local expansions (multilevel FMM: MLFMM [58, 59]). Hereafter, M2M and L2L denote translations between multipole expansions and those between local ones, respectively.

4.3.2.3 Grouping with Hierarchical Cell Structure

A hierarchical cell structure is introduced for grouping of sources and observation points. Multilevel applications of expansions mentioned above are achieved based on grouping with cells and their parent–child relation. Figure 4.10 shows boundaries and hierarchical cell structures in two and three dimensions. A cube (a square in two dimensions) circumscribing the whole boundary, called a root cell (level $l = 0$), is divided into eight child cubes ($l = 1$). Each divided cube is also divided repeatedly until a specified condition is satisfied; for example, until the average number of nodes in a cell at the lowest level, M, is less than a predetermined number ($l = 2, 3, \ldots, L$, where L is the lowest level number). It should be noted that a cube including more than one element is called a cell.

Definition of Cell Relations

Cell sets used in the rest of the chapter are defined as follows (see Fig. 4.11):

- neighbor cell set N_{m_l}: a cell set which consists of m_l itself and its neighbor cells.
- interaction cell set T_{m_l}: a cell set which consists of the cells which are not the neighbor cells of m_l but whose parent cells are neighbor ones of the parent cell of m_l.

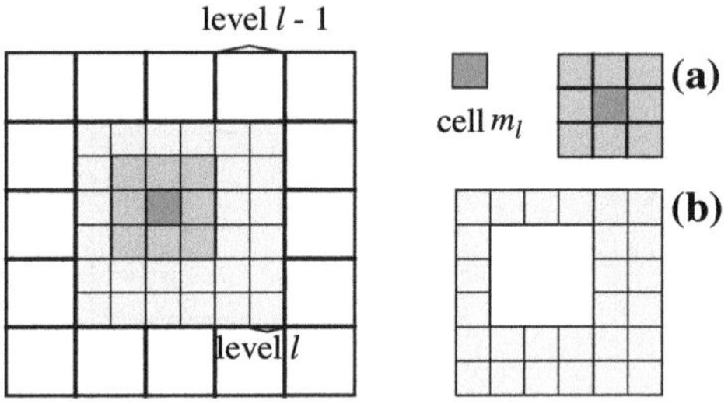

Fig. 4.11 Relations between cells at level l (in 2-D): **a** neighbor cell set N_{m_l} for cell m_l, and **b** interaction cell set T_{m_l} for cell m_l

- child cell set C_{m_l}: a cell set which consists of the child cells of m_l.
- element set G_{m_l}: an element set which consists of boundary elements in cell m_l.

4.3.2.4 Computational Efficiency

When the truncation number for infinite summations (referred to as p) in Eqs. (4.78) and (4.79) is set to be constant independent of N and level l, both the operation count and memory requirement for computation of matrix-vector products using the MLFMM are $O(N)$ [34]. The $O(N)$ efficiency, however, cannot be achieved in a high-frequency range because the number p for keeping computational accuracy increases with the analysis frequency (strictly, the dimensionless wavenumber kD, where D is the representative length of the problem or the diagonal length of a cell for grouping) [42, 59, 60]. In addition, since the number of entries of an expansion coefficient vector is p^2, the translation of expansion coefficients in Eq. (4.80) actually requires multiplication by a $p^2 \times p^2$ dense matrix, the computation of which requires an $O(p^5)$ operation count [61]. This results in huge operation count and memory requirement in total when p is large. Hence, improvement of computational efficiency for translation of expansion coefficients is critical for the FMM for the Helmholtz equation. On these backgrounds, two kinds of the FMM have been proposed: the high-frequency and low-frequency FMMs.

4.3.2.5 Two FMMs

High-Frequency FMM

As mentioned above, in a high-frequency range, the operation count for translation of multipole and local expansion coefficients is huge due to a large truncation number p. An efficient translation technique to overcome this problem has been well known, where the expansion coefficients C_n^m are translated to far-field signature functions $c(\boldsymbol{s})$, a kind of direction-dependent functions, using spherical harmonic expansions or plane wave expansions [62].

$$c(s) = \sum_{nm} j^{-n} C_n^m Y_n^m(s), \tag{4.81}$$

$$C_n^m = j^n \oint c(s) Y_n^{-m}(s)\, ds, \tag{4.82}$$

where $s = (\theta, \phi)$ is the unit vector, and $\oint ds$ represents the integral over the unit sphere. The translation matrices for $c(s)$ is diagonal ones, achieving efficient translations. Both the operation count and memory requirement for computation of matrix-vector products using this technique are $O(N^a \log^b N)$, where $1 \le a \le 2$ and $b \ge 0$, depending on the geometry of problem and implementation. However, this technique has numerical instability at low frequencies, i.e., at small kD. Hence, application of the high-frequency FMM should be limited in a high-frequency range.

Low-Frequency FMM

In a low-frequency range, an $O(N)$ efficiency is achieved because the truncation number p can be set constant. However, the operation count for translation of coefficients is $O\left(p^5\right)$ or $O\left(p^4\right)$, which spoils the whole efficiency. Some techniques have been proposed to overcome this inefficiency: techniques based on plane wave expansions [63, 64], those using Taylor expansions or Legendre polynomials [56], and those using recurrence relations [56, 61]. Many of the techniques are applied in conjunction with coordinate rotation techniques for efficient translation, making use of the characteristics of spherical harmonics. These techniques improve the efficiency of translations of coefficients to $O\left(p^3\right)$.

4.3.3 *Fast Calculation of Matrix-Vector Products*

When the representative element size d is not so small compared to the analysis wavelength λ, for example, $5 \le \lambda/d \le 10$, the high-frequency FMM can give high-accuracy results with high efficiency, independent of the analysis frequency. Here we present the procedures for calculation of matrix-vector products in the basic form, normal derivative form, and Burton–Miller formulation in the BEM, using the high-frequency FMM. Refer to Refs. [50, 54], etc., for the low-frequency FMM and its application to a variety of formulations.

4.3.3.1 Expansions of Matrix Entries

Applying the multipole expansion and spherical harmonic expansion to the fundamental solution Eq. (4.4) for the Helmholtz equation in three-dimensional space gives the following equation [60, 62, 65]:

$$G(\boldsymbol{r}_\mathrm{p}, \boldsymbol{r}_\mathrm{q}) = \frac{jk}{16\pi^2} \oint \underbrace{E_\mathrm{pL}(\boldsymbol{k})}_{\mathrm{L2p}} \underbrace{T_\mathrm{LM}(\boldsymbol{k})}_{\mathrm{M2L}} \underbrace{E_\mathrm{Mq}(\boldsymbol{k})}_{\mathrm{q2M}} \, d\hat{\boldsymbol{k}}, \tag{4.83}$$

where

$$T_\mathrm{LM}(\boldsymbol{k}) = \sum_{l=0}^{N_\mathrm{c}} j^l (2l+1) h_l^{(1)}(kr_\mathrm{LM}) P_l(\hat{\boldsymbol{k}} \cdot \hat{\boldsymbol{r}}_\mathrm{LM}), \tag{4.84}$$

$$E_\mathrm{Mq}(\boldsymbol{k}) = \exp(j\boldsymbol{k} \cdot \boldsymbol{r}_\mathrm{Mq}), \tag{4.85}$$

$\boldsymbol{k}$ is the wave number vector, $k = |\boldsymbol{k}|$, $\hat{\boldsymbol{k}} = \boldsymbol{k}/k$, P_l are the Legendre polynomials, N_c is the truncation number for infinite summation, and $\oint d\hat{\boldsymbol{k}}$ represents the integral over the unit sphere. Here the fundamental solution Eq. (4.4), a function of r_pq, is expressed by using three functions of vectors $\boldsymbol{r}_\mathrm{pL}$, $\boldsymbol{r}_\mathrm{LM}$, and $\boldsymbol{r}_\mathrm{Mq}$, respectively. In addition, Eq. (4.83) can be transformed into the following expression, which corresponds to the computational procedures described in Sect. 4.3.3.2 using the hierarchical cell structure:

$$G(\boldsymbol{r}_\mathrm{p}, \boldsymbol{r}_\mathrm{q}) = \frac{jk}{16\pi^2} \oint \underbrace{E_{\mathrm{p}\lambda_{m_L}}(\boldsymbol{k})}_{\mathrm{L2p}} \underbrace{\prod_{l=I}^{L-1} E_{\lambda_{m_{l+1}}\lambda_{m_l}}(\boldsymbol{k})}_{\mathrm{L2L}}$$
$$\times \underbrace{T_{\lambda_{m_I}\lambda_{m'_I}}(\boldsymbol{k})}_{\mathrm{M2L}} \underbrace{\prod_{l=I}^{L-1} E_{\lambda_{m'_l}\lambda_{m'_{l+1}}}(\boldsymbol{k})}_{\mathrm{M2M}} \underbrace{E_{\lambda_{m'_L}\mathrm{q}}(\boldsymbol{k})}_{\mathrm{q2M}} \, d\hat{\boldsymbol{k}}, \tag{4.86}$$

where $m'_I \in \mathrm{T}_{m_I}$, and I is the level number to execute Step 3 (described in Sect. 4.3.3.2), which is determined by the positions of points p and q. The entries of matrix-vector products $(\boldsymbol{H}+\boldsymbol{D})\boldsymbol{p}$ and $\boldsymbol{G}\boldsymbol{v}$ in Eq. (4.14) and $(\boldsymbol{H}'+\boldsymbol{D}')\boldsymbol{p}$ and $\boldsymbol{G}'\boldsymbol{v}$ in Eq. (4.19) are calculated using the following equations, based on Eq. (4.86):

$$\left\{ \begin{array}{c} (H_{ij}+D_{ij})p_j \\ G_{ij}v_j \end{array} \right\} = \frac{jk}{16\pi^2} \oint E_{i\lambda_{m_L}}(\boldsymbol{k}) \prod_{l=I}^{L-1} E_{\lambda_{m_{l+1}}\lambda_{m_l}}(\boldsymbol{k})$$
$$\times T_{\lambda_{m_I}\lambda_{m'_I}}(\boldsymbol{k}) \prod_{l=I}^{L-1} E_{\lambda_{m'_l}\lambda_{m'_{l+1}}}(\boldsymbol{k}) \left\{ \begin{array}{c} \left[\beta_{\lambda_{m'_L}j}(\boldsymbol{k}) + \gamma_{\lambda_{m'_L}j}(\boldsymbol{k})\right] p_j \\ \alpha_{\lambda_{m'_L}j}(\boldsymbol{k}) v_j \end{array} \right\} d\hat{\boldsymbol{k}}, \tag{4.87}$$

$$\left\{ \begin{array}{c} (H'_{ij}+D'_{ij})p_j \\ G'_{ij}v_j \end{array} \right\} = \frac{-k^2}{16\pi^2} (\boldsymbol{n}_i \cdot \hat{\boldsymbol{k}}) \oint E_{i\lambda_{m_L}}(\boldsymbol{k}) \prod_{l=I}^{L-1} E_{\lambda_{m_{l+1}}\lambda_{m_l}}(\boldsymbol{k})$$

$$\times T_{\lambda_{m_I}\lambda_{m'_I}}(\boldsymbol{k}) \prod_{l=I}^{L-1} E_{\lambda_{m'_l}\lambda_{m'_{l+1}}}(\boldsymbol{k}) \left\{ \begin{matrix} \left[\beta_{\lambda_{m'_L}j}(\boldsymbol{k}) + \gamma_{\lambda_{m'_L}j}(\boldsymbol{k})\right] p_j \\ \alpha_{\lambda_{m'_L}j}(\boldsymbol{k}) v_j \end{matrix} \right\} d\hat{\boldsymbol{k}}, \tag{4.88}$$

where

$$\alpha_{\lambda_{m'_L}j}(\boldsymbol{k}) = a \int_\Gamma E_{\lambda_{m'_L}\mathrm{q}}(\boldsymbol{k}) \varphi_j(\boldsymbol{r}_\mathrm{q})\, d\Gamma, \tag{4.89}$$

$$\beta_{\lambda_{m'_L}j}(\boldsymbol{k}) = jk \int_\Gamma E_{\lambda_{m'_L}\mathrm{q}}(\boldsymbol{k})(\boldsymbol{n}_\mathrm{q}\cdot\hat{\boldsymbol{k}}) \varphi_j(\boldsymbol{r}_\mathrm{q})\, d\Gamma, \tag{4.90}$$

$$\gamma_{\lambda_{m'_L}j}(\boldsymbol{k}) = a \int_\Gamma E_{\lambda_{m'_L}\mathrm{q}}(\boldsymbol{k}) \left[\boldsymbol{\varphi}^T(\boldsymbol{r}_\mathrm{q})\boldsymbol{y}\right] \varphi_j(\boldsymbol{r}_\mathrm{q})\, d\Gamma. \tag{4.91}$$

In the procedures for computation, the integral $\oint d\hat{\boldsymbol{k}}$ is calculated numerically, using the following equation:

$$\oint f(\hat{\boldsymbol{k}})\, d\hat{\boldsymbol{k}} = \sum_{n=1}^{K} w_n f(\hat{\boldsymbol{k}}_n) = \sum_{i=1}^{N_\mathrm{k}} \sum_{j=1}^{2N_\mathrm{k}} w_i^g w_j^c f(\theta_i, \varphi_j), \tag{4.92}$$

where $w_n = w_i^g w_j^c$ is the weight at the quadrature point $\hat{\boldsymbol{k}}_n$, and $K = 2N_\mathrm{k}^2$ is the number of quadrature points. Contributions from/to cells are evaluated using coefficients at $\hat{\boldsymbol{k}}_n$ in Steps 1 to 5. These coefficients correspond to the far-field signature functions expressed as Eq. (4.81), and are called outgoing, interaction, and incoming coefficients, denoted by ξ, τ, and ζ, respectively.

4.3.3.2 Computational Procedures

Here we present a concrete procedure for calculating matrix-vector products $(-\frac{1}{2}\boldsymbol{I} + \boldsymbol{H} + \boldsymbol{D})\boldsymbol{p}$ and $\boldsymbol{G}\boldsymbol{v}$ in Eq. (4.14) for the basic form, $(\boldsymbol{H}' + \boldsymbol{D}' + \frac{a}{2}\boldsymbol{Y})\boldsymbol{p}$ and $(\boldsymbol{G}' + \frac{a}{2}\boldsymbol{I})\boldsymbol{v}$ in Eq. (4.19) for the normal derivative form, and their combined products in Eq. (4.33) for the Burton–Miller formulation. The computational procedures consist of six steps. Steps 1 to 5 correspond to evaluation of far influence using the MLFMM, and Step 6 corresponds to that of near influence in the same manner as the conventional BEM. Refer to Fig. 4.12.

Step 1 (q2M): Compute the outgoing coefficients ξ_{m_L} of each cell m_L at each quadrature point $\hat{\boldsymbol{k}}_n^L$ at the lowest level L, by

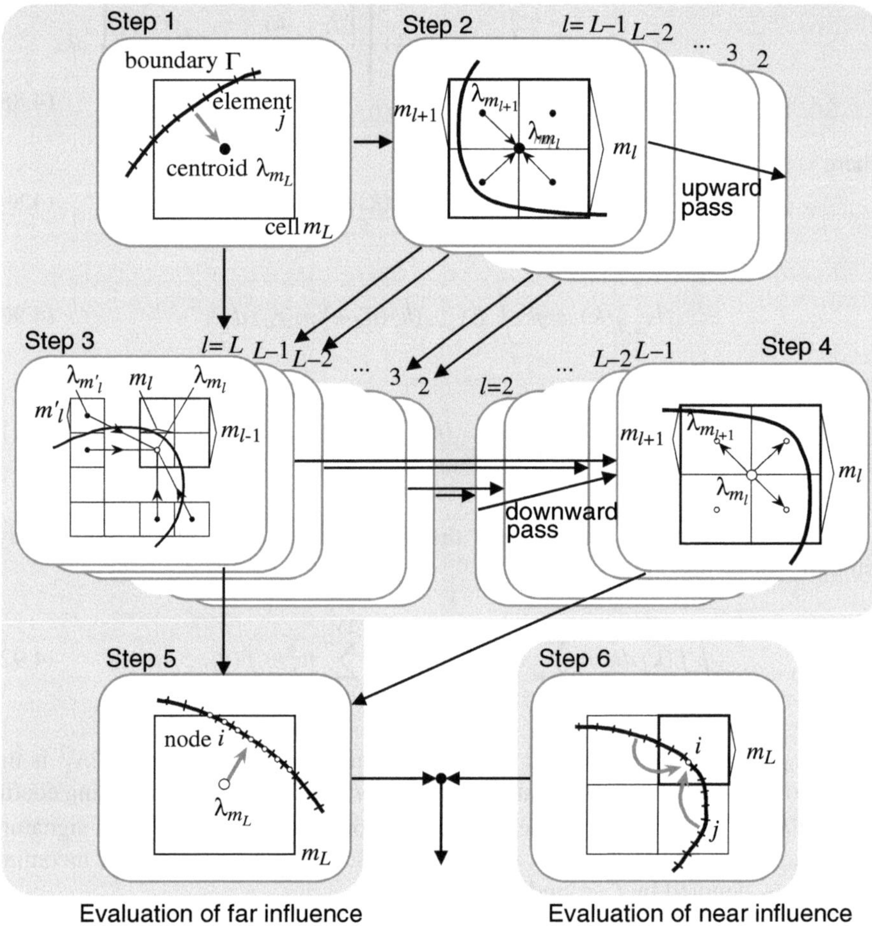

Fig. 4.12 Computational procedures for matrix-vector products in FMBEM

$$\left\{ \begin{array}{c} \xi^{\mathrm{p}}_{m_L}\left(\hat{\boldsymbol{k}}^L_n\right) \\ \xi^{\mathrm{v}}_{m_L}\left(\hat{\boldsymbol{k}}^L_n\right) \end{array} \right\} = \sum_{j \in \mathrm{G}_{m_L}} \left\{ \begin{array}{c} \left[\beta_{\lambda_{m_L} j}\left(\hat{\boldsymbol{k}}^L_n\right) + \gamma_{\lambda_{m_L} j}\left(\hat{\boldsymbol{k}}^L_n\right)\right] p_j \\ \alpha_{\lambda_{m_L} j}\left(\hat{\boldsymbol{k}}^L_n\right) v_j \end{array} \right\}, \tag{4.93}$$

where the superscripts (p) and (v) denote that the coefficients are for the left- and right-hand sides of Eq. (4.14), (4.19), or (4.33), respectively. In the following, we omit these superscripts when the two equations become identical.

Step 2 (M2M): Compute the outgoing coefficients ξ_{m_l} of each cell m_l at each quadrature point $\hat{\boldsymbol{k}}^l_{n'}$ at the next higher level l, by

$$\xi_{m_l}(\hat{\boldsymbol{k}}_{n'}^{l}) = \sum_{m_{l+1} \in \mathrm{C}_{m_l}} E_{\lambda_{m_l}\lambda_{m_{l+1}}}(\hat{\boldsymbol{k}}_{n'}^{l}) \sum_{n=1}^{K_{l+1}} W_{n'n} \xi_{m_{l+1}}(\hat{\boldsymbol{k}}_{n}^{l+1}), \tag{4.94}$$

where K_l is the number of the quadrature points for the spherical integral at level l, and $W_{n'n}$ are the interpolation coefficients. Interpolation for quadrature points is introduced here because the number of the quadrature points for keeping computational accuracy increases at higher levels ($K_l \geq K_{l+1}$). The interpolation method based on FFT [66] and Legendre interpolation [59, 67] are often used. This computation is executed in the upward order at each level ($l = L-1, L-2, \ldots, 2$).

Step 3 (M2L): Compute the interaction coefficients τ_{m_l} of each cell m_l at each quadrature point $\hat{\boldsymbol{k}}_n^{l}$ at each level ($l = 2, 3, \ldots, L$), by

$$\tau_{m_l}\left(\hat{\boldsymbol{k}}_{n}^{l}\right) = \sum_{m'_l \in \mathrm{T}_{m_l}} T_{\lambda_{m_l}\lambda_{m'_l}}(\hat{\boldsymbol{k}}_{n'}^{l}) \xi_{m'_l}\left(\hat{\boldsymbol{k}}_{n'}^{l}\right). \tag{4.95}$$

Step 4 (L2L): Compute the incoming coefficients $\zeta_{m_{l+1}}$ of each cell m_{l+1} at each quadrature point $\hat{\boldsymbol{k}}_n^{l+1}$ at the next lower level $l+1$, by

$$\zeta_{m_{l+1}}\left(\hat{\boldsymbol{k}}_{n}^{l+1}\right) = \sum_{n'=1}^{K_l} \frac{w_{n'}^{l}}{w_{n}^{l+1}} W_{n'n} E_{\lambda_{m_{l+1}}\lambda_{m_l}}\left(\hat{\boldsymbol{k}}_{n'}^{l}\right)\left[\zeta_{m_l}\left(\hat{\boldsymbol{k}}_{n'}^{l}\right) + \tau_{m_l}(\hat{\boldsymbol{k}}_{n'}^{l})\right], \tag{4.96}$$

where $\zeta_{m_2}(\hat{\boldsymbol{k}}_{n'}^{2}) = 0$. Adjoint interpolation [68], a kind of filtering for quadrature points, is introduced here because the number of the quadrature points for keeping computational accuracy decreases at lower levels ($K_{l+1} \leq K_l$). The filtering method based on FFT [66] is also often used. This computation is executed in the downward order at each level ($l = 2, 3, \ldots, L-1$).

Step 5 (L2p): Compute the far influence at each node i, $\phi_{\mathrm{F},i}$ for the basic form, $\varphi_{\mathrm{F},i}$ for the normal derivative form, or $\phi_{\mathrm{F},i} + \alpha\varphi_{\mathrm{F},i}$ for the Burton–Miller formulation, at the lowest level L, by

$$\phi_{\mathrm{F},i} = \frac{jk}{16\pi^2} \sum_{n=1}^{K_L} w_n^L E_{i\lambda_{m_L}}\left(\hat{\boldsymbol{k}}_{n}^{L}\right)\left[\zeta_{m_L}\left(\hat{\boldsymbol{k}}_{n}^{L}\right) + \tau_{m_L}\left(\hat{\boldsymbol{k}}_{n}^{L}\right)\right], \tag{4.97}$$

$$\varphi_{\mathrm{F},i} = \frac{-k^2}{16\pi^2} \sum_{n=1}^{K_L} w_n^L (\boldsymbol{n}_i \cdot \hat{\boldsymbol{k}}_{n}^{L}) E_{i\lambda_{m_L}}\left(\hat{\boldsymbol{k}}_{n}^{L}\right)\left[\zeta_{m_L}\left(\hat{\boldsymbol{k}}_{n}^{L}\right) + \tau_{m_L}\left(\hat{\boldsymbol{k}}_{n}^{L}\right)\right], \tag{4.98}$$

$$\phi_{\mathrm{F},i} + \alpha\varphi_{\mathrm{F},i} = \frac{jk}{16\pi^2}\sum_{n=1}^{K_L} w_n^L \left(1 + \left(\boldsymbol{n}_i \cdot \hat{\boldsymbol{k}}_n^L\right)\right) E_{i\lambda_{m_L}}\left(\hat{\boldsymbol{k}}_n^L\right)\left[\zeta_{m_L}\left(\hat{\boldsymbol{k}}_n^L\right) + \tau_{m_L}\left(\hat{\boldsymbol{k}}_n^L\right)\right], \tag{4.99}$$

respectively.
Step 6: Compute the near influence at each node i, $\phi_{\mathrm{N},i}$ for the basic form, or $\varphi_{\mathrm{N},i}$ for the normal derivative form, as the total effect of the elements in the neighbor cell set at the lowest level L, by

$$\begin{Bmatrix} \phi^{\mathrm{p}}_{\mathrm{N},i} \\ \phi^{\mathrm{v}}_{\mathrm{N},i} \end{Bmatrix} = \sum_{m'_L \in \mathrm{N}_{m_L}} \sum_{j \in \mathrm{G}_{m'_L}} \begin{Bmatrix} \left(-\frac{1}{2}\delta_{ij} + H_{ij} + D_{ij}\right) p_j \\ G_{ij} v_j \end{Bmatrix}, \tag{4.100}$$

$$\begin{Bmatrix} \varphi^{\mathrm{p}}_{\mathrm{N},i} \\ \varphi^{\mathrm{v}}_{\mathrm{N},i} \end{Bmatrix} = \sum_{m'_L \in \mathrm{N}_{m_L}} \sum_{j \in \mathrm{G}_{m'_L}} \begin{Bmatrix} \left(H'_{ij} + D'_{ij} + \frac{a}{2} Y_i \delta_{ij}\right) p_j \\ \left(G'_{ij} + \frac{a}{2}\delta_{ij}\right) v_j \end{Bmatrix}, \tag{4.101}$$

respectively. Regarding the Burton–Miller formulation, the above two equations are simply combined with the combination factor α.

Finally, compute the total influence on each node i by adding the far and near influences, $\phi_i = \phi_{\mathrm{F},i} + \phi_{\mathrm{N},i}$, which gives the matrix-vector products in Eq. (4.14), (4.19), or (4.33).

4.3.4 Implementation Considerations

Here are some considerations when implementing the FMBEM.

4.3.4.1 Parameters for Calculation

Efficient calculation by the FMBEM cannot be achieved with inappropriate setting for calculation parameters. In addition, its computational accuracy may greatly be lower than the standard BEM.

Truncation Number for Infinite Summation, N_c

In the numerical procedures, the infinite summation in Eq. (4.84) should be truncated with a finite number N_c. It should be noted that the summation diverges if the value N_c is too large. The value N_c for high accuracy depends on the wavenumber k, the cell size D, and the distance between centers of cells r_{LM}. Empirical formulae have been proposed [43, 59, 60, 67].

Number of Quadrature Points for Spherical Integral, $K = 2N_k^2$

The value N_k for high accuracy depends on N_c. A formula $N_k = N_c$ is recommended [43, 60].

Number of Quadrature Points for Interpolation, Q

If Legendre interpolation and adjoint interpolation are adopted in Steps 2 and 4, respectively, Q is often set to be constant independent of level [43, 67]. Interpolation and filtering technique based on FFT is known to be highly accurate [66], while its computational cost is larger than the former one.

Lowest Level Number for Hierarchical Cell Structure, L

When the computation time or memory requirement is almost smallest, the average number of nodes, M, in a cell at level L is almost constant, independent of N and geometry of problem [43]. Hence, an appropriate L can be chosen beforehand based on this constant value of M. This value of M depends on the implementation and computer used. In general, the optimum M for computation time and that for memory requirement are different from each other.

4.3.4.2 Iterative Methods

Convergence behavior of an iterative method directly affects the computation time for the FMBEM. GMRes is often referred to as a robust method, whereas it has been reported that the convergence of Restarted GMRes is greatly worse [69, 70]. If an appropriate preconditioner for improvement of convergence is applied, BiCG-type methods, which do not require to be restarted, often give a good convergence [70]. However, a few preconditioners are applicable to the FMBEM, where the system matrix is not computed [71].

4.3.4.3 Non-numerical Operations

In order to optimize the computational efficiency of the FMBEM, special consideration should be required not only on numerical operations but also on non-numerical ones, which include construction of cell data structure, searching interaction cells and neighbor cells [72].

4.3.4.4 Application to a Variety of Problems

Some studies on improvement of applicability of the FMBEM to sound field analyses have been reported, which include efficient techniques for plane-symmetric problems [47, 53, 55] (see Sect. 7.4.2), application to the domain decomposition method [26] (see Sect. 4.3.3), coupling analysis with structural models [73], and analysis of sound fields in porous materials (i.e., complex-wavenumber sound fields) [74, 75].

References

1. C.A. Brebbia, *The Boundary Element Method for Engineers* (Pentech Press, London, 1978)
2. R.D. Ciskowski, C.A. Brebbia (eds.), *Boundary Elements in Acoustics* (Computational Mechanics Publications and Elsevier Applied Science, London, 1991)
3. O. von Estorff (ed.), *Boundary Elements in Acoustics: Advances and Applications* (WIT Press, Southampton, 2000)
4. T.W. Wu (ed.), *Boundary Element Acoustics: Fundamentals and Computer Codes* (WIT Press, Southampton, 2000)
5. S. Marburg, B. Nolte (eds.), *Computational Acoustic of Noise Propagation in Fluids—Finite and Boundary Element Methods* (Springer, Berlin, 2008)
6. D.E. Freund, R. Farrell, A variational principle for the scattered wave. J. Acoust. Soc. Am. **87**, 1847–1860 (1990)
7. Z.S. Chen, G. Hofstetter, H.A. Mang, A symmetric Galerkin formulation of the boundary element method for acoustic radiation and scattering. J. Comput. Acoust. **5**, 219–241 (1997)
8. M. Abramowitz, I.A. Stegun (eds.), *Handbook of Mathematical Functions with Formulas, Graphs, and Mathematical Tables* (Dover Publications, New York, 1965)
9. P.C. Hammer, O.P. Marlowe, A.H. Stroud, Numerical integration over simplexes and cones. Math. Tables Aids Comput. **10**, 130–137 (1956)
10. T. Terai, On calculation of sound fields around three dimensional objects by integral equation methods. J. Sound Vib. **69**, 71–100 (1980)
11. K. Hayami, H. Matsumoto, A numerical quadrature for nearly singular boundary element integrals. Eng. Anal. Bound. Elem. **13**, 143–154 (1994)
12. A.J. Burton, G.F. Miller, The application of integral equation methods to the numerical solution of some exterior boundary value problems. Proc. R. Soc. Lond. Ser. A **323**, 201–210 (1971)
13. H.A. Schenck, Improved integral formulation for acoustic radiation problems. J. Acoust. Soc. Am. **44**, 41–58 (1968)
14. K.A. Cuneface, G. Koopmann, A boundary element method for acoustic radiation valid for all wavenumbers. J. Acoust. Soc. Am. **85**, 39–48 (1989)
15. T. Terai, Y. Kawai, The application of Kirchhoff's formula to the numerical calculation of transient response in an enclosures. J. Acoust. Soc. Jpn. (E) **11**, 1–10 (1990)
16. M.J. Bluck, S.P. Walker, Analysis of three-dimensional transient acoustic wave propagation using the boundary integral equation method. Int. J. Numer. Meth. Eng. **39**, 1419–1431 (1996)
17. A.A. Ergin, B. Shanker, E. Michielssen, Analysis of transient wave scattering from rigid bodies using a burton-miller approach. J. Acoust. Soc. Am. **106**, 2396–2404 (1999)
18. J.A. Hargreaves, T. Cox, A transient boundary element method for acoustic scattering from mixed regular and thin rigid bodies. Acta Acustica United Acustica **95**, 678–689 (2009)
19. Y. Kawai, T. Terai, A numerical method for the calculation of transient acoustic wave scattering from thin rigid plates. J. Sound Vib. **141**, 83–96 (1990)
20. P.J.T. Filippi, Layer potentials and acoustic diffraction. J. Sound Vib. **54**, 473–500 (1977)
21. H. Utsuno, T.W. Wu, A.F. Seybert, T. Tanaka, Prediction of sound fields in cavities with sound absorbing materials. AIAA J **28**, 1870–1876 (1990)

22. A.F. Seybert, C.Y.R. Cheng, T.W. Wu, The solution of coupled interior/exterior acoustic problems using the boundary element method. J. Acoust. Soc. Am. **88**, 1612–1618 (1990)
23. C.Y.R. Cheng, A.F. Seybert, T.W. Wu, A multidomain boundary element solution for silencer and muffler performance prediction. J. Sound Vib. **151**, 119–129 (1991)
24. A.F. Seybert, Z.H. Jia, T.W. Wu, Solving knife-edge scattering problems using singular boundary elements. J. Acoust. Soc. Am. **91**, 1278–1283 (1992)
25. E. Sarradj, Multi-domain boundary element method for sound fields in and around porous absorbers. Acta Acustica United Acustica **89**, 21–27 (2003)
26. Y. Yasuda, S. Sakamoto, T. Sakuma, Application of the fast multipole boundary element method to sound field analysis using a domain decomposition approach, in *Proceedings of Inter-Noise 2006*, no. 624 (Honolulu, 2006)
27. Y. Kosaka, T. Sakuma, Boundary element analysis using thin plate modeling—on calculation of a thin plate with different boundary conditions for both surfaces. in *Proceedings of ASJ Autumn Meeting*, 979–980 (2005) (in Japanese)
28. T. Masumoto, T. Oshima, Y. Yasuda, T. Sakuma, M. Kabuto, M. Akiyama, HRTF caulculation in the full audible frequency range using FMBEM, in *Proceedings of 19th International Congress on Acoustics*, COM-06-012 (Madrid, 2007)
29. N. Kamiya, H. Iwase, E. Kita, Parallel implementation of boundary element method with domain decomposition. Eng. Anal. Bound. Elem. **18**, 209–216 (1996)
30. M.E. Delany, E.N. Bazley, Acoustical properties of fibrous absorbent materials. Appl. Acoust. **3**, 105–116 (1970)
31. Y. Miki, Acoustical properties of porous materials—modifications of Delany-Bazley models. J. Acoust. Soc. Jpn. (E) **11**, 19–24 (1990)
32. P.M. Morse, K.U. Ingard, *Theoretical Acoustics*, Chap. 6 (McGraw-Hill Inc, New York, 1968)
33. V. Rokhlin, Rapid solution of integral equations of classical potential theory. J. Comput. Phys. **60**, 187–207 (1983)
34. L. Greengard, *The Rapid Evaluation of Potential Fields in Paticle Systems* (The MIT Press, Cambridge, 1987)
35. J.K. Salmon, M.S. Warren, G.S. Winckelmans, Fast parallel tree codes for gravitational and fluid dynamical N-body problems. Int. J. Supercomput. Appl. **8**, 124–142 (1994)
36. W.C. Chew, J.M. Jin, C.C. Lu, E. Michelssen, J.M. Song, Fast solution methods in electromagnetics. IEEE Trans. Antennas Propag. **45**, 533–543 (1997)
37. H. Schwichtenberg, G. Winter, H. Wallmeier, Acceleration of molecular mechanic simulation by parallelization and fast multipole techniques. Parallel Comput. **25**, 535–546 (1999)
38. K. Hayami, S.A. Sauter, A formulation of the panel clustering method for the three-dimensional elastostatic problem. JASCOME **13**, 125–130 (1996)
39. Y. Fu, K.J. Klimkowski, G.J. Robin, E. Berger, J.C. Browne, J.K. Singer, R.A. Van De Geijn, K.S. Vemaganti, A fast solution method for three-dimensional many-particle preblems of linear elasticity. Int. J. Numer. Meth. Eng. **42**, 1215–1229 (1998)
40. N. Nishimura, K. Yoshida, S. Kobayashi, A fast multipole boundary integral equation method for crack problems in 3D. Eng. Anal. Bound. Elem. **23**, 97–105 (1999)
41. A. Buchau, W. Rieger, W.M. Rucker, BEM computations using the fast multipole method in combination with higher order elements and the Galerkin method. IEEE Trans. Magn. **37**(5), 3181–3185 (2001)
42. T. Sakuma, Y. Yasuda, Fast multipole boundary element method for large-scale steady-state sound field analysis. Part I: Setup and validation. Acta Acustica United Acustica **88**(4), 513–525 (2002)
43. Y. Yasuda, T. Sakuma, Fast multipole boundary element method for large-scale steady-state sound field analysis. Part II: Examination of numerical items. Acta Acustica United Acustica **89**(1), 28–38 (2003)
44. S. Schneider, Application of fast methods for acoustic scattering and radiation problems. J. Comput. Acoust. **11**, 387–401 (2003)
45. M. Fischer, U. Gauger, L. Gaul, A multipole Galerkin boundary element method for acoustics. Eng. Anal. Bound. Elem. **28**, 155–162 (2004)

46. Y. Yasuda, T. Sakuma, An effective setting of hierarchical cell structure for the fast multipole boundary element method. J. Comput. Acoust. **13**(1), 47–70 (2005)
47. Y. Yasuda, T. Sakuma, A technique for plane-symmetric sound field analysis in the fast multipole boundary element method. J. Comput. Acoust. **13**(1), 71–85 (2005)
48. M. Fischer, L. Gaul, Application of the fast multipole BEM for structural-acoustic simulations. J. Comput. Acoust. **13**(1), 87–98 (2005)
49. H. Cheng, W.Y. Crutchfield, Z. Gimbutas, L.F. Greengard, J.F. Ethridge, J. Huang, V. Rokhlin, N. Yarvin, J. Zhao, A wideband fast multipole method for the Helmholtz equation in three dimensions. J. Comput. Phys. **216**, 300–325 (2006)
50. L. Shen, Y.J. Liu, An adaptive fast multipole boundary element method for three-dimensional acoustic wave problems based on the Burton-Miller formulation. Comput. Mech. **40**, 461–472 (2007)
51. M.S. Tong, W.C. Chew, Multilevel fast multipole algorithm for acoustic wave scattering by truncated ground with trenches. J. Acoust. Soc. Am. **123**, 2513–2521 (2008)
52. N.A. Gumerov, R. Duraiswami, A broadband fast multipole accelerated boundary element method for the three dimensional Helmholtz equation. J. Acoust. Soc. Am. **125**, 191–205 (2009)
53. M.S. Bapat, L. Shen, Y.J. Liu, Adaptive fast multipole boundary element method for three-dimensional half-space acoustic wave problems. Eng. Anal. Bound. Elem. **33**, 1113–1123 (2009)
54. Y. Yasuda, T. Oshima, T. Sakuma, A. Gunawan, T. Masumoto, Fast multipole boundary element method for low-frequency acoustic problems based on a variety of formulations. J. Comput. Acoust. **18**(4), 363–395 (2010)
55. Y. Yasuda, K. Higuchi, T. Oshima, T. Sakuma, Efficient technique in low-frequency fast multipole boundary element method for plane-symmetric acoustic problems. Eng. Anal. Bound. Elem. **36**(10), 1493–1501 (2012)
56. N.A. Gumerov, R. Duraiswami, *Fast Multipole Methods for the Helmholtz Equation in Three Dimensions* (Elsevier, Amsterdam, 2004)
57. Y. Saad, H.A. van der Vorst, Iterative solution of linear systems in the 20th century. J. Comput. Appl. Math. **123**, 1–33 (2000)
58. C.C. Lu, W.C. Chew, A multilevel algorithm for a boundary integral equation of wave scattering. Microw. Opt. Technol. Lett. **7**, 466–470 (1994)
59. J.M. Song, W.C. Chew, Multilevel fast-multipole algolithm for solving combined field integral equations of electromagnetic scattering. Microw. Opt. Technol. Lett. **10**, 14–19 (1995)
60. R. Coifman, V. Rokhlin, S. Wandzura, The fast multipole method for the wave equation: a pedestrian prescription. IEEE Antennas Propag. Mag. **35**(3), 7–12 (1993)
61. N.A. Gumerov, R. Duraiswami, Recursions for the computation of multipole translation and rotation coefficients for the 3-D Helmholtz equation. SIAM J. Sci. Comput. **25**, 1344–1381 (2003)
62. V. Rokhlin, Diagonal forms of translation operators for the Helmholtz equation in three dimensions. Appl. Comput. Harmon. Anal. **1**, 82–93 (1993)
63. L. Greengard, J. Huang, V. Rokhlin, S. Wandzura, Accelerating fast multipole methods for the Helmholtz equation at low frequencies. IEEE Comput. Sci. Eng. **5**, 32–38 (1998)
64. E. Darve, P. Havé, Efficient fast multipole method for low-frequency scattering. J. Comput. Phys. **197**, 341–363 (2004)
65. M.A. Epton, B. Dembart, Multipole translation theory for the three-dimensional Laplace and Helmholtz equations. SIAM J. Sci. Comput. **16**(4), 865–897 (1995)
66. E. Darve, The fast multipole method I: error analysis and asymptotic complexity. SIAM J. Numer. Anal. **38**(1), 98–128 (2000)
67. S. Koc, W.C. Chew, Calculation of acoustical scattering from a cluster of scatterers. J. Acoust. Soc. Am. **103**(2), 721–734 (1997)
68. A. Brandt, Multilevel computations of integral transforms and particle interactions with oscillatory kernels. Comput. Phys. Commun. **65**, 24–38 (1991)

69. S. Schneider, S. Marburg, Performance of iterative solvers for acoustic problems. Part II. Acceleration by ILU-type preconditioner. Eng. Anal. Bound. Elem. **27**, 751–757 (2003)
70. Y. Yasuda, S. Sakamoto, Y. Kosaka, T. Sakuma, N. Okamoto, T. Oshima, Numerical analysis of large-scale sound fields using iterative methods. Part I: Application of Krylov subspace methods to boundary element analysis. J. Comput. Acoust. **15**(4), 449–471 (2007)
71. S. Marburg, B. Nolte (eds.), *Computational Acoustics of Noise Propagation in Fluids—Finite and Boundary Element Methods*, Chap. 12 (Springer, Berlin, 2008)
72. T. Oshima, Y. Yasuda, T. Sakuma, A high efficiency implementation of fast multipole boundary element method, in *Proceedings of 18th International Congress on Acoustics*, vol. I, pp. 487–490 (Kyoto, 2004)
73. T. Masumoto, A. Gunawan, T. Oshima, Y. Yasuda, T. Sakuma, Coupling analysis between FMBEM-based acoustic and modal-based structural models -convergence behavior of iterative solutions, in *Proceedings of Inter-Noise 2008*, no. 0308 (Shanghai, 2008)
74. Y. Yasuda, T. Sakuma, Analysis of sound fields in porous materials using the fast multipole BEM, in *Proceedings of Inter-Noise 2008*, no. 0531 (Shanghai, 2008)
75. Y. Yasuda, K. Eda, T. Sakuma, T. Oshima, A fast multipole BE analysis of a small room with sound absorbers using domain decomposition approach, in *Proceedings of Inter-Noise 2010*, no. 0503 (Lisbon, 2010)

Chapter 5
Alternative Time-Domain Methods

Takuya Oshima, Takashi Ishizuka and Kan Okubo

Abstract The finite-difference time-domain (FDTD) method has been the only choice of time-domain methods for practical applications with its simplicity and efficiency. However, the simple discretization of the simple wave equation model in which the method has its basis is not sufficient for modeling more complex wave propagation phenomena, high-accuracy simulations, or acoustic fields with complex geometries. In this chapter, alternative time-domain methods that may be applied to such situations are discussed as follows: the linearized Euler equation (LEE) method, the constrained interpolation profile (CIP) method, and the finite-volume time-domain (FVTD) method. The LEE method is applicable to wave propagation phenomena under the influence of arbitrary background flows. The main application of the method is sound propagation simulations outdoors where wind effects are not negligible. The CIP method is characteristic in that the method is in principle free from numerical dispersion. The characteristic allows simulations with high phase accuracy. The FVTD method is constructed on an unstructured grid system. The method thus has an advantage in modeling complex geometries compared to the FDTD method where orthogonal structured grid is used.

T. Oshima (✉)
Faculty of Engineering, Niigata University, 8050 Ikarashi-Ninocho, Nishi-ku, Niigata-shi, Niigata 950-2181, Japan
e-mail: oshima@eng.niigata-u.ac.jp

T. Ishizuka
Institute of Technology, Shimizu Corporation, 3-4-17 Etchujima, Koto-ku, Tokyo 135-8530, Japan
e-mail: ishiduka@shimz.co.jp

K. Okubo
Graduate School of System Design, Tokyo Metropolitan University, 6-6 Asahigaoka, Hino-shi, Tokyo 191-0065, Japan
e-mail: kanne@tmu.ac.jp

T. Sakuma et al. (eds.), *Computational Simulation in Architectural and Environmental Acoustics*, DOI: 10.1007/978-4-431-54454-8_5,

Keywords Linearized Euler equation (LEE) method · Wind effects · Constrained interpolation profile (CIP) method · Method of characteristics (MOC) · Finite-volume time-domain (FVTD) method · Unstructured grid

5.1 Linearized Euler Equation Method

The linearized Euler equation (LEE) method is an approach to solve sound propagation under an arbitrary background flow by an addition of the advection effect of the medium to the basic equations of the finite-difference time-domain (FDTD) method. The main application of the method is sound propagation simulations outdoors with wind effects.

5.1.1 Governing Equations

The medium velocity U is represented as a composition of the background flow component $\bar{U}$ and the acoustic particle velocity component U' as

$$U = \bar{U} + U'.$$

Other physical quantities are similarly represented. For example, pressure deviation from the average medium air pressure is represented by decomposing into the airflow component $\bar{p}$ and the acoustic pressure component p' as

$$p = \bar{p} + p'.$$

For outdoor acoustic propagation problems, a simplified linearized Euler model under adiabatic, barotropic, and non-buoyant conditions can be employed. In the simplified model, the fundamental equations of fluid flow, the equations of motion, and continuity, are linearized by the background flow components of velocity, pressure, and density, $\bar{U}$, $\bar{p}$ and $\bar{\rho}$ respectively [1, 2].

$$\frac{\partial U'}{\partial t} = -\left(\bar{U} \cdot \nabla\right) U' - \left(U' \cdot \nabla\right) \bar{U} - \frac{1}{\bar{\rho}} \nabla p', \tag{5.1}$$

$$\frac{\partial p'}{\partial t} = -\bar{\rho}\bar{c}^2 \nabla \cdot U' - \bar{U} \cdot \nabla p', \tag{5.2}$$

where $\bar{\rho}$ and $\bar{c}$ are the medium air density and the speed of sound respectively.

5.1.2 Discretization

Equations (5.1), (5.2) are discretized using a three-dimensional orthogonal staggered grid where p' is located at the cell centroid and each component of U' is located at the cell face of the respective coordinate direction. The pressure and velocity variables are located at staggered positions in time as well. Hereafter, the components of the background flow vector and the acoustic velocity vector are denoted as follows:

$$\bar{U} = \{\bar{u}\ \bar{v}\ \bar{w}\}^{\mathrm{T}},\ U' = \{u'\ v'\ w'\}^{\mathrm{T}}.$$

where ^{T} denotes the transpose operator.

5.1.2.1 Temporal Discretization

Time advancement is performed using the prediction-step staggered-in-time (PSIT) technique [3], which allows a simple implementation and smaller memory footprints. The technique calculates firstly a predicted value of the acoustic velocity at time step n, $\tilde{U}'^n$, using the acoustic velocity at time step $n - 1/2$, $U'^{n-1/2}$, and the acoustic pressure at time step n, p'^n, with the following equation

$$\tilde{U}'^n = U'^{n-1/2} - \frac{\Delta t}{2\bar{\rho}} \nabla p'^n, \tag{5.3}$$

which is a backward Euler discretized form of Eq. (5.1) between time step $n - 1/2$ to n, but without the contribution of the background flow velocity $\bar{U}$

$$\frac{\partial U'}{\partial t} = -\frac{1}{\bar{\rho}} \nabla p'. \tag{5.4}$$

The predicted value is used to calculate the acoustic velocity at the next time step $n + 1/2$, $U'^{n+1/2}$, using a timewise discretized form of Eq. (5.1).

$$U'^{n+1/2} = U'^{n-1/2} - \frac{\Delta t}{\bar{\rho}} \nabla p'^n - \Delta t \left(\bar{U} \cdot \nabla\right) \tilde{U}'^n - \Delta t \left(\tilde{U}'^n \cdot \nabla\right) \bar{U}. \tag{5.5}$$

Similarly, for Eq. (5.2) the predicted value of the acoustic pressure at time step $n+1/2$ is calculated by using the acoustic pressure at time step n, p'^n, and the acoustic velocity at time step $n+1/2$, $U'^{n+1/2}$. The predicted value in turn is used to calculate the acoustic pressure at the next time step $n + 1$, p'^{n+1}.

$$\tilde{p}'^{n+1/2} = p'^n - \frac{\bar{\rho}\bar{c}^2 \Delta t}{2} \nabla \cdot U'^{n+1/2}, \tag{5.6}$$

$$p'^{n+1} = p'^n - \bar{\rho}\bar{c}^2 \Delta t\, \nabla \cdot U'^{n+1/2} - \Delta t\, \bar{U} \cdot \nabla \tilde{p}'^{n+1/2}. \tag{5.7}$$

5.1.2.2 Spatial Discretization

The spatial derivatives in the prediction-step equations, namely Eqs. (5.3), (5.6), are discretized by the second-order central finite difference scheme that are widely used in usual FDTD simulations. For example, the x component of $\nabla p'^n$ in the right hand side of Eq. (5.3) is written as

$$\left.\frac{\partial p'^n}{\partial x}\right|_{i+1/2,j,k} \approx \frac{p'^n_{i+1,j,k} - p'^n_{i,j,k}}{\Delta x}, \tag{5.8}$$

where Δx is the x directional grid spacing.

On the other hand, Eqs. (5.5), (5.7) are simplified assuming that the background flow $\bar{U}$ is a uniform x-directional flow of speed $\bar{u}$

$$\bar{U} = \{\bar{u}\ 0\ 0\}^{\mathrm{T}}.$$

By substituting this equation to Eq. (5.5), only the x-directional derivative remains in the third term in the right hand side of Eq. (5.5). Besides, the fourth term in the right hand side is cancelled out due to the omnidirectional zero gradient of $\bar{U}$. Hence, by representing each component of $\tilde{U}'$ by

$$\tilde{U}' = \{\tilde{u}'\ \tilde{v}'\ \tilde{w}'\}^{\mathrm{T}},$$

each component of Eq. (5.5) and Eq. (5.7) are written as follows.

$$u'^{n+1/2}_{i+1/2,j,k} = u'^{n-1/2}_{i+1/2,j,k} - \frac{\Delta t}{\bar{\rho}} \left.\frac{\partial p'^n}{\partial x}\right|_{i+1/2,j,k} - \Delta t\, \bar{u} \left.\frac{\partial \tilde{u}'^n}{\partial x}\right|_{i+1/2,j,k}, \tag{5.9}$$

$$v'^{n+1/2}_{i,j+1/2,k} = v'^{n-1/2}_{i,j+1/2,k} - \frac{\Delta t}{\bar{\rho}} \left.\frac{\partial p'^n}{\partial y}\right|_{i,j+1/2,k} - \Delta t\, \bar{u} \left.\frac{\partial \tilde{v}'^n}{\partial x}\right|_{i,j+1/2,k}, \tag{5.10}$$

$$w'^{n+1/2}_{i,j,k+1/2} = w'^{n-1/2}_{i,j,k+1/2} - \frac{\Delta t}{\bar{\rho}} \left.\frac{\partial p'^n}{\partial z}\right|_{i,j,k+1/2} - \Delta t\, \bar{u} \left.\frac{\partial \tilde{w}'^n}{\partial x}\right|_{i,j,k+1/2}, \tag{5.11}$$

$$p'^{n+1}_{i,j,k} = p'^n_{i,j,k} - \bar{\rho}\bar{c}^2 \Delta t \left(\left.\frac{\partial u'^{n+1/2}}{\partial x}\right|_{i,j,k} + \left.\frac{\partial v'^{n+1/2}}{\partial y}\right|_{i,j,k} + \left.\frac{\partial w'^{n+1/2}}{\partial z}\right|_{i,j,k} \right) - \Delta t\, \bar{u} \left.\frac{\partial \tilde{p}'^{n+1/2}}{\partial x}\right|_{i,j,k}. \tag{5.12}$$

The spatial derivatives in Eqs. (5.9)–(5.12) are discretized by two strategies: one is to apply the second-order central finite difference scheme, which is denoted hereafter as Type FD-2nd. Another is to apply a compact difference scheme to the third term in the right hand side of each equation and to apply the second-order central finite difference scheme to all other terms, which is denoted hereafter as Type FD-Compact.

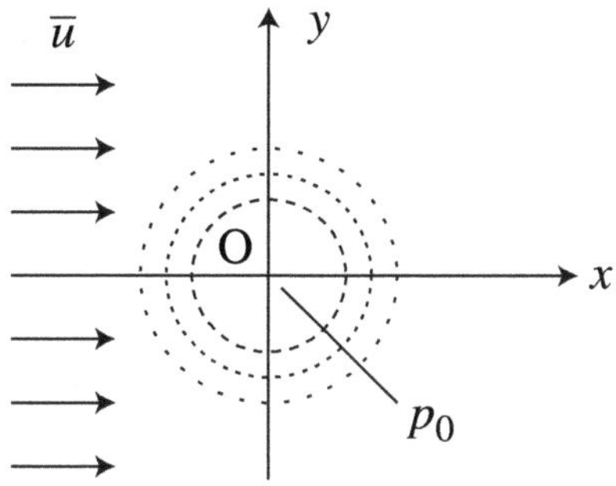

Fig. 5.1 Acoustic pulse in a uniform flow

Type FD-2nd. Taking the x-directional momentum equation Eq. (5.9) as example, the spatial derivative of the second term in the right hand side is discretized by Eq. (5.8). The third term is discretized using the finite difference of double grid spacing $2\Delta x$ as follows:

$$\left.\frac{\partial \tilde{u}'^n}{\partial x}\right|_{i+1/2,j,k} \approx \frac{\tilde{u}'^n_{i+3/2,j,k} - \tilde{u}'^n_{i-1/2,j,k}}{2\Delta x}.$$

Type FD-Compact. Taking Eq. (5.9) as example, the spatial derivative of the second term in the right hand side is identically discretized to Type FD-2nd by Eq. (5.8). The third term, on the other hand, is discretized with the Padé scheme [4], which is a fourth-order compact difference scheme that represents the approximation of $\partial \tilde{u}'^n/\partial x$, ϕ'^n, as follows:

$$\frac{1}{4}\phi'^n_{i-1/2,j,k} + \phi'^n_{i+1/2,j,k} + \frac{1}{4}\phi'^n_{i+3/2,j,k} = \frac{3}{4}\frac{\tilde{u}'^n_{i+3/2,j,k} - \tilde{u}'^n_{i-1/2,j,k}}{\Delta x}.$$

The matrix equation of ϕ'^n obtained by applying the Padé scheme to the entire grid is solved by a tridiagonal matrix algorithm (TDMA) solver [5].

5.1.3 Computational Setup

A sound propagation problem in a uniform medium under a uniform background flow within a two-dimensional free space is solved in order to compare the formulations and the analytical solutions.

An x-directional uniform background flow of speed $\bar{u}$ is given as the background flow. The initial acoustic pressure and acoustic velocity are given as

$$p' = p_0 \exp\left\{-\alpha\left(x^2 + y^2\right)\right\},$$
$$U' = 0,$$

as shown in Fig. 5.1. Under the initial conditions, the analytical solution of the acoustic pressure is given by Ref. [6] as

$$p' = \frac{p_0}{2\alpha}\int_0^\infty \exp\left(-\frac{\xi^2}{4\alpha}\right)\cos\left(\bar{c}\xi t\right) J_0\left(\xi\eta\right)\xi d\xi,$$
$$\eta = \sqrt{(x - \bar{u}t)^2 + y^2},$$
$$\alpha = \frac{\log 2}{b^2},$$

where p_0 is the initial pressure amplitude, b is the half width at half maximum of the initial pressure amplitude, and J_0 is the Bessel function of the first kind of order 0. In the tests, b is set to 3 and p_0 and $\bar{c}$ are set to unity. The Mach number is assumed to be 0.1, namely

$$M = \frac{\bar{u}}{\bar{c}} = 0.1.$$

The computational domain is a square of diagonal points $(x, y) = (-100, -100)$ and $(100, 100)$, which is divided by an orthogonal uniform grid of $201 \times 201 \times 1$ cells in $x \times y \times z$ directions. The time step Δt is set to 0.5.

5.1.4 Results

The contours of instantaneous pressure amplitudes $|p'|$ at time $t = 40$ and 80 for the analytical solutions and the discretized formulations are shown in Fig. 5.2. Slow advection in x-direction following the uniform background flow of $M = 0.1$ is seen in all cases. At $t = 40$, all computed results show good agreements with the analytical case. However, as the propagation distance grows, at $t = 80$ we see differences. At the downstream direction, a numerical dispersion is seen in Type FD-2nd. On the contrary, Type FD-Compact shows no such error and agrees well with the analytical case.

5.2 Constrained Interpolation Profile Method

The constrained interpolation profile (CIP) method was developed in the field of fluid dynamics as a kind of the method of characteristics (MOC) [7–9]. This method has been applied to numerical simulations of sound field in time domain [10–14] as it

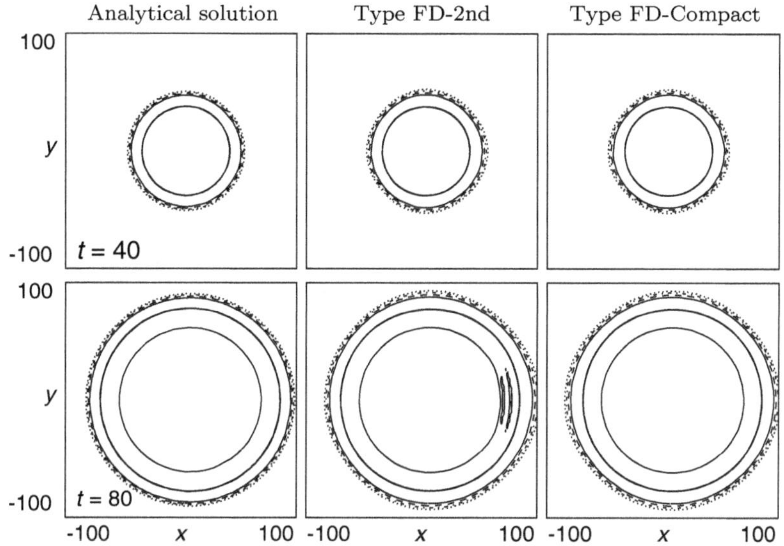

Fig. 5.2 Propagation of acoustic pulse in uniform flow at $t = 40$ and 80. *Dotted, dash-dotted, dashed* and *solid iso-contours* denote $|p'|$ values defined by a geometric ratio of 4 (which corresponds to about 12 dB in relative SPL) from 6.25×10^{-5} to 0.004

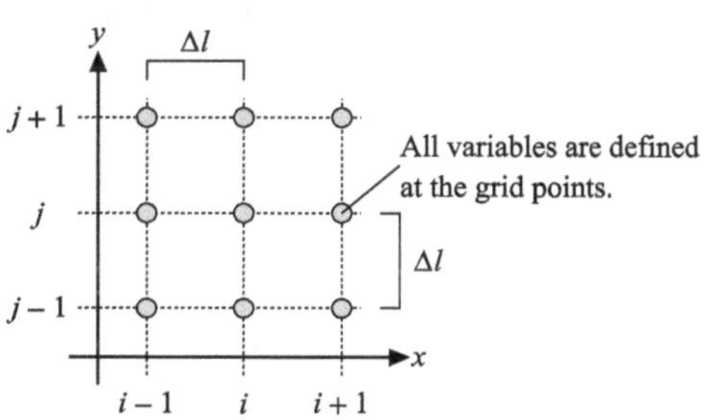

Fig. 5.3 Collocated grid system in the CIP method

has an advantage of low numerical dispersion over the FDTD method. Additionally, a spatial grid size, which the CIP method requires for adequate accuracy, is larger than that the FDTD method requires. High accuracy is achieved by a technique called "multi-moments" where not only physical values, such as the sound pressure and the particle velocity, but also their spatial derivatives are explicitly and simultaneously calculated at all grid points. In the MOC, including the CIP method, a time step size is free from Courant condition. The CIP method does not require a small time step size other time-domain methods require for accuracy and stability, resulting in a reduction of total computational time.

A collocated grid system is employed in the CIP method. All discrete physical values and their spatial derivatives are defined at the same grid point as shown in Fig. 5.3.

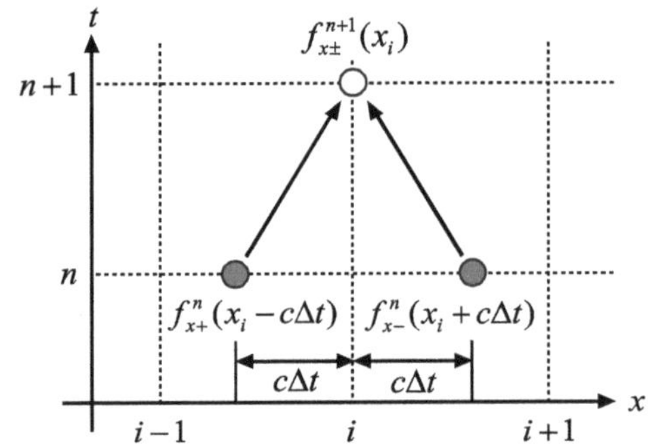

Fig. 5.4 Advection of values in the method of characteristics

5.2.1 Formulation

The CIP method for acoustic simulation computes advection equations derived from the governing equations of sound fields. The equation of continuity and the equation of motion for a lossless linear sound field are transformed into

$$\partial_t p + cZ\partial_x v_x = 0, \quad Z\partial_t v_x + c\partial_x p = 0, \tag{5.13}$$

where ∂_α represents an operator $\partial/\partial\alpha$, and p, v_x, c, and Z are the sound pressure, the particle velocity in the x-direction, the speed of sound, and the characteristic impedance of a medium, respectively. It should be noted that these equations express one-dimensional wave propagation in the x-direction. Addition and subtraction of the two equations in Eq. (5.13) lead to the following advection equations:

$$\partial_t f_{x\pm} \pm c\partial_x f_{x\pm} = 0, \quad \text{for} \quad f_{x\pm} = p \pm Zv_x. \tag{5.14}$$

f_{x+} and f_{x-}, which are called "characteristic curves", represent forward and backward components of the sound field in the x-direction and Eq. (5.14) expresses propagation of them with the velocity of c.

In the MOC, the advection equations expressing wave propagation are computed by advection calculations schematically illustrated in Fig. 5.4. As shown in the diagram, advecting $f_{x\pm}$ at points $\mp c\Delta t$ distant from a grid point x_i at a time step n gives those at x_i at the next time step:

$$f_{x\pm}^{n+1}(x_i) = f_{x\pm}^{n}(x_i \mp c\Delta t), \tag{5.15}$$

where Δt is the time step size, and the superscripts n and $n+1$ denote time steps. When the points $x_i \mp c\Delta t$, called advection sources, are not at grid points, $f_{x\pm}^{n}(x_i \mp c\Delta t)$ are obtained by using interpolations.

In the CIP method, a kind of the MOC, values at advection sources are interpolated with high accuracy by using the Hermite interpolation [7, 15], normally the third-order Hermite interpolation. These calculations need spatial derivatives of $f_{x\pm}$ at each

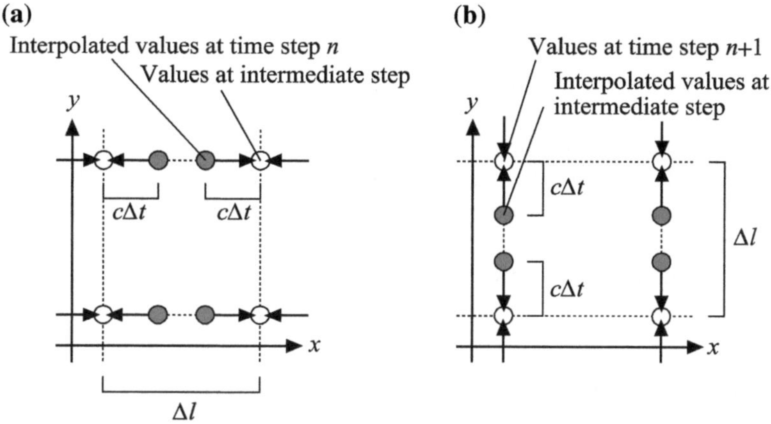

Fig. 5.5 Directionally separated advection of values in the multi-dimensional CIP method. **a** x-directional advection. **b** y-directional advection

grid. Advection equations for the derivatives are derived from the differentiation of Eq. (5.14):

$$\partial_t g_{x\pm} \pm c\partial_x g_{x\pm} = 0, \quad \text{for} \quad g_{x\pm} = \partial_x p \pm Z\partial_x v_x. \tag{5.16}$$

The CIP method simultaneously computes Eq. (5.14) and (5.16) using the third-order Hermite interpolation.

Multi-dimensional wave propagation is computed by the directionally separated advection formulation, where one-dimensional advections for each axis are alternately computed as shown in Fig. 5.5, for a two-dimensional simulation for example. Lettig results of x-directional advections be values at an intermediate step, values at the next time step is obtained by advecting the values at the intermediate step in the y-direction. This technique requires additional advection calculations for spatial derivatives with respect to the direction perpendicular to the advection direction. For the x-directional advection in a two-dimensional field, for example, the CIP method simultaneously computes the following advection equations as well as Eqs. (5.14) and (5.16):

$$\partial_t \eta_{x\pm} \pm c\partial_x \eta_{x\pm} = 0\,, \quad \text{for} \quad \eta_{x\pm} = \partial_y p \pm Z\partial_y v_x, \tag{5.17}$$
$$\partial_t \mu_{x\pm} \pm c\partial_x \mu_{x\pm} = 0\,, \quad \text{for} \quad \mu_{x\pm} = \partial_x\partial_y p \pm Z\partial_x\partial_y v_x, \tag{5.18}$$

where $\eta_{x\pm}$ and $\mu_{x\pm}$ are derivatives of $f_{x\pm}$ and $g_{x\pm}$ with respect to the y-direction, respectively. A kind of the CIP method, called the "type-C" CIP method, employs the third-order Hermite interpolation to obtain the perpendicular derivatives $\eta_{x\pm}$ and $\mu_{x\pm}$ at the advection sources [9, 11, 12]. Additionally, in a three-dimensional simulation, advection equations for the derivatives of $f_{x\pm}$, $g_{x\pm}$, $\eta_{x\pm}$, and $\mu_{x\pm}$ with respect to the z-direction have to be taken into calculations. We can calculate the y-directional (and the z-directional in three-dimensional simulations) advections in a similar procedure to that for the x-direction described above.

Although the type-C CIP method employing the Hermite interpolation for the perpendicular derivatives has high accuracy, it is computationally expensive. A simplified scheme, called the "type-M" CIP method, has been proposed [8, 10, 12]. The type-M CIP method calculates advection of the perpendicular derivatives only for Eq. (5.17) using the first-order Lagrange interpolation, i.e., the upwind scheme. Omitting calculations of Eq. (5.18), this scheme saves the computer memory and computational time, but has lower accuracy than that the type-C CIP method has.

5.2.2 Implementation

The followings explain an implementation procedure of the type-C CIP method for a two-dimensional sound field simulation. We assume here that the sound field is discretized in the x- and the y-direction by the same grid size of Δl and i and j denote the grid point number in each direction as shown in Fig. 5.3.

Firstly, we carry out advection calculations in the $\pm x$-direction. At a time step n, the characteristic curves and their derivatives at a grid point (i,j) are derived from the physical values and their derivatives at the same point:

$$f^n_{x\pm}(i,j) = p^n(i,j) \pm Z v^n_x(i,j), \tag{5.19}$$

$$g^n_{x\pm}(i,j) = \partial_x p^n(i,j) \pm Z \partial_x v^n_x(i,j), \tag{5.20}$$

$$\eta^n_{x\pm}(i,j) = \partial_y p^n(i,j) \pm Z \partial_y v^n_x(i,j), \tag{5.21}$$

$$\mu^n_{x\pm}(i,j) = \partial_x \partial_y p^n(i,j) \pm Z \partial_x \partial_y v^n_x(i,j). \tag{5.22}$$

The third-order Hermite interpolation simultaneously calculates $f_{x\pm}$ and $g_{x\pm}$ at an intermediate step, which are values at the advection sources at the time step n:

$$f^*_{x\pm}(i,j) = a_\pm \xi^3_\pm + b_\pm \xi^2_\pm + g^n_{x\pm}(i,j)\xi_\pm + f^n_{x\pm}(i,j), \tag{5.23}$$

$$g^*_{x\pm}(i,j) = 3a_\pm \xi^2_\pm + 2b_\pm \xi_\pm + g^n_{x\pm}(i,j), \tag{5.24}$$

where superscript $*$ represents the intermediate step and

$$a_\pm = \frac{g^n_{x\pm}(i,j) + g^n_{x\pm}(i \mp 1,j)}{(\mp\Delta l)^2} + \frac{2\left(f^n_{x\pm}(i,j) - f^n_{x\pm}(i \mp 1,j)\right)}{(\mp\Delta l)^3}, \tag{5.25}$$

$$b_\pm = \frac{3\left(f^n_{x\pm}(i \mp 1,j) - f^n_{x\pm}(i,j)\right)}{(\mp\Delta l)^2} - \frac{2g^n_{x\pm}(i,j) + g^n_{x\pm}(i \mp 1,j)}{\mp\Delta l}, \tag{5.26}$$

$$\xi_\pm = \mp c\Delta t. \tag{5.27}$$

Equations (5.23)–(5.27), which are polynomial forms of the third-oder Hermite interpolation, can be transformed into simple sum-of-product forms for efficient computation:

$$f^*_{x\pm}(i,j) = C_{1\pm} f^n_{x\pm}(i \mp 1, j) + C_{2\pm} f^n_{x\pm}(i,j) + C_{3\pm} g^n_{x\pm}(i \mp 1, j) + C_{4\pm} g^n_{x\pm}(i,j), \tag{5.28}$$

$$g^*_{x\pm}(i,j) = C'_{1\pm} f^n_{x\pm}(i \mp 1, j) + C'_{2\pm} f^n_{x\pm}(i,j) + C'_{3\pm} g^n_{x\pm}(i \mp 1, j) + C'_{4\pm} g^n_{x\pm}(i,j). \tag{5.29}$$

The coefficients will be obtained before starting computations:

$$C_{1\pm} = -2\chi^3 + 3\chi^2, \tag{5.30}$$
$$C_{2\pm} = 2\chi^3 - 3\chi^2 + 1, \tag{5.31}$$
$$C_{3\pm} = \xi_\pm(\chi^2 - \chi), \tag{5.32}$$
$$C_{4\pm} = \xi_\pm(\chi^2 - 2\chi + 1), \tag{5.33}$$
$$C'_{1\pm} = 6(-\chi^3 + \chi^2)/\xi_\pm, \tag{5.34}$$
$$C'_{2\pm} = 6(\chi^3 - \chi^2)/\xi_\pm, \tag{5.35}$$
$$C'_{3\pm} = 3\chi^2 - 2\chi, \tag{5.36}$$
$$C'_{4\pm} = 3\chi^2 - 4\chi + 1, \tag{5.37}$$

where $\chi = c\Delta t/\Delta l$. Advection of the perpendicular derivatives $\eta_{x\pm}$ and $\mu_{x\pm}$ are also calculated in the same way:

$$\eta^*_{x\pm}(i,j) = C_{1\pm} \eta^n_{x\pm}(i \mp 1, j) + C_{2\pm} \eta^n_{x\pm}(i,j) + C_{3\pm} \mu^n_{x\pm}(i \mp 1, j) + C_{4\pm} \mu^n_{x\pm}(i,j), \tag{5.38}$$

$$\mu^*_{x\pm}(i,j) = C'_{1\pm} \eta^n_{x\pm}(i \mp 1, j) + C'_{2\pm} \eta^n_{x\pm}(i,j) + C'_{3\pm} \mu^n_{x\pm}(i \mp 1, j) + C'_{4\pm} \mu^n_{x\pm}(i,j). \tag{5.39}$$

Results of advection calculations in the x-direction are restored to the physical values and their derivatives at the intermediate step:

$$p^*(i,j) = \left(f^*_{x+}(i,j) + f^*_{x-}(i,j)\right)/2, \tag{5.40}$$
$$v^*_x(i,j) = \left(f^*_{x+}(i,j) - f^*_{x-}(i,j)\right)/2Z, \tag{5.41}$$
$$\partial_x p^*(i,j) = \left(g^*_{x+}(i,j) + g^*_{x-}(i,j)\right)/2, \tag{5.42}$$
$$\partial_x v^*_x(i,j) = \left(g^*_{x+}(i,j) - g^*_{x-}(i,j)\right)/2Z, \tag{5.43}$$
$$\partial_y p^*(i,j) = \left(\eta^*_{x+}(i,j) + \eta^*_{x-}(i,j)\right)/2, \tag{5.44}$$
$$\partial_y v^*_x(i,j) = \left(\eta^*_{x+}(i,j) - \eta^*_{x-}(i,j)\right)/2Z, \tag{5.45}$$
$$\partial_x \partial_y p^*(i,j) = \left(\mu^*_{x+}(i,j) + \mu^*_{x-}(i,j)\right)/2, \tag{5.46}$$
$$\partial_x \partial_y v^*_x(i,j) = \left(\mu^*_{x+}(i,j) - \mu^*_{x-}(i,j)\right)/2Z. \tag{5.47}$$

The particle velocity in the x-direction and their derivatives are directly assigned to those at the next time step: $v_x^{n+1}(i, j) = v_x^*(i, j)$, for example.

Secondly, advection calculations in the $\pm y$-direction are carried out in a similar procedure described above. The y-directional characteristic curves and their derivatives are derived from the physical values and their derivatives at the intermediate step:

$$f_{y\pm}^*(i, j) = p^*(i, j) \pm Z v_y^*(i, j), \tag{5.48}$$
$$g_{y\pm}^*(i, j) = \partial_y p^*(i, j) \pm Z \partial_y v_y^*(i, j), \tag{5.49}$$
$$\eta_{y\pm}^*(i, j) = \partial_x p^*(i, j) \pm Z \partial_x v_y^*(i, j), \tag{5.50}$$
$$\mu_{y\pm}^*(i, j) = \partial_x \partial_y p^*(i, j) \pm Z \partial_x \partial_y v_y^*(i, j). \tag{5.51}$$

The third-order Hermite interpolation gives the values at the next time step $n+1$:

$$f_{y\pm}^{n+1}(i, j) = C_{1\pm} f_{y\pm}^*(i, j \mp 1) + C_{2\pm} f_{y\pm}^*(i, j) + C_{3\pm} g_{y\pm}^*(i, j \mp 1) + C_{4\pm} g_{y\pm}^*(i, j), \tag{5.52}$$
$$g_{y\pm}^{n+1}(i, j) = C'_{1\pm} f_{y\pm}^*(i, j \mp 1) + C'_{2\pm} f_{y\pm}^*(i, j) + C'_{3\pm} g_{y\pm}^*(i, j \mp 1) + C'_{4\pm} g_{y\pm}^*(i, j), \tag{5.53}$$
$$\eta_{y\pm}^{n+1}(i, j) = C_{1\pm} \eta_{y\pm}^*(i, j \mp 1) + C_{2\pm} \eta_{y\pm}^*(i, j) + C_{3\pm} \mu_{y\pm}^*(i, j \mp 1) + C_{4\pm} \mu_{y\pm}^*(i, j), \tag{5.54}$$
$$\mu_{y\pm}^{n+1}(i, j) = C'_{1\pm} \eta_{y\pm}^*(i, j \mp 1) + C'_{2\pm} \eta_{y\pm}^*(i, j) + C'_{3\pm} \mu_{y\pm}^*(i, j \mp 1) + C'_{4\pm} \mu_{y\pm}^*(i, j). \tag{5.55}$$

Results of advection calculations in the y-direction are restored to the physical values and their derivatives at the time step $n+1$:

$$p^{n+1}(i, j) = \left(f_{y+}^{n+1}(i, j) + f_{y-}^{n+1}(i, j)\right)/2, \tag{5.56}$$
$$v_y^{n+1}(i, j) = \left(f_{y+}^{n+1}(i, j) - f_{y-}^{n+1}(i, j)\right)/2Z, \tag{5.57}$$
$$\partial_y p^{n+1}(i, j) = \left(g_{y+}^{n+1}(i, j) + g_{y-}^{n+1}(i, j)\right)/2, \tag{5.58}$$
$$\partial_y v_y^{n+1}(i, j) = \left(g_{y+}^{n+1}(i, j) - g_{y-}^{n+1}(i, j)\right)/2Z, \tag{5.59}$$
$$\partial_x p^{n+1}(i, j) = \left(\eta_{y+}^{n+1}(i, j) + \eta_{y-}^{n+1}(i, j)\right)/2, \tag{5.60}$$
$$\partial_x v_y^{n+1}(i, j) = \left(\eta_{y+}^{n+1}(i, j) - \eta_{y-}^{n+1}(i, j)\right)/2Z, \tag{5.61}$$
$$\partial_x \partial_y p^{n+1}(i, j) = \left(\mu_{y+}^{n+1}(i, j) + \mu_{y-}^{n+1}(i, j)\right)/2, \tag{5.62}$$
$$\partial_x \partial_y v_y^{n+1}(i, j) = \left(\mu_{y+}^{n+1}(i, j) - \mu_{y-}^{n+1}(i, j)\right)/2Z. \tag{5.63}$$

In the type-M CIP method, the third-oder Hermite interpolation of the perpendicular derivatives, Eqs. (5.38) and (5.54), are replaced with the first-order Lagrange interpolation as follows:

$$\eta^{*}_{x\pm}(i,j) = C^{L}_{1\pm}\eta^{n}_{x\pm}(i \mp 1, j) + C^{L}_{2\pm}\eta^{n}_{x\pm}(i,j) \tag{5.64}$$

$$\eta^{n+1}_{y\pm}(i,j) = C^{L}_{1\pm}\eta^{*}_{y\pm}(i, j \mp 1) + C^{L}_{2\pm}\eta^{*}_{y\pm}(i,j), \tag{5.65}$$

where

$$C^{L}_{1\pm} = \chi, \tag{5.66}$$

$$C^{L}_{2\pm} = 1 - \chi. \tag{5.67}$$

The above equations correspond with the upwind scheme. Advection calculations for Eqs. (5.39) and (5.55) are omitted and the second derivatives of the physical values, $\partial_x\partial_y p$, $\partial_x\partial_y v_x$, and $\partial_x\partial_y v_y$, are not defined at the grid points, resulting in reduction of required computer memory and computational time.

5.2.3 Boundary Conditions

In the CIP method, boundary conditions at the interface between two media are given using the reflection coefficient [16]. When f_{x+} and g_{x+} are incident on the boundary at x_b, for example, the boundary conditions are represented as

$$f_{x-}(x_b) = \Gamma f_{x+}(x_b), \tag{5.68}$$

$$g_{x-}(x_b) = \Gamma' g_{x+}(x_b), \tag{5.69}$$

where Γ and Γ' denote the reflection coefficient of p and v_x and that of the spatial derivatives $\partial_x p$ and $\partial_x v_x$, respectively.

Here, we designate a medium in the $-x$ side of the boundary as medium-1 and that in the opposite side as medium-2. The reflection coefficient Γ is given as

$$\Gamma = \frac{Z_2 - Z_1}{Z_2 + Z_1}, \tag{5.70}$$

where Z_1 and Z_2 are the characteristic impedance of each medium (hereafter, values with a subscript 1 and 2 denote those in each medium). Eq. (5.70) is derived from Dirichlet conditions of the sound pressure and the particle velocity on the boundary:

$$p_1 = p_2, \tag{5.71}$$

$$v_{x1} = v_{x2}. \tag{5.72}$$

Meanwhile, Neumann conditions on the boundary are

$$\frac{1}{\rho_1}\partial_x p_1 = \frac{1}{\rho_2}\partial_x p_2, \tag{5.73}$$

$$\rho_1 c_1^2 \partial_x v_{x1} = \rho_2 c_2^2 \partial_x v_{x2}, \tag{5.74}$$

where ρ_1 and ρ_2 are the density of each medium. These conditions lead to the following relation:

$$\Gamma' = -\frac{Z_2 - Z_1}{Z_2 + Z_1} = -\Gamma. \tag{5.75}$$

We can explicitly give the reflection coefficients Γ and Γ' as constants as long as they satisfy Eq. (5.75). For example, $\Gamma = 1$ and $\Gamma' = -1$ are set to a perfectly reflective surface.

In multi-dimensional simulations, boundary conditions for the perpendicular derivatives are given in similar forms to Eqs. (5.68) and (5.69). In a two-dimensional simulation, for example, the boundary conditions for the perpendicular derivatives are represented as

$$\eta_{x-}(x_b) = \Gamma \eta_{x+}(x_b), \tag{5.76}$$

$$\mu_{x-}(x_b) = \Gamma' \mu_{x+}(x_b). \tag{5.77}$$

The equations above are derived from the differentiation of Eqs. (5.71)–(5.74).

5.2.4 Perfectly Matched Layer

The CIP method computes a sound field formed into a finite discrete domain. When we apply the CIP method to a simulation of an infinite sound filed, such as an outdoor sound field, an absorption treatment is required to suppress reflection waves from outer boundaries of the finite simulation domain as the FDTD method requires. The CIP method computes wave propagation dividing it into forward and backward components. Consequently, the CIP method naturally provides an automatic-absorbing boundary condition (A-ABC) on the outer boundaries letting incoming components be zero. However, the A-ABC is similar to the Mur's ABC of the first kind in the FDTD method [17]. In multi-dimensional simulations, the A-ABC has insufficient absorption performance for oblique-incident waves on the outer boundaries although it is effective for a normal-incident wave.

The perfectly matched layer-absorbing boundary condition (PML-ABC) [18] is often employed in multi-dimensional FDTD simulations to achieve high absorption performance on the outer boundaries for oblique-incident waves as well as a normal-incident wave. The followings explain formulation to implement the PML-ABC in multi-dimensional sound field simulations using the type-C CIP method [19, 20].

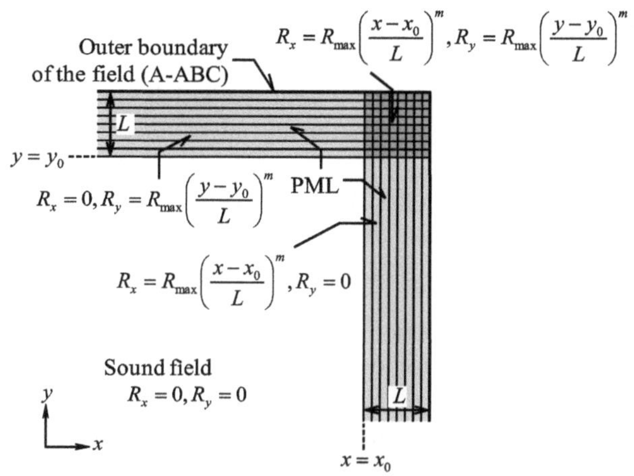

Fig. 5.6 Perfectly matched layer surrounding a sound field

Although, for simplicity, two-dimensional formulation is shown below, it can be simply extended to three-dimensional simulations.

Figure 5.6 schematically illustrates the PML, which surrounds a calculated sound field to suppress reflection waves from the outer boundaries. In the PML region, nonzero attenuation parameter reduces sound waves and results in suppression of incoming waves. Discontinuous change of the attenuation parameter causes reflection. Therefore, increasing attenuation parameter is given as a function of the distance from the PML surface, that is the interface between the sound field and the PML:

$$R = R_{\max}\left(\frac{d}{L}\right)^m, \tag{5.78}$$

where $R_{\max}$ is the maximum value of the attenuation parameter, L is the thickness of the PML, and m decides distribution of the attenuation parameter. d denotes the distance from the PML surface and can be expressed by coordinates as shown in Fig. 5.6. It should be noted that nonzero attenuation parameter is given only in the direction where the PML intends to suppress reflection waves; $R_x = R$, $R_y = 0$ in the PML for the x-direction, $R_x = 0$, $R_y = R$ in that for the y-direction, and $R_x = R$, $R_y = R$ in that at a corner of calculated sound field.

In the PML region with nonzero attenuation parameter R_x, the governing equations for acoustic fields corresponding to Eq. (5.13) are [21]

$$\partial_t p + cZ\partial_x v_x = -\frac{R_x}{\rho}p, \quad Z\partial_t v_x + c\partial_x p = -\frac{R_x}{\rho}Zv_x. \tag{5.79}$$

Addition and subtraction of them lead to the following equations letting $r_x = R_x/\rho$:

$$\partial_t f_{x\pm} \pm c\partial_x f_{x\pm} = -r_x f_{x\pm}. \tag{5.80}$$

Although Eq. (5.80) has a non-advection term, that is a dissipation term, in the right-hand side, it can be computed by separating into advection and non-advection phases [8]:

$$\partial_t f_{x\pm} \pm c\partial_x f_{x\pm} = 0, \tag{5.81}$$

$$\partial_t f_{x\pm} = -r_x f_{x\pm}. \tag{5.82}$$

Equation (5.81) is a normal advection equation. Therefore, it can be calculated by the procedure previously described for lossless fields. We let here results of the advection calculations at a time step n be $f_{x\pm}^{n(A)}$, $g_{x\pm}^{n(A)}$, $\eta_{x\pm}^{n(A)}$, and $\xi_{x\pm}^{n(A)}$.

In the non-advection phase, the analytical solution of Eq. (5.82) gives $f_{x\pm}$ at the next time step using $f_{x\pm}^{n(A)}$:

$$f_{x\pm}^{n+1} = f_{x\pm}^{n(A)} \exp(-r_x \Delta t). \tag{5.83}$$

The derivatives $g_{x\pm}$ at the next time step are also analytically obtained by the spatial differentiation of Eq. (5.83) with respect to x:

$$g_{x\pm}^{n+1} = \left\{ g_{x\pm}^{n(A)} - \Delta t (\partial_x r_x) f_{x\pm}^{n(A)} \right\} \exp(-r_x \Delta t), \tag{5.84}$$

where the spatial derivative of r_x is given by

$$\partial_x r_x = (\partial_x d) \frac{m R_{\max}}{\rho L} \left(\frac{d}{L} \right)^{m-1}. \tag{5.85}$$

Furthermore, the differentiations of Eqs. (5.83) and (5.84) with respect to y derive the solutions of the perpendicular derivatives $\eta_{x\pm}$ and $\mu_{x\pm}$:

$$\eta_{x\pm}^{n+1} = \eta_{x\pm}^{n(A)} \exp(-r_x \Delta t), \tag{5.86}$$

$$\mu_{x\pm}^{n+1} = \left\{ \mu_{x\pm}^{n(A)} - \Delta t (\partial_x r_x) \eta_{x\pm}^{n(A)} \right\} \exp(-r_x \Delta t). \tag{5.87}$$

As described above, the values at the next time step, that is the results of the non-advection phase, can be straightly calculated by using those previously obtained in the advection phase.

We can compute y-directional wave propagations in the PML region in a similar procedure to the above.

Figure 5.7 exhibits sound pressure distributions in a two-dimensional free field with and without the PML over the lower outer boundary. The PML consists of 64 layers and the outer boundaries including the lower one behind the PML employ the A-ABC. The illustrated area is a part of an adequately large calculated field, which

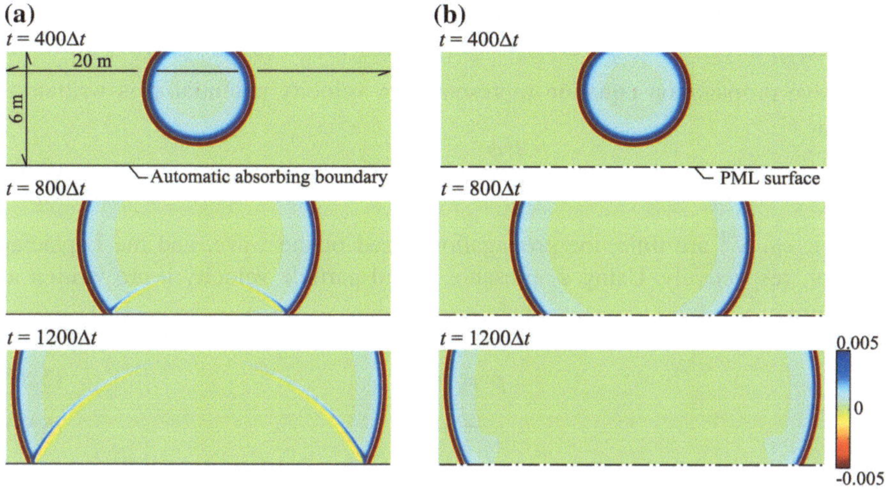

Fig. 5.7 Comparison of sound pressure distributions in the two-dimensional free field with and without the PML over the lower outer boundary. Calculation parameters are: $c = 340.5$ m/s, $\Delta t = 0.25 \times 10^{-4}$ s, $\Delta l = 0.025$ m, $R_{\max} = 2.0\rho c$, $m = 2$, and $N = 64$. **a** Without PML. **b** With 64-layer PML

has dimensions of 65.6 × 27.2 m. Therefore, no reflection waves from the outer boundaries except the lower one appear in the illustrated area in a time range shown in Fig. 5.7. In the results without the PML, we can observe the reflection wave from the A-ABC and find that the amplitude of it increases as the incident angle increases. In contrast, in the results with the PML, the reflection wave almost disappears.

5.3 Finite-Volume Time-Domain Method

The FDTD approach has widely been accepted as a simple, fast, and proven method for numerical sound propagation prediction. However, the method requires an orthogonal structured grid system for a discretized representation of an acoustic field. The use of the orthogonal grid system is recognized as one of shortcomings of the FDTD method because it is generally not suitable for the representation of complex geometries. Meshing (preprocessing) and postprocessing complex geometries with orthogonal structured grids often require extensive human works.

The issue is addressed here through an introduction of a full finite-volume time-domain (FVTD) method meant as a replacement for the FDTD method. The main strength of the full FVTD approach is the capability to utilize unstructured grids, which contain arbitrary polyhedral cells. The strength opens possibility of using a vast variety of general-purpose pre- and post-processors designed for finite volume or finite element grids.

5.3.1 Finite-Volume Formulation

The wave propagation equation represented in velocity potential ϕ is written as follows.

$$\frac{\partial^2 \phi}{\partial t^2} = c_0^2 \nabla^2 \phi, \tag{5.88}$$

where t, c_0, ∇^2 are time, the propagation speed of the wave, and the Laplacian operator, respectively. Using ϕ, pressure p and particle velocity u are written as follows.

$$p = \rho \frac{\partial \phi}{\partial t}, \tag{5.89}$$

$$u = -\nabla \phi. \tag{5.90}$$

Equation (5.88) is discretized under an unstructured grid system as shown in Fig. 5.8 where the definition point of physical quantities are taken at the barycenter of each control volume (CV). For the left hand side of Eq. (5.88), by integrating over the CV with time-invariant volume V and applying central time-differential scheme, we get

$$\frac{\partial^2}{\partial t^2} \int_V \phi \, dV \approx \frac{\phi^{n+1} - 2\phi^n + \phi^{n-1}}{\Delta t^2} V,$$

where ϕ^{n-1}, ϕ^n, ϕ^{n+1} denote the values of ϕ at the $(n-1)$th, nth, $(n+1)$th steps of time step Δt. For the right hand side, by integrating Eq. (5.88) within a CV and applying divergence theorem, we get

$$\int_V c_0^2 \nabla^2 \phi dV = c_0^2 \int_S dS \cdot \nabla \phi$$
$$\approx c_0^2 \sum_{\mathrm{f}} S_{\mathrm{f}} \cdot (\nabla \phi)_{\mathrm{f}}, \tag{5.91}$$

where S_{f} denotes the face area vector of the f-th face that constitutes the polyhedral CV in question as follows:

$$S_{\mathrm{f}} = S_{\mathrm{f}} n_{\mathrm{f}}, \tag{5.92}$$

where S_{f} and n_{f} are the area and the unit outward normal vector of the face f, respectively.

If a vector connecting the centers of the CV P and its adjacent CV N, d_{PN}, is parallel to S_f, $S_{\mathrm{f}}\,(\nabla \phi)_{\mathrm{f}}$ is written in terms of $\partial \phi / \partial n_{\mathrm{f}}$, the surface-normal gradient

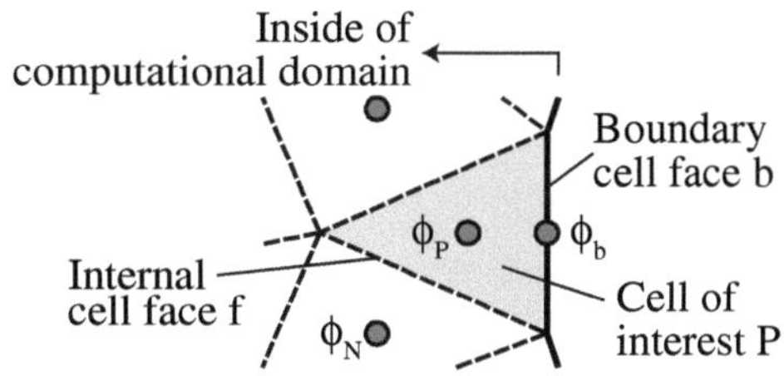

Fig. 5.8 Unstructured mesh system

of ϕ. Thus, the term within the summation in the rightmost hand side of Eq. (5.91) is discretized as follows.

$$S_{\mathrm{f}} \cdot (\nabla\phi)_{\mathrm{f}} = S_{\mathrm{f}} \frac{\partial \phi}{\partial n_{\mathrm{f}}} \approx S_{\mathrm{f}} \frac{\phi_{\mathrm{N}} - \phi_{\mathrm{P}}}{|d_{\mathrm{PN}}|}. \tag{5.93}$$

However, if d_{PN} is nonorthogonal to S_{f}, S_{f} has to be decomposed into its orthogonal part Δ_{f} and nonorthogonal part k_{f}.

$$S_{\mathrm{f}} \cdot (\nabla\phi)_{\mathrm{f}} = \Delta_{\mathrm{f}} \cdot (\nabla\phi)_{\mathrm{f}} + k_{\mathrm{f}} \cdot (\nabla\phi)_{\mathrm{f}}.$$

The first term of the right hand side of the equation above, the orthogonal part, is discretized similarly to Eq. (5.93) as follows:

$$\Delta_{\mathrm{f}} \cdot (\nabla\phi)_{\mathrm{f}} \approx |\Delta_{\mathrm{f}}| \frac{\phi_{\mathrm{N}} - \phi_{\mathrm{P}}}{|d_{\mathrm{PN}}|}.$$

The nonorthogonal part, $(\nabla\phi)_{\mathrm{f}}$ in the second term, is given by interpolating the gradient of ϕ at the centers of CVs P and N.

$$(\nabla\phi)_{\mathrm{f}} = f_x (\nabla\phi)_{\mathrm{P}} + (1 - f_x)(\nabla\phi)_{\mathrm{N}}. \tag{5.94}$$

Here, the interpolation coefficient f_x and the gradient $(\nabla\phi)_{\mathrm{P}}$ are given as follows.

$$f_x = \frac{\overline{\mathrm{fN}}}{|d_{\mathrm{PN}}|},$$

$$(\nabla\phi)_{\mathrm{P}} = \frac{1}{V} \int_S dS\, \phi \approx \frac{1}{V} \sum_{\mathrm{f}} S\phi_{\mathrm{f}},$$

where ϕ_{f} is a face-interpolated value of ϕ at the center of CVs.

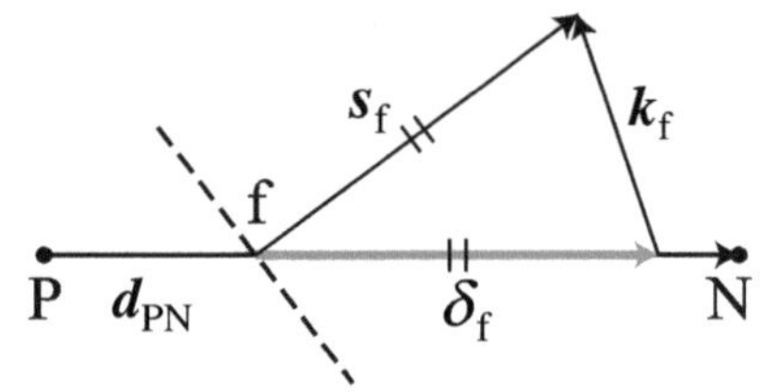

Fig. 5.9 Nonorthogonal mesh correction vectors

The orthogonal and nonorthogonal component vectors Δ_f and k_f can be calculated arbitrarily [22]. An orthogonal correction as follows is applied here.

$$\Delta_f = \frac{d_{PN}}{|d_{PN}|} S_f, \tag{5.95}$$

$$k_f = S_f - \Delta_f. \tag{5.96}$$

The vectors are schematically drawn in Fig. 5.9.

5.3.2 Rigid Boundary Conditions

On acoustically rigid boundaries b, the normal component of particle velocity u_b is fixed to zero.

$$u_b = n_b \cdot u_b = 0. \tag{5.97}$$

Substituting the relationship above to Eqs. (5.90) and (5.92) leads to the equation below, which represents the surface normal gradient of ϕ being zero.

$$S_b \cdot (\nabla\phi)_b = 0.$$

5.3.3 Computational Setup

In order to validate the proposed FVTD technique under unstructured meshes, a comparative test using a sound propagation problem in a closed cube of 1 m × 1 m × 1 m, one of the AIJ-BPCA (Benchmark Platform on Computational Methods for Architectural and Environmental Acoustics) [23] problems, was carried out. The detail of the tested cases are shown in Table 5.1.

Case 1 The problem was solved using a conventional FDTD code written in Fortran 77 employing a pressure-particle velocity leapfrog scheme. The case is meant to be the benchmark case to which the results obtained by the proposed technique is compared for validation. Each edge of the cube was divided to

Table 5.1 Computational setup

Case	1	2	3	4
Approach	FDTD	FVTD		
Type of mesh	–	Hexahedral	Unstructured tetrahedral	
Number of cells/CVs	$81^3 = 531\,441$		531 333	
Δx [m]	0.0123		–	
l_c [m]	–		0.025 (40 elements per edge)	
Δt [ms]	0.02		0.0049	
c_0 [m/s]	343.7			
Courant number	0.96		0.99 (max)	
Nonorth. correction	–	Uncorrected		Orthogonal
Initial condition	A single wave of offset cosine (Eqs. (5.98), (5.99))			

81 subedges to create a mesh of cell width $\Delta x = 0.0123$ m and the number of cells 531,441 (Fig. 5.10b). The time step Δt and the Courant number C_o were set to 0.02 ms and 0.96, respectively.

Case 2 The problem was solved with the proposed technique under a hexahedral orthogonal mesh and setup both identical to the ones for Case 1.

Case 3 The problem was solved with the proposed technique under a nonuniform tetrahedral unstructured mesh automatically generated by an open-source mesher, Gmsh [24]. The characteristic length l_c (the length with which each edge of the cube is divided) is set to 0.025 m, to make a mesh with the number of CVs 531,333 (roughly the same as Cases 1 and 2). The ratio of maximum and minimum CV edge lengths of the generated mesh was 6.32. The time step Δt was set to 0.0049 ms to keep the maximum Courant number to 0.99. In this case no nonorthogonal techniques was applied.

Case 4 The setups are same as Case 3, except that the orthogonal correction technique was applied.

For all cases, the initial values of ϕ were set to represent the pressure and particle velocity conditions of

$$p^{-1/2}(r) = \begin{cases} \dfrac{\cos 8\pi r + 1}{2} & (r < 0.125) \\ 0 & (\text{otherwise}) \end{cases} \text{[Pa]}, \tag{5.98}$$

$$u^0(r) = 0, \tag{5.99}$$

where r [m] is the distance from the center of the cube. All cases were run up to $t = 0.04$s.

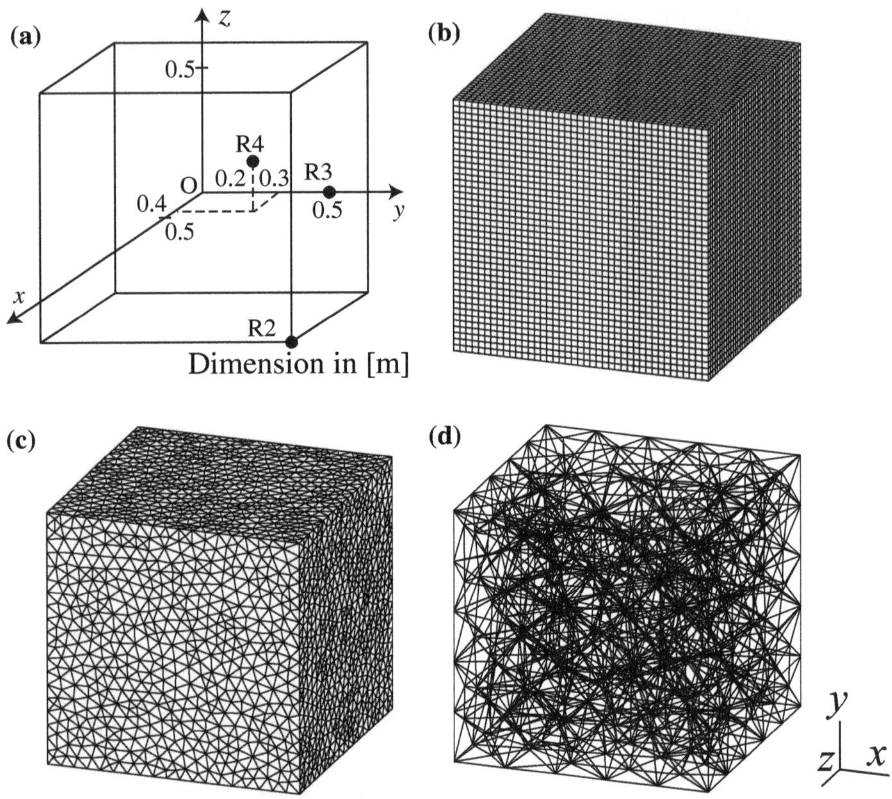

Fig. 5.10 **a** Problem geometry of the benchmark problem AIJ-BPCA B0-1T, **b** Surface mesh for Cases 1 and 2 (coarsened by factor of 2 for visibility), **c** Surface mesh for Cases 3 and 4 (coarsened by factor of 2) and **d** Surface and internal mesh (coarsened by factor of 8) for Cases 3 and 4

5.3.4 Results and Discussion

The transient sound pressure waveforms at the receiving point R2 shown in Fig. 5.10a are plotted in Fig. 5.11, using the result of Case 1 as the benchmark case for comparison with other cases. From Fig. 5.11a, one can see that the results of FDTD and FVTD techniques agree so precisely that they can be regarded as virtually identical results. From the results one can verify the proposed FVTD technique has the same accuracy as the conventional FDTD under identical geometry, mesh, and computational setups.

On the other hand, from the comparison of Cases 1 and 3 in Fig. 5.11b, the waveform obtained by the FVTD technique under the tetrahedral mesh is phasing forward in about 1.5 %. Also, the overall waveform is gradually dispersing over time. In addition, the results of Case 4 is shown in Fig. 5.11c. Despite the employment of the

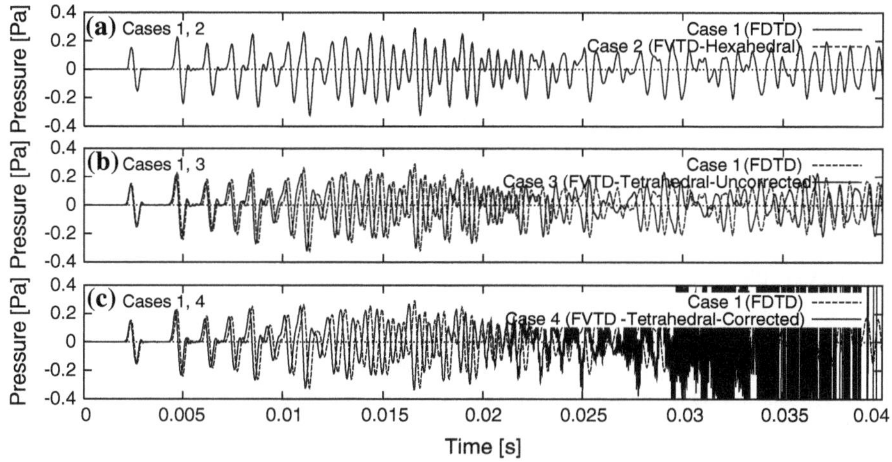

Fig. 5.11 Transient sound pressure waveforms at the receiving point R2

Table 5.2 Processor and memory usages

Case	1	2	3	4
Processor [s]	28.0	343	865	2013
Per time step [s]	0.0140	0.172	0.106	0.247
Memory [MB]	18	301	260	260

correction technique, the drift of the phase did not improve. Even worse, waveforms started to oscillate and diverged eventually.

From the results one can conclude that while one can expect identical results between FDTD and FVTD techniques under identical setups, there remains works for the FVTD technique in reducing the phase error coming from nonorthogonalities of unstructured grids.

5.3.5 Computational Loads

To compare the proposed technique with the conventional FDTD from the standpoint of computational loads, processor times and memory usages were instrumented for Cases 1–4, as shown in Table 5.2. The instrumentations were carried out on an Opteron 2.4 GHz 64-bit Linux platform. The FVTD computations turned out to require more than ten times of processor and memory usages. Hence it should be noted that, from the computational load of view, the proposed FVTD technique is not meant to completely replace FDTD especially in large cases, but should rather be used for small to medium cases where rapid preprocessing (case setup) and postprocessing have of particular importance.

Although Cases 2 and 3 have roughly the same number of cells, one may notice that Case 3 requires smaller amount of computational time per time step. This is because the computational load required in calculating the Laplacian term is determined mostly by the number of faces per CV, as shown in Eq. (5.91). It is also shown that, from Cases 3 and 4, applying nonorthogonal correction technique more than doubles the processor usage.

References

1. D. Heimann, R. Blumrich, Time-domain simulations of sound propagation through screen-induced turbulence. Appl. Acoust. **65**(6), 561–582 (2004)
2. V.E. Ostashev, D.K. Wilson, L. Liu, D.F. Aldridge, N.P. Symons, D. Marlin, Equations for finite-difference, time-domain simulation of sound propagation in moving inhomogeneous media and numerical implementation. J. Acoust. Soc. Am. **117**(2), 503–517 (2005)
3. T. van Renterghem, D. Botteldooren, Prediction-step staggered-in-time FDTD: An efficient numerical scheme to solve the linearized equations of fluid dynamics in outdoor sound propagation. Appl. Acoust. **68**(2), 201–216 (2007)
4. S.K. Lele, Compact finite difference schemes with spectral-like resolution. J. Comput. Phys. **103**, 16–42 (1992)
5. C. Arakawa, in *Computational Fluid Dynamics for Engineering* (University of Tokyo Press, Tokyo, 1994) (in Japanese)
6. J.C. Hardin, J.R. Ristorcelli, C.K.W. Tam, in *ICASE/LaRC Workshop on Benchmark Problems in Computational Aeroacoustics (CAA). NASA Conference Publications*, pp. 3300 (1995)
7. T. Yabe, F. Xiao, T. Utsumi, Constrained interpolation profile method for multiphase analysis. J. Comput. Phys. **169**, 556–593 (2001)
8. T. Yabe, T. Utsumi, Y. Ogata, *The Constrained Interpolation Profile Method* (Morikita, Tokyo, 2003). (in Japanese)
9. T. Aoki, Multi-dimensional advection of CIP (cubic interpolated propagation) scheme. Comput. Fluid Dynamics J. **4**, 279–291 (1995)
10. K. Okubo, S. Oh, T. Tsuchiya, N. Takeuchi, The examination on the three-dimensional acoustic field analysis using the CIP method. IEICE Technical Report, US2006-52:25–30, (2006) (in Japanese)
11. S. Oh, K. Okubo, T. Tsuchiya, N. Takeuchi, Analysis of acoustic field using the type-C CIP method. IEICE Technical Report, US2006-98:17–22 (2006) (in Japanese)
12. M. Konno, K. Okubo, T. Tsuchiya, N. Tagawa, Performance of various types of constrained interpolation profile method for two-dimensional numerical acoustic simulation. Jpn. J. Appl. Phys. **47**, 3962–3963 (2008)
13. Y. Tachioka, Y. Yasuda, T. Sakuma, Application of the constrained interpolation profile method to room acoustic problems: Examination of boundary modeling and spatial/time discretization. Acoust. Sci. Tech. **33**, 22–32 (2012)
14. T. Ishizuka, K. Okubo, Y. Ara, Variable-grid technique for sound field analysis using the constrained interpolation profile method. Acoust. Sci. Tech. **33**, 387–390 (2012)
15. G. Szego, in *Orthogonal Polynomials* (American Mathematical Society, Providence, 1939)
16. K. Okubo, S. Oh, T. Tsuchiya, N. Takeuchi, Consideration of the boundary condition between two media in acoustic field analysis using the constrained interpolation profile (CIP) method. IEICE Trans. Fundam. **E90-A**, 2000–2005 (2007) (in Japanese)
17. G. Mur, Absorbing boundary conditions for the finite difference approximation of the time domain electromagnetic-field equation. IEEE Trans. Antennas Propag. **45**, 377–382 (1981)
18. J.P. Berenger, A perfectly matched layer for the absorption of electromagnetic waves. J. Comput. Phys. **114**, 185–200 (1994)

19. Y. Ando, M. Hayakawa, Implementation of the perfectly matched layer to the CIP method. IEICE Trans. Electron. **E98-C**, 645–648 (2006)
20. T. Ishizuka, K. Okubo, Formulation and examination of the perfectly matched layer for sound field analyses using the constrained interpolation profile method. Acoust. Sci. Tech. **34**, 378–381 (2013)
21. Q. Qui, T.L. Geers, Evaluation of the perfectly matched layer for computational acoustics. J. Comput. Phys. **139**, 166–183 (1998)
22. H. Jasak, Error analysis and estimation for the finite volume method with applications to fluid flows. PhD thesis, Imperial College, June 1996
23. Architectural Institute of Japan, Benchmark platform on computational methods for architectural/environmental acoustics. http://gacoust.hwe.oita-u.ac.jp/AIJ-BPCA/
24. C. Geuzaine, J.-F. Remacle, Gmsh: a three-dimensional finite element mesh generator with built-in pre- and post-processing facilities. Int. J. Numer. Methods Eng. **79**, 1309–1331 (2009)

Part II
Applications to Architectural and Environmental Acoustic Problems

Chapter 6
Room Acoustics Simlation

Reiji Tomiku, Shinichi Sakamoto, Noriko Okamoto, Yosuke Yasuda, Yoshinari Horinouchi and Kazuma Hoshi

Abstract This chapter shows examples of numerical simulation results on various sound fields in the interior of buildings and vehicles, such as auditoria, large space with seating rows, reverberation rooms, vehicle cabins, and small space with partitions. In each section, the features and problems of each sound field are introduced, and numerical modeling schemes of the simulation and the calculated results of the sound fields are described and illustrated. Some of the calculated results are compared with those obtained by other methods or measured ones to discuss the applicability and efficiency of the analysis method.

R. Tomiku (✉)
Faculty of Engineering, Oita University, 700 Dannoharu, Oita-shi, Oita 870-1192, Japan
e-mail: tomiku-reiji@oita-u.ac.jp

S. Sakamoto
Institute of Industrial Science, University of Tokyo, 4-6-1 Komaba, Meguro-ku,
Tokyo 153-8505, Japan
e-mail: sakamo@iis.u-tokyo.ac.jp

N. Okamoto
Department of Architecture, Ariake National College of Technology, 150 Higashihagio-machi,
Omuta-shi, Fukuoka 836-8585, Japan
e-mail: okamoto@ariake-nct.ac.jp

Y. Yasuda
Faculty of Engineering, Kanagawa University, 3-27-1 Rokkakubashi,
Kanagawa-ku, Yokohama 221-8686, Japan
e-mail: yyasuda@kanagawa-u.ac.jp

Y. Horinouchi
Faculty of Engineering, Kyoto University, C1-4-384 Kyoto-Daigaku-Katsura,
Nishikyo-ku, Kyoto 615-8540, Japan
e-mail: horinch@archi.kyoto-u.ac.jp

K. Hoshi
Junior college, Nihon University,7-24-1 Narashinodai, Funabashi-shi, Chiba 274-8501,
Japan
e-mail: hoshi@arch.jcn.nihon-u.ac.jp

T. Sakuma et al. (eds.), *Computational Simulation in Architectural and Environmental Acoustics*, DOI: 10.1007/978-4-431-54454-8_6,

Keywords Room acoustics · Auditoria · Seating rows · Reverberation room · Vehicle cabin · Partition

6.1 Auditoria

Concert hall acoustics is the most attractive topic to many acousticians from both physical and psycho-acoustical viewpoints. Also, in computational acoustics an auditorium has been being an interesting object, but the computation of an auditorium in wide frequency range has been difficult due to the requirement of huge computational resource. In recent years, rapid advances in hardware/software technologies are promoting applications of wave-based numerical methods to analysis of acoustical phenomena in auditoria. In this section, several applications of the finite-difference time-domain method (FDTD method), the finite element method (FEM), and the boundary element method (BEM) to numerical analyses of auditoria are introduced.

6.1.1 Application of FDTD Method

Figure 6.1 shows a small hall under investigation. As seen in plan of the figure, this hall with about 260 seats mainly for lectures and conference, has a unique room shape consisting of concave curved boundary. Generally, the concave boundary is apt to raise the concentration of sound and echoes, and so this hall could be acoustically hazardous. In acoustical design, such concave curved boundaries are substituted by convex boundaries in order to raise scatter and diffusion of sound. Regarding this hall, triangular diffusers were attached to side walls and the rear wall was treated absorptive to avoid the concentration of sound and echoes.

Treatment of absorptive boundary conditions of building walls, floor, ceiling, and other surface of furniture has become a serious problem to be solved. In many cases, locally reacting on surfaces is assumed and normal acoustical impedance is taken into account. In order to treat frequency characteristics of the acoustic impedance, some contrivances should be done for time-domain solvers such as the FDTD method, whereas a complex impedance can be introduced for respective frequency for frequency domain solvers. As such contrivances to treat frequency-dependent acoustic impedance in time-domain solvers, the following methods are proposed; methods using of z transform [1] and IIR filters [2] to treat transient characteristics of the acoustic impedance, a method in which absorptive layers are directly discretized and the internal propagation is calculated based on the Rayleigh model [3]. In this example, continuous physical phenomenon of sound absorption and sound reflection by an absorptive boundary was substituted by an equivalent discrete mechanical system consisting of masses, springs, and resistances [4]. In this calculation model, hypothetical mechanical model with two degrees of freedom which moves by pressure of sound is set on the boundaries and their parameters of the mass, spring, and resistance,

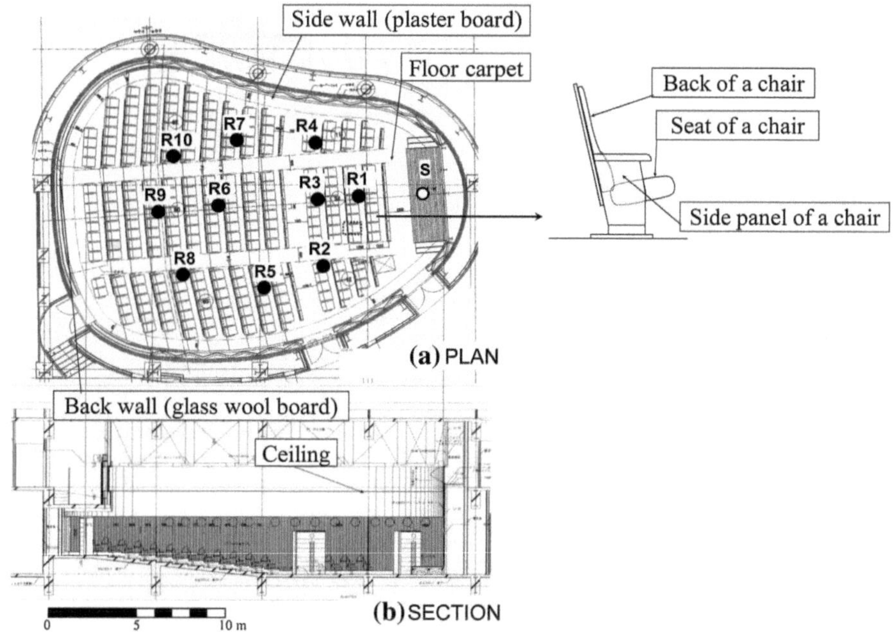

Fig. 6.1 Plan and section of a hall under investigation

$m_1, m_2, k_1, k_2, c_1, c_2$, are appropriately determined so that the vibration velocity of the mass, m_1, is equal to particle velocity at the surface of the boundary. More many degrees of freedom were possible to fit the acoustic impedance characteristics, but in this example two degrees of freedom was adopted because of convenience of the parameters' determination (Fig. 6.2).

Among the materials consisting of the hall, seven kinds of materials shown in Fig. 6.1 were picked up as acoustically principal materials and their normal acoustic impedance were measured in order to determine the parameters of the hypothetical mechanical systems. Among them, correspondence of measured acoustic impedance and modeled ones based on the determined parameters for three kinds of the materials are shown in Fig. 6.3. For (a) floor carpet and (c) seat of chair, both characteristics were fairly in good agreement. For (b) side wall, however, the agreement was poor, especially in real part of the acoustic impedance because the characteristics could not be appropriately modeled by the hypothetical model having two degrees of freedom. As indicated in the discrepancy, all of the acoustic impedance characteristics cannot be always appropriately approximated.

Impulse responses at 10 receiving points shown in Fig. 6.1 were calculated by the FDTD method. In the calculation, spatial grid size, Δh, and discrete time interval, Δt, in the FDTD analysis were made to be 0.06 m and 0.05 ms, respectively, and transient responses of 30,000 steps (1.5 s) were calculated. The calculated responses were corrected so that their spectral characteristics became flat in the frequency range

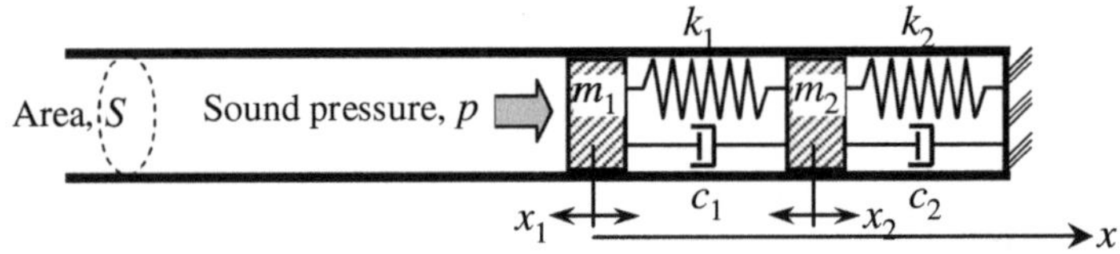

Fig. 6.2 Hypothetical mechanical system which simulates surface acoustic impedance of the boundary with two degrees of freedom

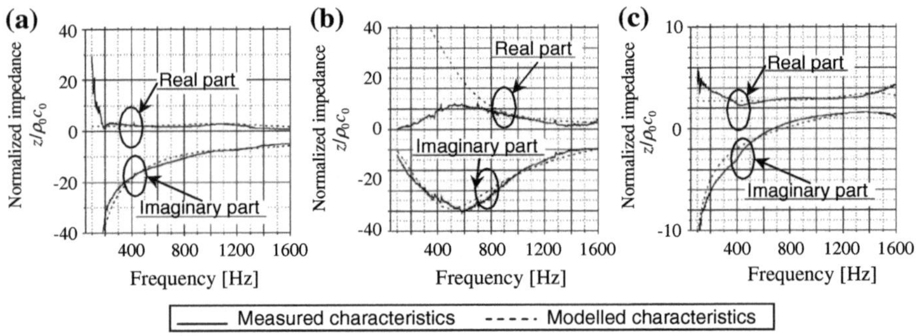

Fig. 6.3 Comparisons between measured and modeled surface acoustic impedance

between 88 Hz and 1,414 Hz (4 octave bands from 125 Hz to 1 kHz). To validate the calculation results, the impulse responses were measured in the hall for the same sound source and receiving positions by applying the swept-sine method. For this measurement, a dodecahedral loudspeaker was placed on the stage at a height of 1.5 m and a (1/2)-inch omni-directional microphone was placed at each receiving point at a height of 1.2 m above the floor. Four examples of the comparison between the echo diagrams of the calculated and the measured responses—front seat, R3; back seat, R9; middle seats, R5 and R7—are shown in Fig. 6.4. These echo diagrams were obtained by passing the impulse response signal through a numerical RMS detector with a time constant of 1 ms. In these figures, 'Corr.' means the cross-correlation coefficient between the echo diagrams of the calculated and measured responses calculated for the range of 1 s of the responses. The coefficient is approximately 0.8 for all the results shown in Fig. 6.4. Similar tendencies were observed in the results obtained for the other receiving points.

To further compare the impulse responses obtained by the FDTD calculation and the actual measurement more quantitatively, major room acoustic parameters calculated from the impulse responses were compared. First, the reverberation time (RT) in each octave band was calculated by the integrated impulse response method for all the receiving points and the results were averaged for each octave band. The results for the calculation and the measurement are shown in Fig. 6.5. In the figure, it is observed that the calculated and measured values are in fairly good agreement in the 250, 500 Hz, and 1 kHz bands, whereas a large discrepancy is observed in the 125 Hz

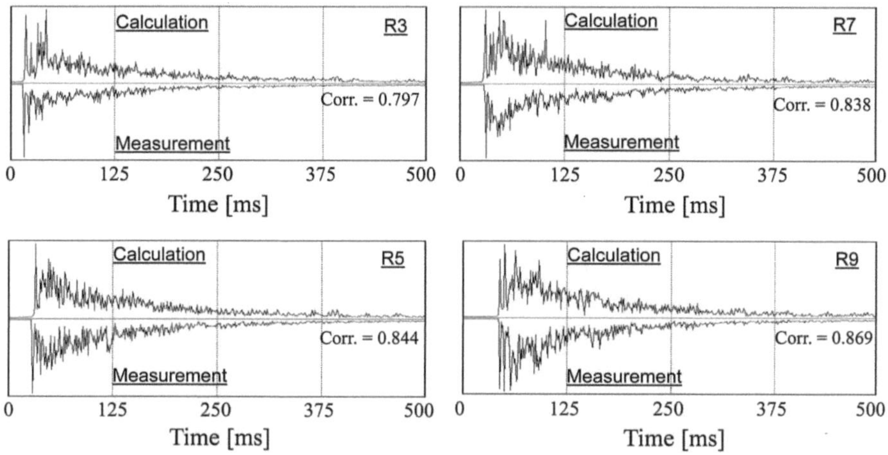

Fig. 6.4 Comparison of impulse responses between calculation and measurement

Fig. 6.5 Comparison of reverberation time obtained from impulse responses between calculation and experiment (averaged value of 10 receiving points)

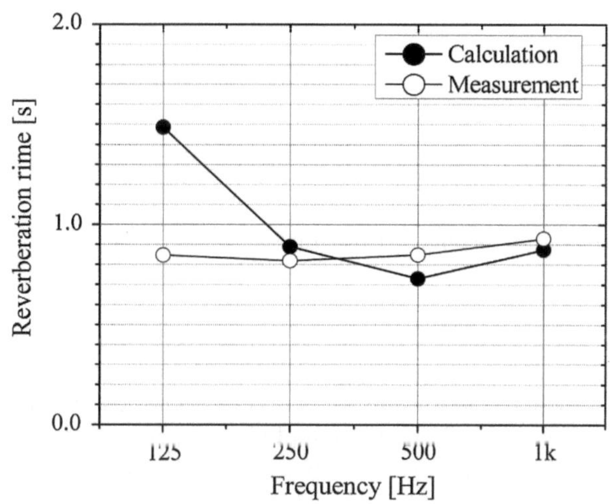

band. This discrepancy might be attributed to the setting of the boundary condition at low frequencies in the FDTD calculation. In other words, the sound absorption effects of panel vibration and Helmholtz resonance in the rib wall at low frequencies were not considered in the calculation because these effects cannot be measured by the impedance-tube method. The other room acoustic parameters obtained from the impulse response-definition D_{50}; clarity C_{80}; and center time T_s (specified in ISO 3382) were calculated for the frequency range including the 2 octave bands of 500 Hz and 1 kHz in order to reduce the interference effects. For both calculation and measurement, the parameters were deduced from the impulse responses for 1 s. The calculations and the measured values of each of these parameters are shown in Fig. 6.6. Although some discrepancies are observed in the results of C_{80} at the

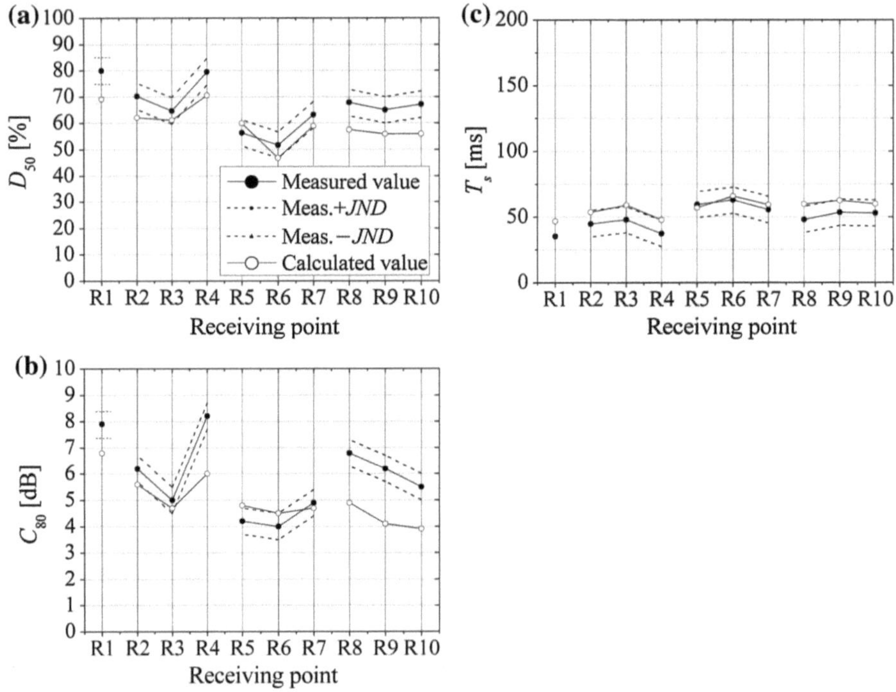

Fig. 6.6 Comparison of monaural acoustic parameters calculated from impulse responses. **a** D_{50}; **b** C_{80}; **c** T_s for 500 Hz–1 kHz band

receiving points R1, R4, R8, R9, and R10, the calculations and the measured values are in fair agreement on the whole. In the figures, the dotted lines indicate the range of the measured value $\pm JND$ (just noticeable difference limen) [5].

Figure 6.7 shows a result of visualization of instantaneous sound pressure distribution in the hall. We can observe sound propagation process in the hall where sound wave is scattered by sidewall diffusers and series of chairs.

6.1.2 Application of FEM and BEM

This section is an introduction of sound simulation on a small hall by FEM and BEM using fast multipole method (FMBEM). The hall to be analyzed is shown in Fig. 6.8. It is a multi-purpose event hall which can change the stage form and has a room volume of 3,000 m^3 and from 220 to 310 audience seats. Generally, elliptic geometry is attractive one from the viewpoint of architectural design, however, it frequently causes room acoustic problems such as sound focus at specific location. In this hall, many diffusers are installed on the wall to avoid those problems.

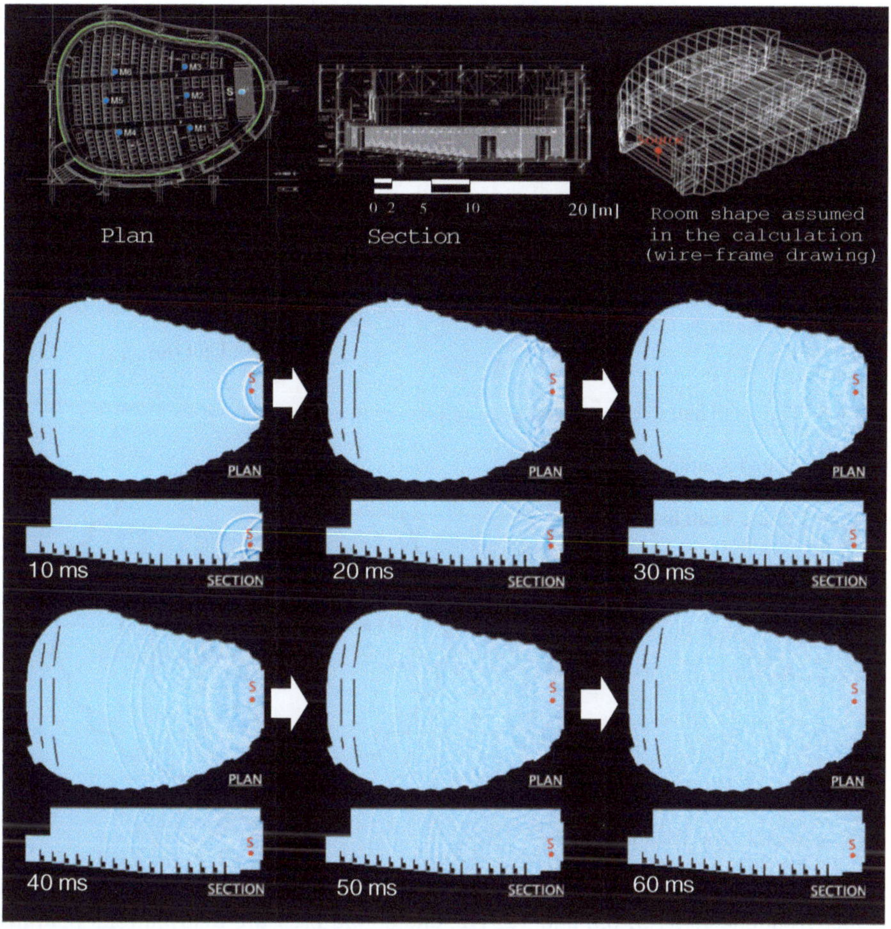

Fig. 6.7 Snapshots of instantaneous sound pressure distribution

The hall is listed as "B1-1F, Task A" of a benchmark platform, i.e., 'Benchmark Platform on Computational Methods for Architectural/Environmental Acoustics' [6], which has been provided by a committee in Architectural Institute of Japan. The "B1-1F, Task A" is a problem of computing the sound pressure amplitude at 201 points (Fig. 6.9), on the line from S:(−7.558, 0, 1.5) to (12.65, 0, 1.5), located at 0.1 m intervals, in the small hall shown in Fig. 6.8. Computing frequencies are set to the center frequencies of octave-band from 31.5 Hz to 4 kHz. A point source, S, is located at (−7.558, 0.0, 1.50), assuming its stationary vibration with volume acceleration amplitude 1.0 m^3/s^2. The absorption coefficient of the ceiling is 1.0. Other surfaces are rigid boundaries. Values c and ρ are set to 343.7 m/s and 1.205 kg/m^3, respectively.

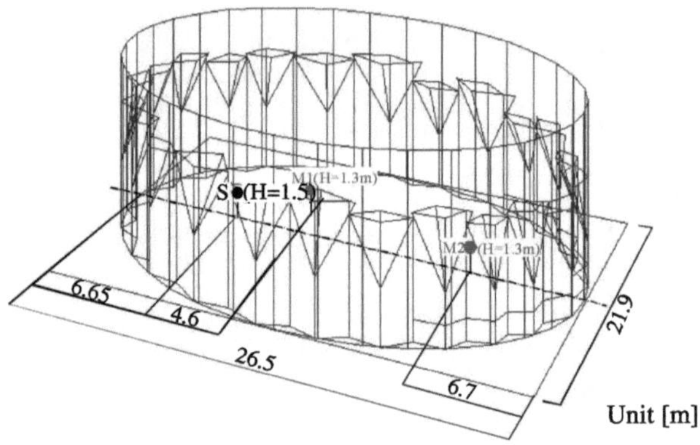

Fig. 6.8 The small hall to be analyzed

Fig. 6.9 Sound source location and receiving points on Line-1 of the small hall

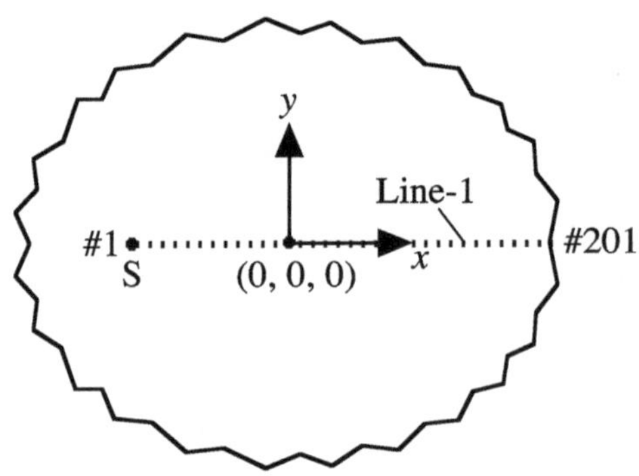

In this section, analyses by FEM and FMBEM are, respectively, carried out at center frequencies 31.5, 63, and 125 Hz, and sound pressure distributions obtained by both methods are compared.

In the FE analysis, DOF is 8,926,001 regardless of frequency. The elemental division was determined to satisfy that λ/d exceeds 4.5. The COCG method (stopping criterion: $\varepsilon = 10^{-4}$; initial value: $\boldsymbol{x}_0 = \boldsymbol{0}$), which performs good convergence for sound field analysis in rooms [7], is adopted. Incidentally, as for the geometry used in the FE analysis, diffusers on a wall are more simplified than those of FMBEM as shown in Fig. 6.10.

In the BE analysis, constant elements were employed, and DOF is 6,110 at both 31.5 Hz and 63 Hz, and it is 24,514 at 125 Hz. The elemental division was determined to satisfy that λ/d exceeds 9.0. The GPBiCG method ($\varepsilon = 10^{-3}$; $\boldsymbol{x}_0 = \boldsymbol{0}$; initial shadow residual: $\boldsymbol{r}_0^* = \boldsymbol{r}_0$) [8], which performs good convergence, was adopted. The ILUT preconditioning (the parameter in regard to the number of fill-in element: $p = 10$; the relative minimum value of element in a matrix: $\tau = 10^{-5}$) [8] was implemented.

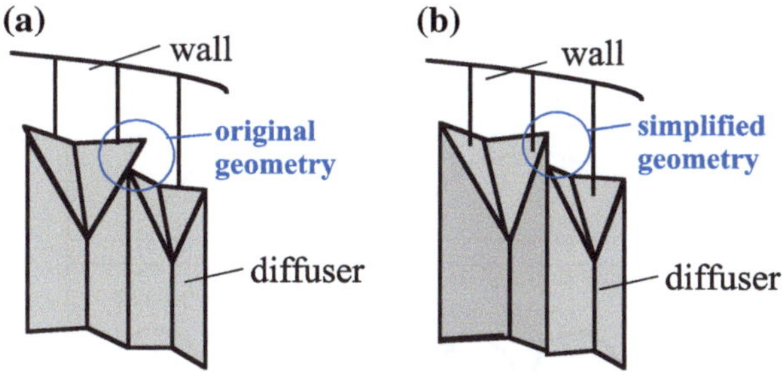

Fig. 6.10 Geometry of diffusers used in analyses by **a** FMBEM; by **b** FEM

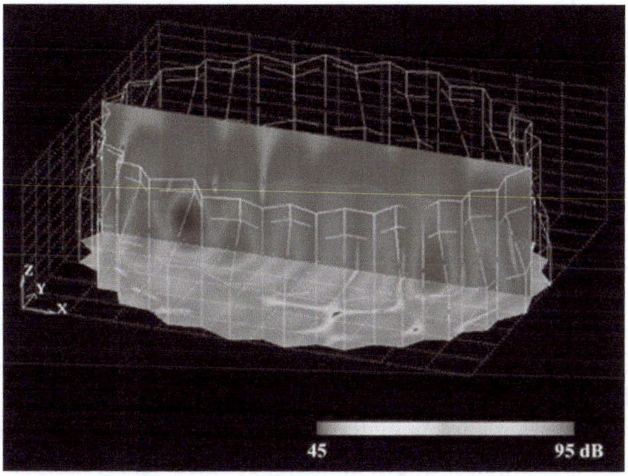

Fig. 6.11 The contour map of sound pressure levels obtained by FEM (63 Hz)

Numerical analyses were carried out using a VT64 Opteron Workstation (CPU: Opteron (2 GHz), Number of Processors: 2). For numerical experiments of this section, single processor was used.

Figure 6.11 shows a contour map of relative sound pressure levels at 63 Hz obtained by FEM, where sound pressure level at the source point is assumed as 100 dB. From the figure, strong influence of room's mode can be confirmed. Figure 6.12 shows a comparison of absolute sound pressure levels between FEM and FMBEM. The FEM shows good agreement with FMBEM at both 31.5 and 63 Hz. The correlation coefficients between FEM and FMBEM are, respectively, 0.99 at 31.5 Hz, and 0.97 at 63 Hz. However, in 125 Hz, the sound pressure levels obtained by FEM are considerably different from those of FMBEM at several dips of those distributions. Then, the correlation coefficient became 0.81. That difference should be caused by the difference in diffusers modeling between FEM and FMBEM, as described previously.

For reference, when the sound field of 125 Hz was analyzed, the computation time and memory required for FEM were 635,687 s and 9,426 MB, respectively,

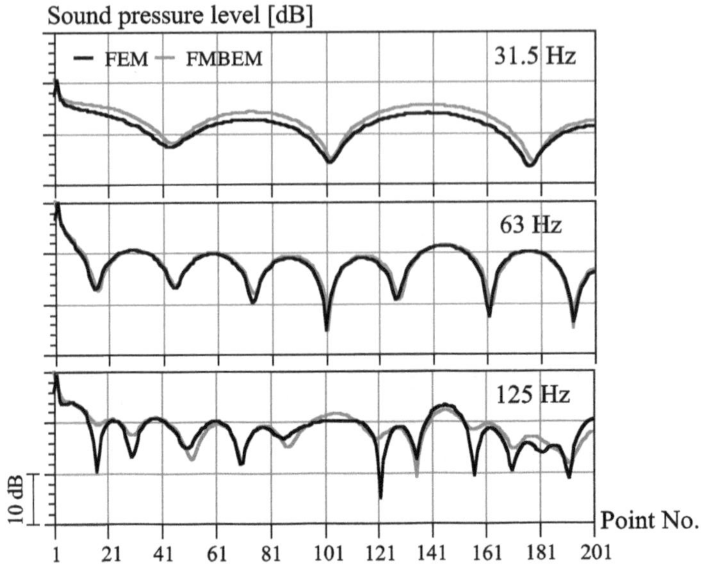

Fig. 6.12 Comparison of sound pressure levels between FEM and FMBEM

and those of FMBEM were 846 s and 571 MB. The computation time and memory required in FMBEM are smaller than those of FEM. In the case of a problem that has receiving points located at narrow intervals for wavelengths such as 'B1-1F, Task A', the DOF of FEM that requires a split domain becomes numerous. It is important that investigations are carried out in detail, including assembling characteristics among numerical analyses for various acoustic problems.

6.2 Seating Rows

The seating in concert halls has been an interesting subject of research in the field of room acoustics for a considerable time. The seat dip effect (SDE) [9, 10] is considered to trigger excessive attenuation, particularly in the low-frequency acoustic range, and has been examined by numerous researchers.

Seating rows were initially the subject of direct analysis using numeric simulations, with numerous studies conducted to examine their acoustic characteristics and the occurrence mechanisms of SDE [11, 12]. In recent years, however, the analysis of space in the dimensional scales of concert halls has become possible with progress in computing. The boundary surface aspect of seating rows for the analysis of the space as a whole has become more prominent than the seating rows being an independent subject of analysis. In other words, the issue of setting the boundary conditions for the floor area is being raised.

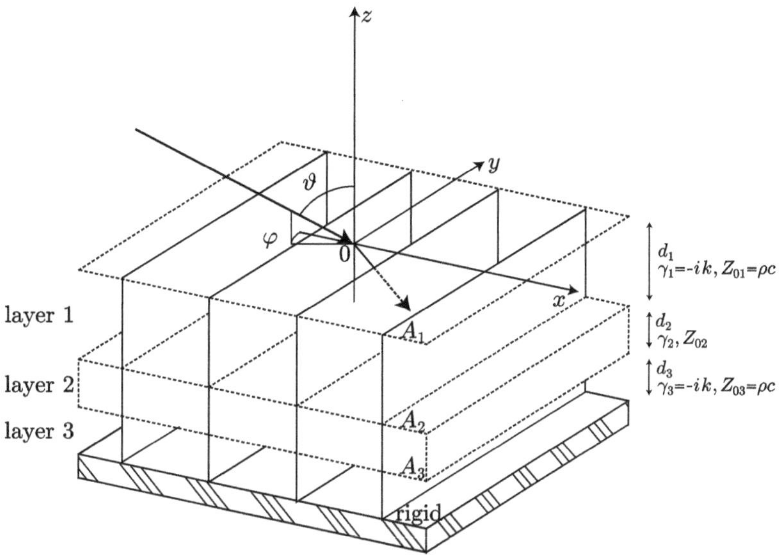

Fig. 6.13 Layer model of seating rows

In this section, a method for considering the seating rows as a uniform imaginary sound absorbing surface, for which localized effects can be assumed, will be introduced in order to set boundary conditions for the floor area for the purpose of analyzing a large-scale space. In addition, a case study, in which a two-dimensional analysis was performed, is presented.

6.2.1 Analysis with Layer Model Admittance

In performing an acoustic field analysis of concert halls, the seating sections have complex forms and are made of different materials with various acoustic characteristics. There are many practical difficulties in breaking them down into finer segments to understand the boundary conditions for each such segment. For this reason, there have been analytical methods that replace the seating rows (seat surfaces) with an imaginary boundary that can be assumed to have a uniform impedance and local reaction [13]. In this section, we shall first present a case study of an analysis performed using the layer model admittance proposed by Osa et al. [13].

This method sets up the admittance in an imaginary boundary surface, which is set up with the seat top sections that form a ridge across the rows, and the lower sections of the imaginary boundary surface is considered to be divided into three layers (Layer 1: an air layer assumed to extend from the upper section of the seat to knee height; Layer 2: an acoustic absorption layer equivalent to the seat surfaces;

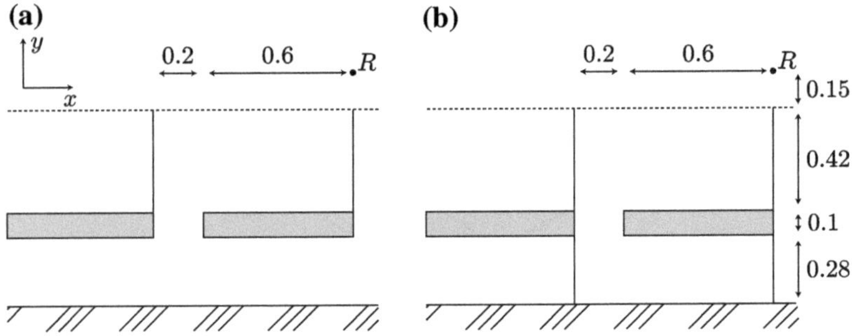

Fig. 6.14 Seat models (seat surface length: 0.6 m). **a** With underpass; **b** Without underpass

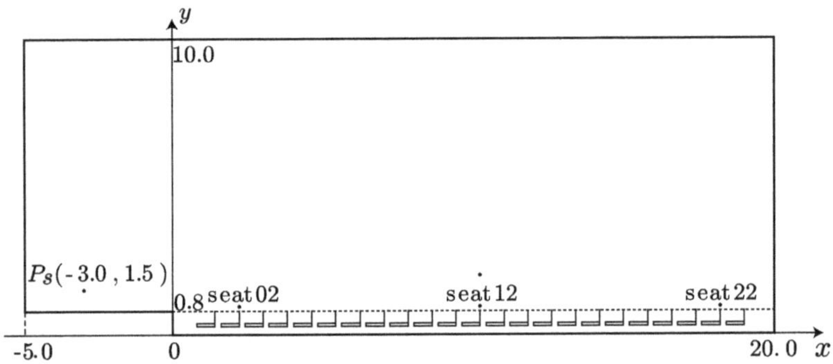

Fig. 6.15 Cross-sectional diagram of the closed space in which the two-dimensional analysis was performed

and Layer 3: an air layer extending from the lower seat surfaces to the floor), with the particle velocities restricted to the z direction (Fig. 6.13).

A case study in which layer model admittance was applied to a two-dimensional closed spatial acoustic field is shown below. The shape of the seats used for the analysis is shown in Fig. 6.14, while the cross-sectional diagram and co-ordinate axes of the space are shown in Fig. 6.15. The length of the floor sections was given by two patterns of 0.4 and 0.6 m for the purpose of comparison, while the spacing between seats was 0.8 m, and the height of the top line of the seats from the floor was 0.8 m. The height of the seat backs was considered to be 0.42 m, while the lower sections of the seat backs and seat surfaces were considered to be rigid. The acoustic absorption coefficient for the walls, ceiling, and stage floor was considered to be 0.2.

Seats in Figs. 6.14a and 6.15 were without legs, resulting in acoustic fields in which the underpass existed. The imaginary boundary surfaces were set uniformly from the stage to the rear walls along the peak positions of the seats that form the top ridge lines, as indicated by the dotted lines in Fig. 6.15.

A situation in which the seating row models shown in Fig. 6.14 are arranged is notated as the "seating row model," while the case in which the imaginary boundary surfaces were used is notated as the "imaginary surface model," with the results from

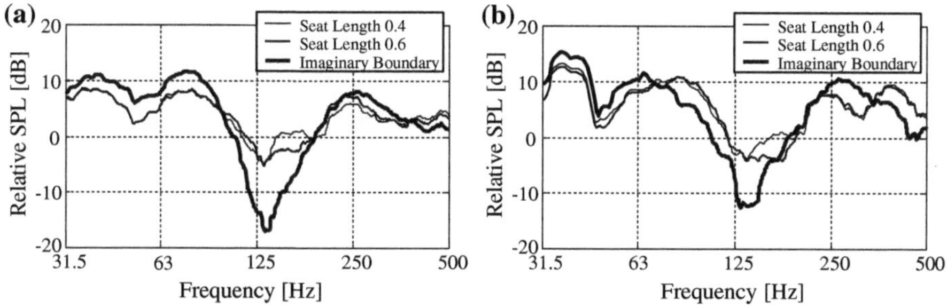

Fig. 6.16 Frequency response. **a** A seat in the 2nd row; **b** A seat in the 12th row

each model compared to examine the effectiveness of the methods. Furthermore, all analyses described hereafter were conducted using the BEM of the normal derivative form.

Furthermore, the method developed by Terai and Kawai [11, 14] was used for processing the singular term in the two-dimensional analysis.

Direct sound was used to normalize the seat in the 2nd row and the seat in the 12th row (seats 02 and 12 in Fig. 6.15).

The analysis results of the frequency response are shown in Fig. 6.16. The results of the seating row model and imaginary surface models, one each for the cases of a seat surface length of 0.4 and 0.6 m, are shown together in the figure. The sound receiving point was set at a height of 0.95 m from the floor. The sound receiving point was located roughly at the center of seating the 12th row in Fig. 6.16b with the results for the seating row model and the imaginary surface model, including that of the position of the dip, matching up quite well. In the comparison of the cases in which the seat surface lengths were 0.4 and 0.6 m, the case of 0.6 m showed excessive attenuation in the low-frequency range, i.e., this case exhibited SDE quite well. This coincides with the research result [12] indicating that the SDE is more strongly affected by the seat surface than by the floor surface.

Furthermore, the frequency response at a point on the seat in the 12th row, at a height of 2 m, is shown in Fig. 6.17. A comparison with Fig. 6.16b reveals that there is less of an effect due to the seating row. The excessive attenuation also decreased. It also shows how the attenuation is overestimated in the vicinity of 125 Hz in the imaginary surface model for the front area of the seating rows. This is believed to be due to the effect arising from setting the imaginary boundary surface between the front of the seating rows and the stage.

The number of elements and the CPU time in the two-dimensional analysis described earlier are shown in Fig. 6.18. Since more element divisions must be made in the higher frequency ranges, the method of increasing the number of elements in stages according to the frequency was adopted for our analysis. This meant that Fig. 6.18 was plotted using the center of the frequency range for the analysis performed using the same number of elements. The figure indicates that the required

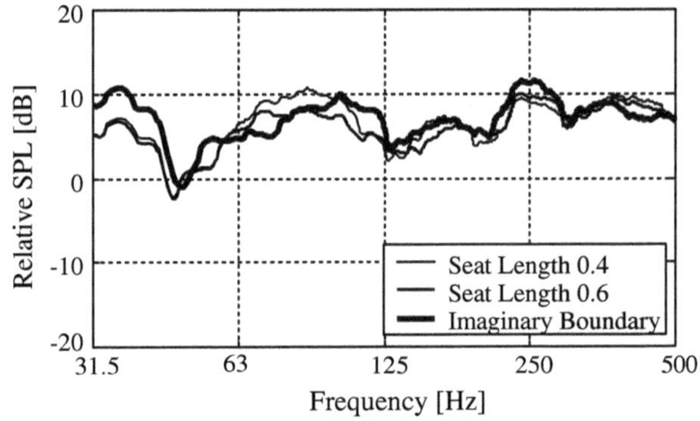

Fig. 6.17 Frequency response: a point on the seat in the 12th row at a height of 2 m

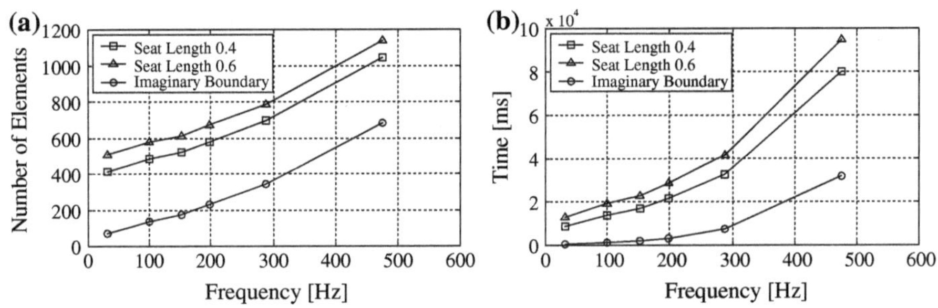

Fig. 6.18 Calculation load. **a** Number of elements; **b** CPU time

number of elements and the CPU time both decrease significantly with the implementation of the imaginary surface model.

Furthermore, the results for the case in which no underpass exists, i.e., where the seats have legs, are shown in Fig. 6.19. This is a case in which the seats in Fig. 6.15 were replaced by the seats in Fig. 6.14b. The imaginary surface model in this instance, however, had the imaginary acoustic absorption surface set only on the seat at the very front and the seat at the very back as well as between the top parts of the respective seats (it was set uniformly from the stage to the rear walls for the case with underpass).

The results of Fig. 6.19 indicate that the imaginary surface model is a good approximation of the seating row model for frequencies in the vicinity of 125 Hz. The attenuation that occurs in the vicinity of 63 Hz with the seating row model, however, was not reproduced with the imaginary surface model. Furthermore, it is evident that the excessive attenuation shown by the imaginary surface model has been suppressed in comparison with the case where underpass exists.

Furthermore, an analysis of a case in which the shape of the space was changed was also performed in order to examine the validity of the imaginary surface model. The shape includes a reflector plate installed on the stage ceiling, as shown in Fig. 6.20.

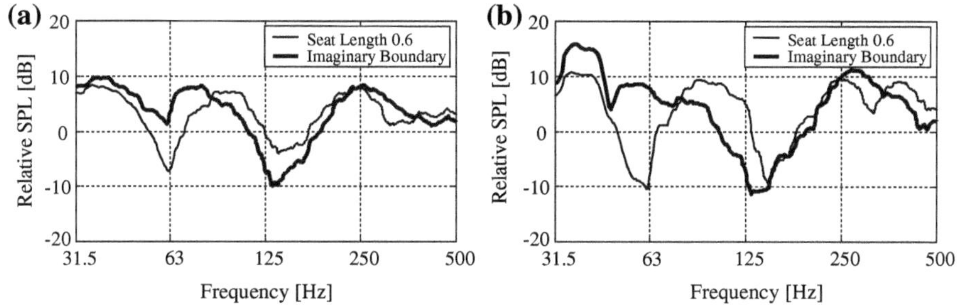

Fig. 6.19 Frequency response (no underpass). **a** A seat in the 2nd row; **b** A seat in the 12th row

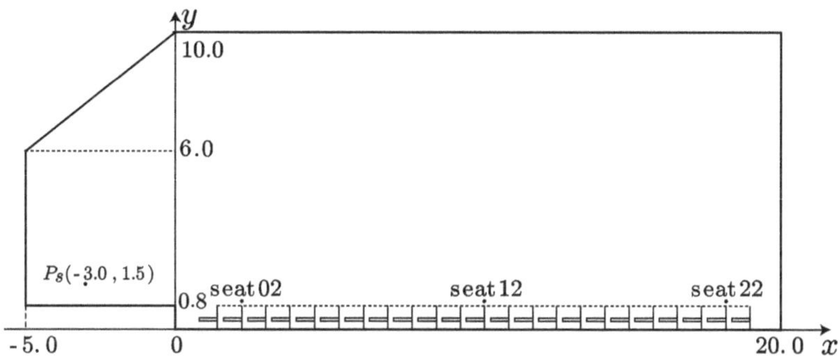

Fig. 6.20 Cross-sectional diagram of space subject to two-dimensional analysis (reflector plate present on the stage ceiling)

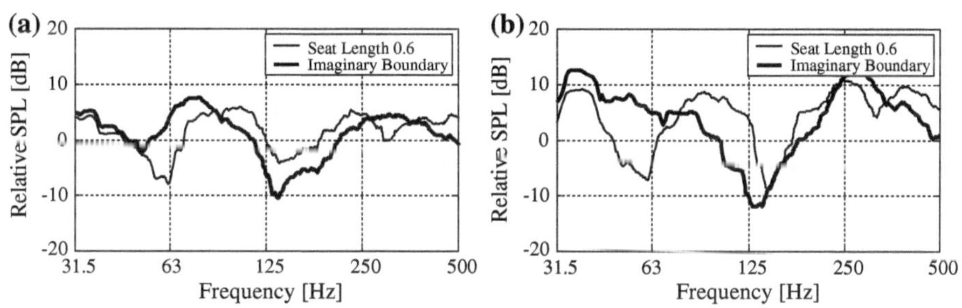

Fig. 6.21 Frequency response. **a** A seat in the 2nd row; **b** A seat in the 12th row

A comparison of results for the seating row model with a seat length of 0.8 and no underpass and the acoustic absorption surface model with an imaginary acoustic absorption surface set between the top sections of seats for the seat at the very front and the seat at the very back is shown in Fig. 6.21. The spatial acoustic pressure distribution for the case of 31.5 Hz is shown in Fig. 6.22, while the calculated number of elements and the CPU time are shown in Fig. 6.23. A review of the figure for the frequency response in Fig. 6.21 reveals that the analysis results for both models vary

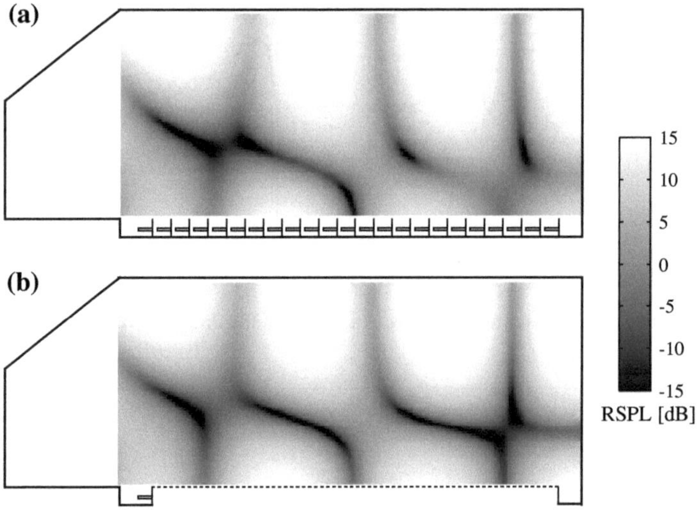

Fig. 6.22 Spatial distribution of acoustic pressure (3.15 Hz). **a** Seating row model; **b** Imaginary surface model

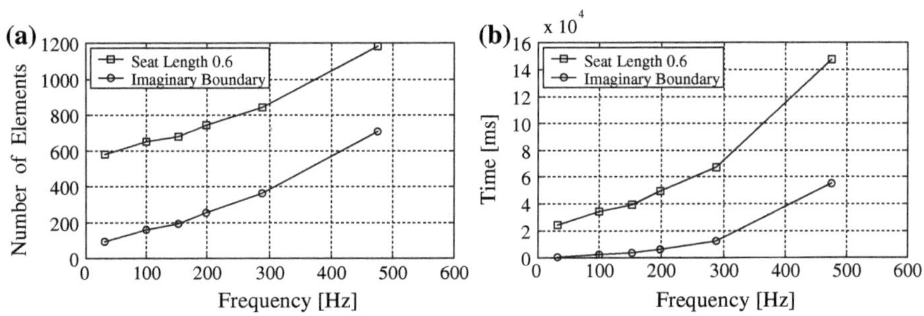

Fig. 6.23 Calculation load. **a** Number of elements; **b** CPU time

significantly depending on the positions of the sound receiving points. When the spatial acoustic pressure distribution is calculated as shown in Fig. 6.22, however, there is not much difference between the two models.

6.2.2 Analysis with Seat Top Section Admittance

The analysis performed with the layer model admittance described in the previous section indicated a rough match in the trends, but inconsistencies arose in the results for the seating row model and the imaginary surface model, particularly when no

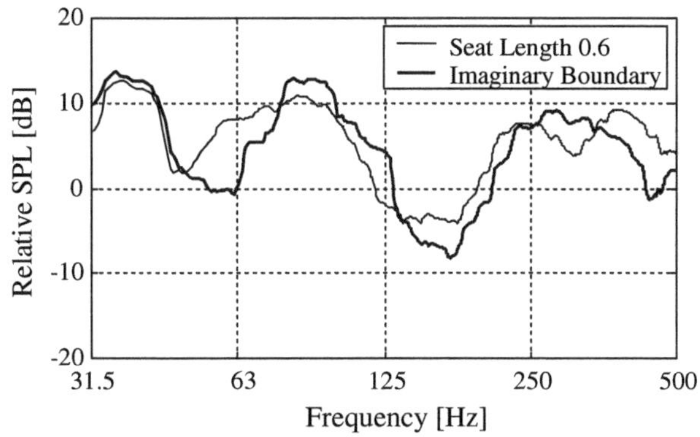

Fig. 6.24 Frequency response for the seat top section admittance model (with underpass, at the 12th seat)

underpass was present. The seat top section admittance described below was therefore implemented.

We considered a case in which seating rows existed on a rigid floor of infinite size, i.e., a field in which only seating rows exist in a semi-free space. We calculated the velocity potential ϕ and the differential coefficient to the incident direction $\frac{\partial \phi}{\partial y}$ for the point in the seat top regions in the case of a plane wave perpendicularly incident on the seating rows. A, as given by the equation below, was set as the specific acoustic admittance for the imaginary boundary surface.

$$A = -\frac{\frac{\partial \phi}{\partial y}}{jk\phi}, \qquad (6.1)$$

where y is the axis perpendicular to the rigid plane.

In reality, A was calculated at multiple sound receiving points near the center of the seating rows, and the average value thereof was used as the specific acoustic admittance for the imaginary boundary surface

This method involves analyzing only the seating rows in advance; the resulting admittance is given as the boundary condition for the imaginary surface model being analyzed. This means that more calculation is required in comparison with the case in which the layer model admittance is used as described in the previous section; however, there is a reduction in the calculation load in comparison with the method for simultaneously analyzing the entire space, including seats.

The comparison of the results from the seating row model and the results from the calculation for the frequency characteristics at the sound receiving point at the 12th row in the imaginary surface model, for which the seat top section admittance was given as the boundary condition as described previously, is shown in Fig. 6.24.

The tendency of the seating row frequency characteristics is considered to have been reproduced quite well in the calculated admittance. The attenuation of the SDE in the vicinity of 125 Hz, in particular, was revealed to have been reproduced with

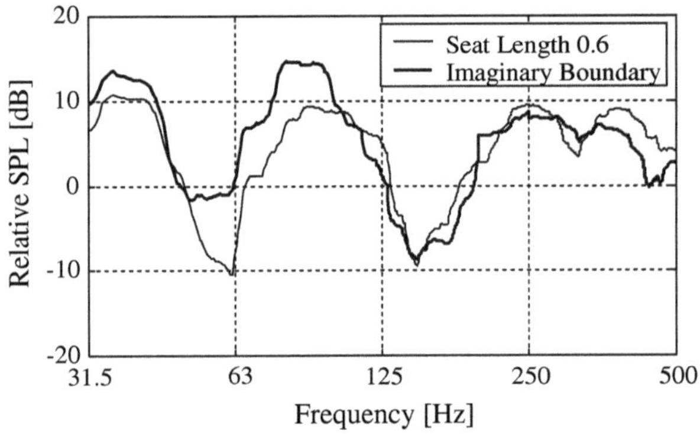

Fig. 6.25 Frequency response for seat top section admittance model (without underpass, at the 12th seat)

Table 6.1 Calculation time comparison between the seating row model and the imaginary surface model

	Underpass present		Underpass not present	
	Seating row model	Imaginary surface model	Seating row model	Imaginary surface model
Seating row potential		0.277		0.327
Admittance calculation		0.091		0.091
Hall boundary potential	1.000	0.273	1.000	0.237
Total CPU time	1.000	0.641	1.000	0.655

much higher accuracy than in the case of the layer model admittance shown in Fig. 6.16.

Similar calculations were performed for the cases where no underpass was present. As with the previous section, the acoustic absorption surface was set only between the top section of the seat backs in the very first row and the very last row.

A comparison of the frequency characteristic calculation results for the sound receiving point located at the 12th row from the seat arrangement model and the imaginary surface model is shown in Fig. 6.25. The seat top section admittance was given as the boundary condition for the imaginary boundary surface, which led to a good reproduction of the acoustic characteristics for the case in which there was no underpass; this was not accomplished by the layer model admittance in Fig. 6.19.

A comparison of the calculation time for the seating row model and the imaginary surface model is shown in Table 6.1. Values in the table represent ratios, and the total time for seat arrangement models with and without underpass was considered to be 1.0.

The seat top section admittance calculation method, in comparison with the seating row model that requires a large amount of calculation elements, can be handled by distributing the calculation loads, and it is evident that the total calculation time was also reduced. The impact of load reduction becomes more effective as the complexity of the seat shape increases; thus, it would be reasonable to apply the admittance calculation method for seat shapes with no underpass and the layer model admittance to simplified seat shapes with underpass. A case study of the analysis method involving the replacement of the seating rows with a uniform imaginary boundary surface in two dimensions was presented. However, further examination may be required in the future, especially for cases in which three dimensions are involved.

6.3 Reverberation Rooms

Reverberation rooms are essential for measurements in many kinds of building acoustics: for instance, the absorption coefficient in reverberation room (ISO 354, JIS A 1409), sound power of noise source, and transmission loss (ISO 140-3, JIS A 1416). All measurements in reverberation rooms presume a diffuse sound field. There remain issues caused by differences in sound fields or diffuseness in rooms used for measurements. In particular, although the Japanese Industrial Standard (JIS) approves use of both regularly and irregularly shaped reverberation rooms for these measurements (e.g., JIS A 1416, 2000), it cannot be inferred that their characteristics of diffuseness are the same.

Diffuseness of sound fields in reverberation rooms has been widely investigated by experiments and statistical considerations. In these studies, various descriptors were applied: sound intensity [15, 16], modal density [17], the coherence function [18], the spatial correlation function [19], spatial uniformity of sound fields [20], and so on. On the other hand, numerical analyses based on the wave equation have been intensively used to explore many kinds of acoustic problems [16, 21–23]. Among the analyses, FEM can be applied to such sound fields as those with complex absorbent walls and materials, those with temperature distribution, and so on. In this section, data obtained by the FE-analysis are applied to calculation of descriptors of diffuseness of sound fields in regularly and irregularly shaped reverberation rooms.

6.3.1 Comparison with Measurement Results

At first, to confirm accuracy of FE analysis, computed sound pressure levels by the FEM are compared with those of a measurement on the condition that a source is located in the middle of the room. Figure 6.26 shows schematic drawings of the irregularly shaped reverberation room (Volume = 165 m^3, Surface area = 178 m^2) to be analyzed and finite element division. Sound source location and 378 receiving points are illustrated in Fig. 6.27. A point source is assumed to be located in the field, which radiates white noise filtered by 200 Hz in 1/3 octave band and in a steady-state condition. An omnidirectional speaker is employed for measurement.

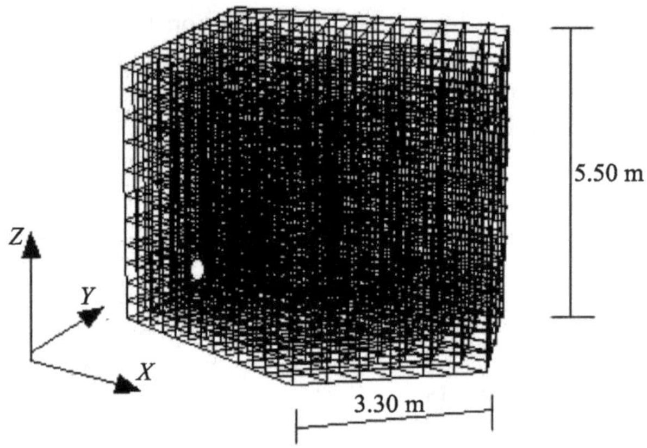

Fig. 6.26 Schematic drawing of an irregularly shaped reverberation room and finite element division (13 × 13 × 11)

Fig. 6.27 Measured and computed points and a sound source location

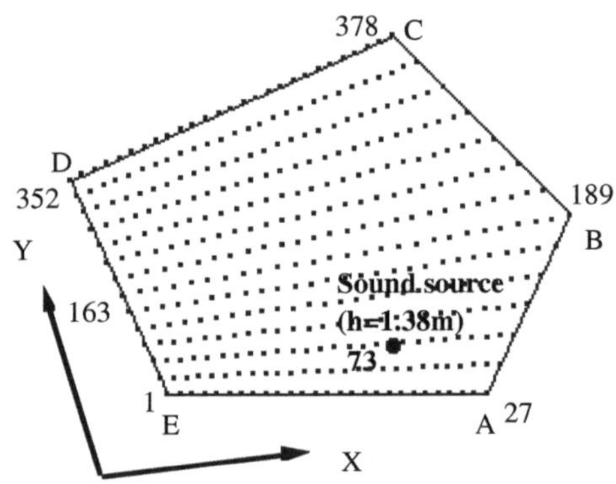

Figure 6.28 shows a comparison of measured values with computed sound pressure level distribution obtained at 378 points. The average residual of computed sound pressure values to measured values is 0.83 dB. Thus, it can be said that results obtained by the FE sound field analysis in frequency domain agree well with measured data in this case.

6.3.2 Descriptors for Diffuseness of Sound Fields

In this section, data obtained by the authors' FE-analysis are applied to calculate descriptors of diffuseness of sound fields in irregularly shaped reverberation rooms analyzed in the former section and regularly shaped reverberation room (Fig. 6.29, Volume = 220 m^3, Surface area = 227 m^2). Spatial correlation function of the sound field and the standard deviation of sound pressure levels are utilized as descriptors

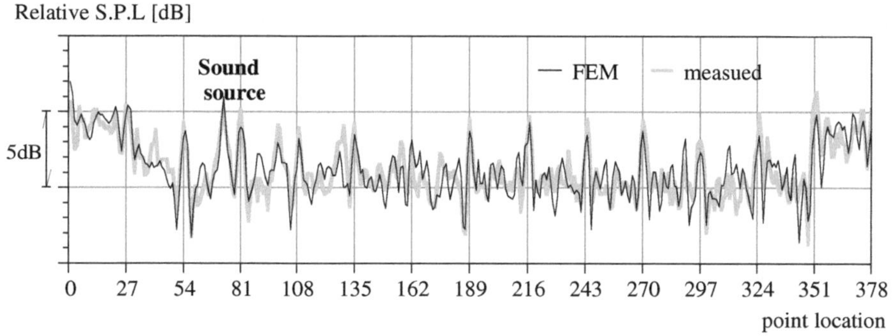

Fig. 6.28 Comparison of relative sound pressure levels; measurement versus FEM ($h = 1.20$ [m], 378 points)

Fig. 6.29 Schematic drawing of a regularly shaped reverberation room and finite element division (10 × 16 × 17)

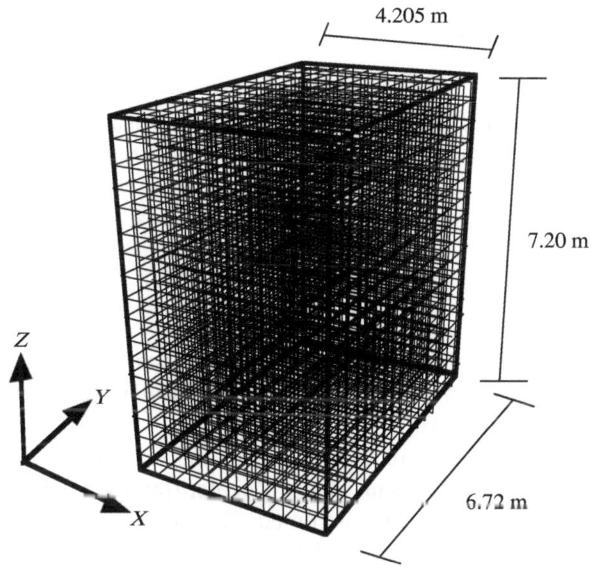

because these are widely employed in many studies and are easily calculated from results obtained by FEM.

Cook et al. [19] derived the spatial correlation function of a perfectly diffuse sound field, $(\sin kr)/kr$, and showed good agreement between the function and spatial correlation values of sound fields in actual reverberation rooms for high-frequency. It has been noted that if one or some correlation function values of the sound field correspond with $(\sin kr)/kr$, the field is not ensured to be perfectly diffused.

Then, assuming that sound pressures at two arbitrary points of the sound field obtained by FEM are expressed as $p_{1\_FEM}(f)$ and $p_{2\_FEM}(f)$; spatial correlation function values can be calculated as follows:

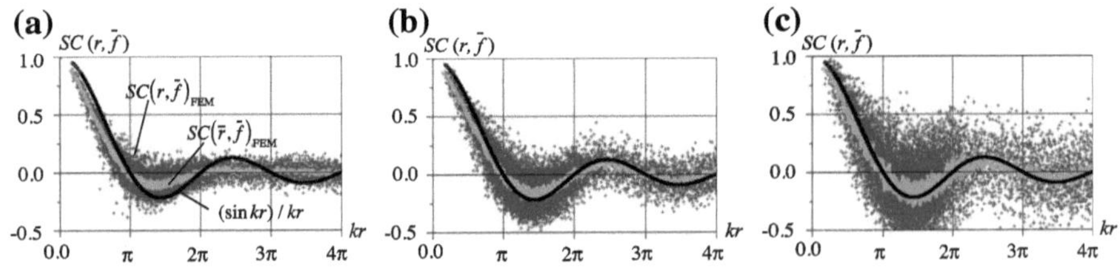

Fig. 6.30 Comparisons of spatial cross correlation in IR, theoretical values ($\sin kr/kr$) versus FEM (center frequency = 250 Hz). **a** 1 octave band; **b** 1/3 octave band; **c** 1/12 octave band

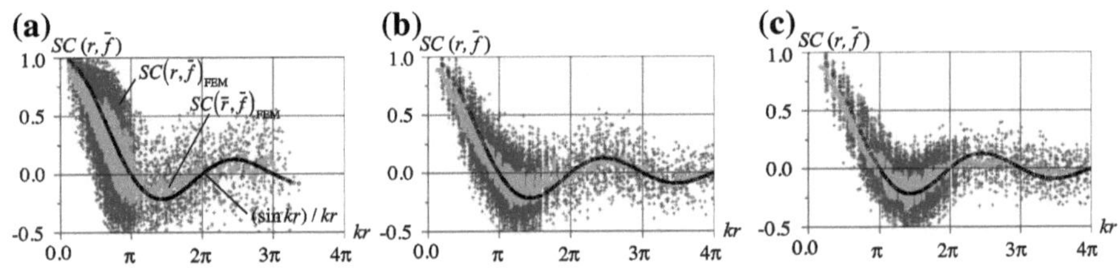

Fig. 6.31 Comparisons of spatial cross correlation in RR, theoretical values ($\sin kr/kr$) versus FEM (1/3 octave band). **a** Center frequency 100 Hz; **b** Center frequency 160 Hz; **c** Center frequency 200 Hz

$$SC_{\text{FEM}}(r, f) = \frac{\text{Re}[p_{1\_\text{FEM}}(f) \cdot p^*_{2\_\text{FEM}}(f)]}{|p_{1\_\text{FEM}}(f)| \cdot |p_{2\_\text{FEM}}(f)|}. \tag{6.2}$$

In this equation, *, Re[ ] and Im[ ] denote complex conjugate, real and imaginary parts, respectively. If the sound field has a band of frequency range, FEM values are averaged in several octave bands.

Figure 6.30a to c shows comparisons between ($\sin kr$)/kr and $SC_{\text{FEM}}(r, f)$, within 1/1, 1/3 and 1/12 octave band at center frequency 250 Hz in IR when one of the sound sources in the field is assumed. To make comparisons easier, these figures include spatially averaged data of $SC_{\text{FEM}}(r, f)$, which are calculated by a moving average of the same number regardless of frequencies or bandwidths. Among the three, best agreement is shown in the 1/3 octave band between ($\sin kr$)/kr and $SC_{\text{FEM}}(r, f)$.

Figure 6.31a to c shows the same comparison within 1/3 octave band at center frequencies of 100, 160, and 200 Hz in RR when one of the sound sources in the field is assumed. It is confirmed that $SC_{\text{FEM}}(r, f)$ at higher frequency agrees better with ($\sin kr$)/kr rather than $SC_{\text{FEM}}(r, f)$ at lower frequency.

Koyasu and Yamashita [24] noted that it is important to observe correlation function values for all directions in the field. It is also pointed out that quantitative assessments of sound fields are difficult by the correlation function [18, 25]. Then, to characterize sound fields in rooms using the spatial correlation function, difference between ($\sin kr$)/kr and $SC_{\text{FEM}}(r, f)$ can be calculated as

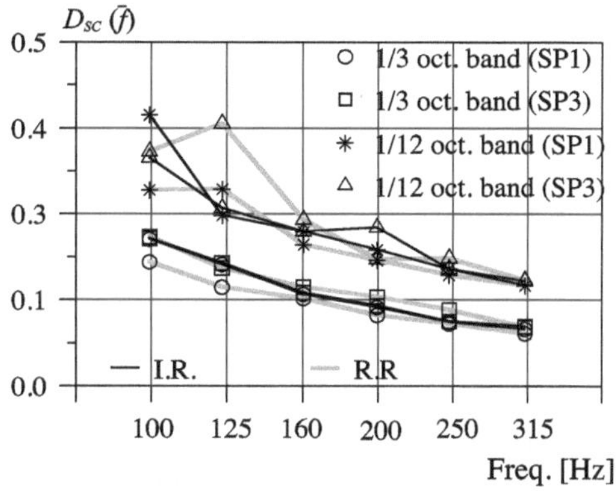

Fig. 6.32 $D_{SC}(f)$ values when room shapes, sound source locations ("SP1": a point sound source at one corner; "SP3": a point source in the middle of the room), frequencies, or bandwidths are changed

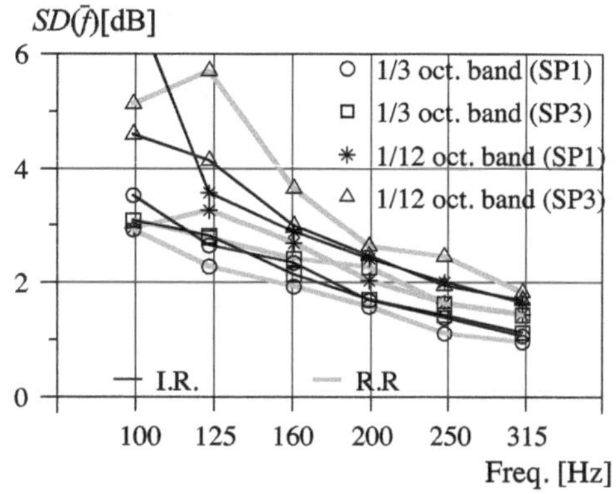

Fig. 6.33 $SD(f)$ values when room shapes, sound source locations ("SP1": a point sound source at one corner; "SP3": a point source in the middle of the room), frequencies, or bandwidths are changed

$$D_{SC}(\bar{f}) = \frac{1}{M}\sum_{\iota}^{M} |SC(r_i, \bar{f})_{\mathrm{FEM}} - (\sin kr_i)/kr_i|. \tag{6.3}$$

In this equation, M denotes combination number of $SC_{\mathrm{FEM}}(r, f)$. The descriptor includes several kr in addition to several directions.

The $D_{SC}(f)$ are given to show diffuseness difference in Fig. 6.32 when room shapes, sound source locations ("SP1": a point sound source at one corner; "SP3": a point source in the middle of the room), frequencies, or bandwidths are changed. For comparison, standard deviation of sound pressure levels in a room ($SD(f)$) are calculated and shown in Fig. 6.33.

Both results correspond to each other in the following aspects: (1) Clear difference can be found between results using SP1 and those using SP3 in RR while little difference can be found in the IR; (2) When bandwidth is identical, IR results fall on intermediate points between results using SP1 and results using SP3 in RR; (3) Worst characteristics of diffuseness in each room are shown when using SP3 within 125 Hz 1/12 octave band in RR and using SP1 within 100 Hz 1/12 octave band in IR. Pearson's correlation coefficient between $D_{SC}(f)$ and $SD(f)$ in 108 conditions (2 rooms $\times$ 3 sound source locations $\times$ 3 frequency bandwidths $\times$ 6 center frequencies) is 0.92.

6.4 Vehicle Cabins

Quietness, as acoustic environmental property in a vehicle cabin, has been required so far. in recent years, however, the quietness in a car cabin is highly improving and the interest will be changing to comfortness such as communication ability between passengers and intelligibility of electrical audible information from car audio and car navigation systems in future. Especially, various audible information from car navigation and collision alarm system is essential for safety of drive. Much detailed acoustical design technology is indispensable to realize such a comfortable car cabin environment. The wave-based numerical analysis is one of the promising tools as such acoustical design technologies. In this section, from such a viewpoint, an example of sound field analysis in a car cabin using the FDTD method is introduced.

6.4.1 Boundary Conditions

In wave-based numerical analysis of a space having small volume and complicated boundary shape and condition such as a car cabin, responses with wide frequency range (from low frequency to high frequency) can be obtained because the mesh size can be very small. Therefore, the simulation accuracy depends on the modeling of the boundary shape and the boundary condition. A car cabin consists of variety of parts such as window glasses, resin boards, leather or cloth seats, floor mats, ceiling panels, and so on, and their acoustical boundary conditions should be suitably modeled and the boundary conditions should be quantitatively set at correct values. In this example, the boundary conditions of many kinds were set as surface acoustic impedance under the assumption of locally reacting. In detail, the frequency characteristics of the acoustic impedance were reflected on the calculation by adopting the IIR filtering technique. For passenger seats, which occupy the largest areas in an inner space of a car cabin, sound propagation inside the seat materials—leather or cloth sheets and urethane foams—was treated in detail under the assumption of the extended reacting because transmitting sound through soft materials of seats might affect the sound field properties.

6.4.1.1 Expression of Boundary Conditions by Using IIR Filtering Technique

At a surface, its normal particle velocity pointing to outside of the domain is expressed by a sound pressure p and a normal acoustic impedance z_{n} as,

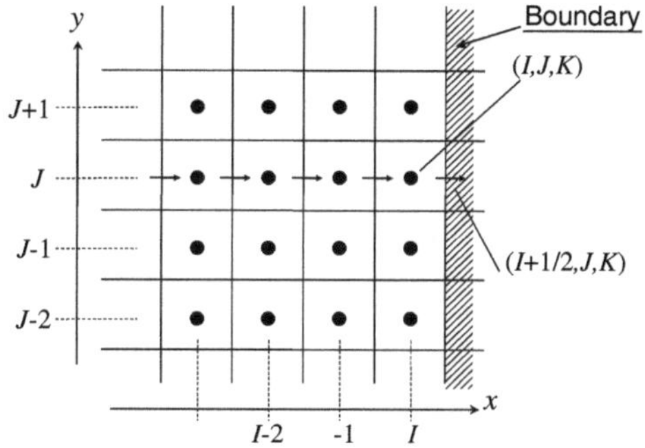

Fig. 6.34 Definition of boundary

$$u_{\mathrm{n}} = \frac{p}{z_{\mathrm{n}}}. \tag{6.4}$$

Absorbing characteristics of boundaries depend on the frequency f, and the acoustic impedance also depends on the frequency and is expressed as function of f. In case of steady-state condition with an angular frequency of ω, relationship among sound pressure p, particle velocity u_{n} and acoustical admittance $a_{\mathrm{n}}(\omega)$ which is the inverse value of acoustic impedance is

$$u_{\mathrm{n}}(\omega) = a_{\mathrm{n}}(\omega) p(\omega). \tag{6.5}$$

Based on the convolution theorem, the product in the frequency domain corresponds to the convolution integral in the time domain. Therefore, the particle velocity in x direction at a point (I,J,K) can be calculated using the acoustic admittance $a^n_{I,J,K}$ by a digital form as

$$u^n_{I+1/2,J,K} = \sum_{m=0}^{M_o} a^m_{I,J,K} \, p^{n-m}_{I,J,K}, \tag{6.6}$$

where M_o is the number of the order of the admittance filter (Fig. 6.34). Equation (6.6) means that detailed boundary absorption characteristics can reflect numerical solutions by acquiring transient characteristics of the boundary admittance as IIR digital filter coefficients.

Among the variety of parts consisting of a car cabin, several kinds of parts which have relatively large area, such as a door panel, a floor mat, a ceiling panel, and a instrumental panel, were picked up and their acoustic impedance measured by transfer function method using an impedance tube (B&K 4206). Based on the measurement results, the coefficients of the IIR filters were determined. Figures 6.35 and 6.36 show the comparisons of the acoustic impedance and normal absorption coefficients calculated from the acoustic impedance, respectively, between the measurement results and set values in the FDTD analysis by determined IIR filters.

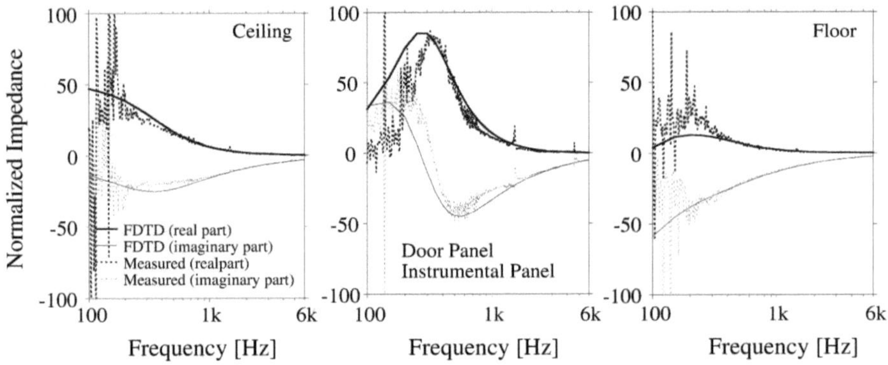

Fig. 6.35 Comparison of acoustic impedance

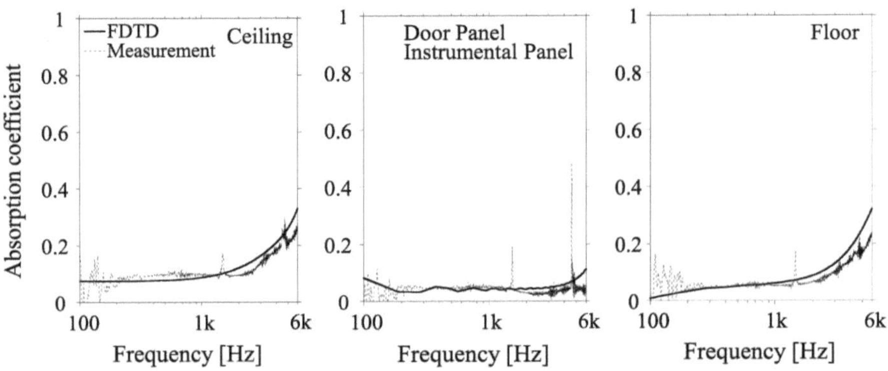

Fig. 6.36 Comparison of normal absorption coefficients calculated from the acoustic impedance

6.4.1.2 Expression of Boundary Conditions Considering Propagation Inside Material

Regarding the boundary condition of seats, internal propagation inside the material of the seats is taken into consideration. As shown in Fig. 6.37, a surface and an internal material of a seat are modeled as a membrane without tension having area density of m [kg/m^2] and a lossy media with a flow resistance of σ [Ns/m^4], respectively. Then, finite-difference schemes for particle velocity in x direction inside and on the seat are as follows.

$$u_{i+1/2,j,k}^{n+1} = \frac{\rho_0 - \sigma\Delta t}{\rho_0 + \sigma\Delta t} u_{i+1/2,j,k}^{n} - \frac{\Delta t}{(\rho_0 + \sigma\Delta t)\Delta h}\left(p_{i+1,j,k}^{n+1/2} - p_{i,j,k}^{n+1/2}\right), \quad (6.7)$$

$$u_{i+1/2,j,k}^{n+1} = u_{i+1/2,j,k}^{n} + \frac{\Delta t}{m}\left(p_{i+1,j,k}^{n+1/2} - p_{i,j,k}^{n+1/2}\right). \quad (6.8)$$

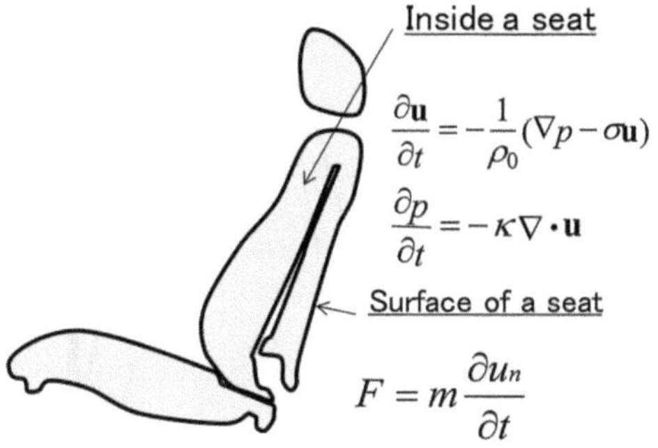

Fig. 6.37 Concept of analysis of seat

Sound propagation is calculated according to the above schemes and the schemes in the air described in Sect. 2.1 simultaneously. For the values of the parameters of the flow resistance of the material of the seat and the area density of the surface sheet of the seat were determined experimentally. That is, the former was determined from propagation constants of the material measured using an impedance tube, the latter was determined from inverse analysis based on frequency characteristics of normal absorption coefficient. Figure 6.38 shows comparisons of normal absorption coefficients of the seats in cases where its surface sheet is a leather (left) and a cloth (right) between their measurement results and calculated values based on the set parameters in the FDTD analysis. Whereas the respective values vary between the measurement and calculation, the tendency of the frequency characteristics of the measurements and of the calculations are similar. Comparing Figs. 6.36 with 6.38 shows that the absorption coefficients of the seats are much higher than the other materials and therefore the seats are dominant materials determining absorption characteristics of the car cabin sound field.

6.4.2 Numerical Analysis of Impulse Responses

Figure 6.39 shows an outline of a vehicle under investigation. An omni-directional sound source was set at a position of a driver's head and two receiving points were located at positions of passenger's heads at a assistant driver's seat and a rear seat, and impulse responses between a source and two receiving points were obtained for both measurement and numerical analysis. In numerical analysis, spatial grid size and discrete time interval were made 5 mm and 5.2 μs, respectively. Figures 6.40, 6.41, and 6.42 show snapshots of instantaneous sound pressure distribution in a car cabin, waveforms in time, and frequency characteristics of impulse responses, respectively, for the case of cloth seats. Regarding correspondence of impulse responses between calculation and measurement, fair agreement is recognized in transient waveforms. From frequency characteristics shown in Fig. 6.42, there exists discrepancy of sound pressure level in higher frequency than 4 kHz. This discrepancy is caused by disagree-

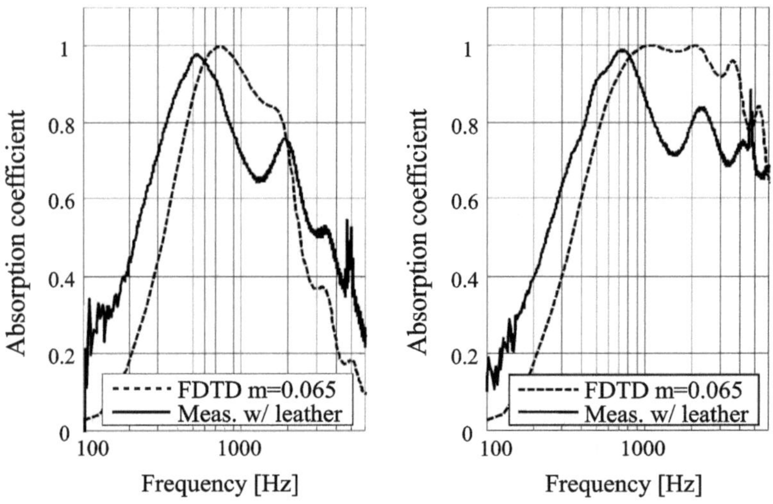

Fig. 6.38 Comparisons of normal absorption coefficients between measurement and set values in the FDTD analysis. *Left* figure is for a leather seat and right figure is for a cloth seat

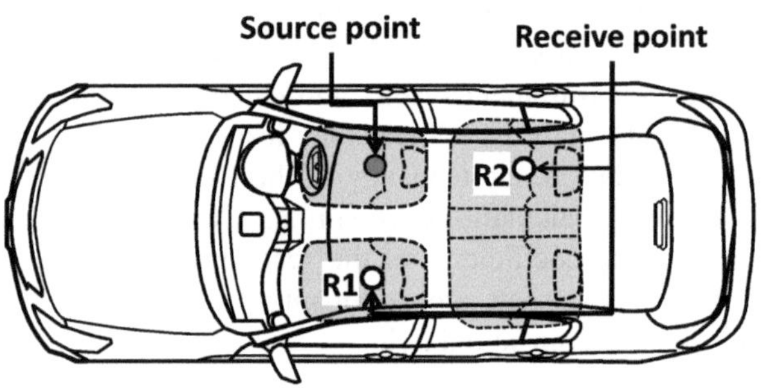

Fig. 6.39 A car cabin under investigation, sound source and receiving points

ment of sound absorbing condition between measurement and the FDTD analysis as shown in Fig. 6.38.

6.5 Partitions

When a person talks over a counter in a small space such as a pharmacy or a bank located in a city, the contents of the conversation often leak to the waiting area behind the counter. This problem is known as speech privacy (confidential privacy) problems. In Japan, partitions such as shown in Fig. 6.43 are often installed on a counter on both side of a speaker and a listener. These partitions seem to be installed for visual privacy and focus on the conversation. These partitions have potential to

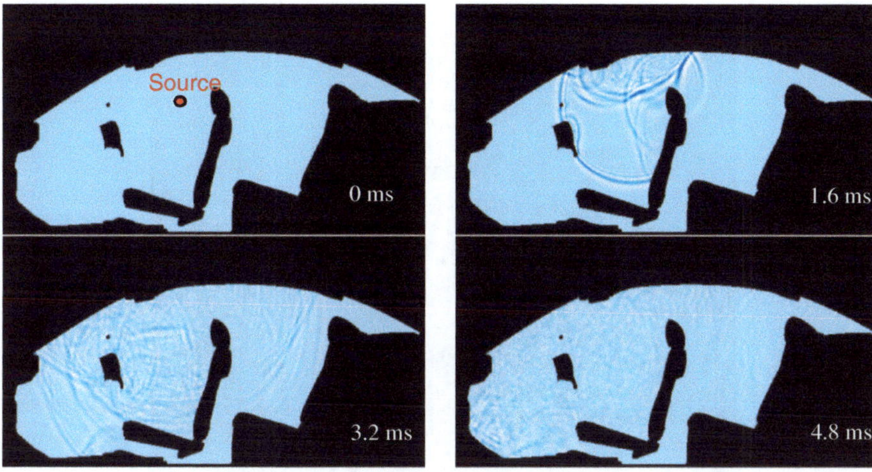

Fig. 6.40 Snapshots of sound pressure distribution inside car cabin

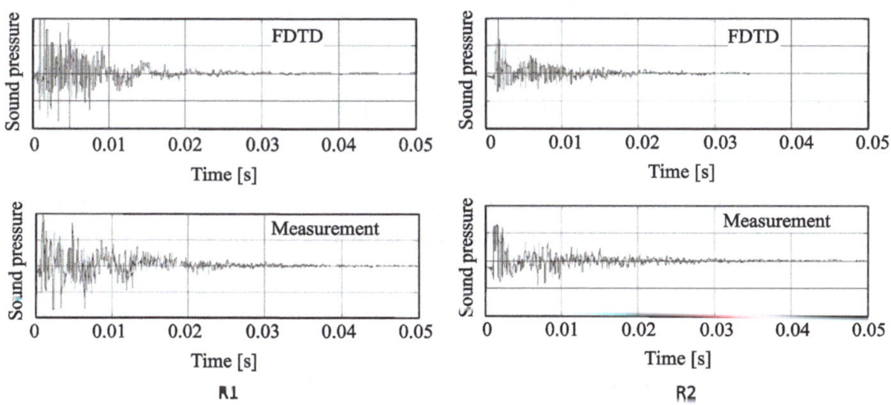

Fig. 6.41 Comparison of impulse response—transient wave form

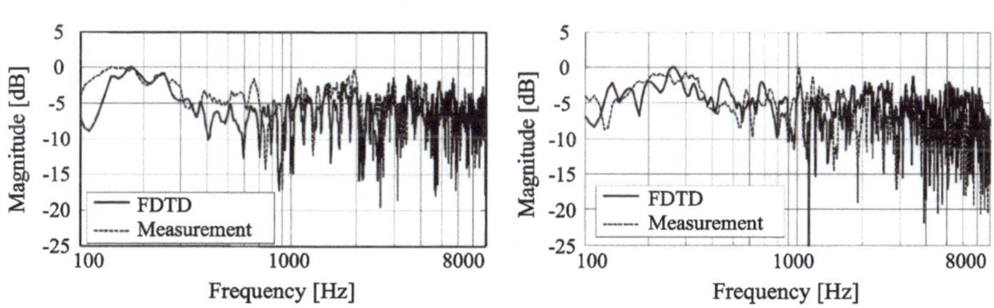

Fig. 6.42 Comparison of impulse response—frequency characteristics

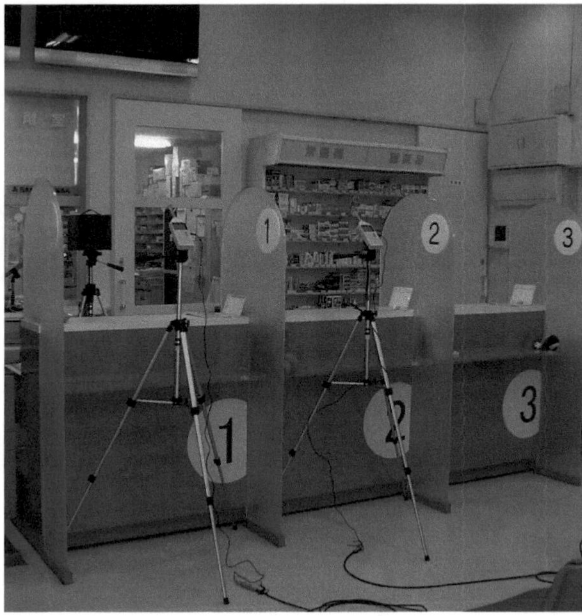

Fig. 6.43 Parallel partitions on a counter in a pharmacy, Japan

insulate the sound of conversation if the partitions had been designed appropriately. In this section, numerical studies to obtain the knowledge of these series of partitions are introduced.

6.5.1 Insulation Effect of Arrayed Partitions

When a counter is separated by a an array of partitions, we should check the variety of sound pressure level at a listening position in a conversation booth, at listening positions in other booths, and at waiting areas.

6.5.1.1 Configurations

Figure 6.44 shows the floor plan for the examination. For checking effects of partitions, the edge of the room drawn in dotted lines is set to absorption boundary. Two types of partitions arrangement are ready for comparison, one is set to only one booth using two partitions drawn black partitions in Fig. 6.44, the other is set to five booths using six partitions drawn black and gray partitions in it. Additionally, two conditions of absorption coefficients of partition are ready. Table 6.2 shows four examination types that are mentioned above. Figure 6.44 also shows areas to obtain

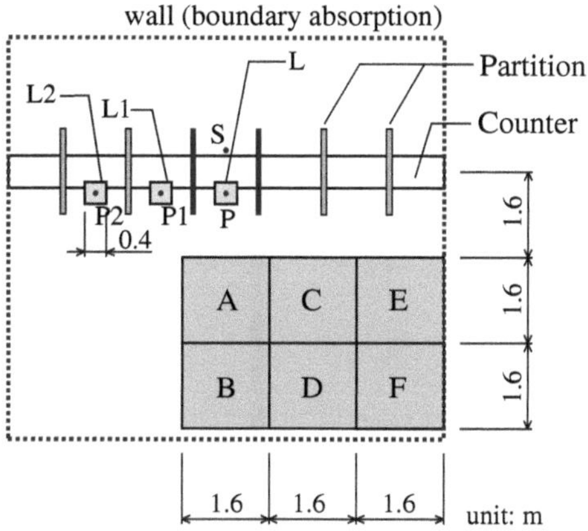

Fig. 6.44 Floor plan and sampling areas to obtain averaging level

Table 6.2 Partition types for examination

Type	Number of booths	Abs. coeff. of partition
b1α0	1	0
b1α0.6	1	0.6
b5α0	5	0
b5α0.6	5	0.6

averaging sound pressure level. L is the area of listener in a conversation booth, L1 and L2 are the areas of listener in the other booths. A to F separate the waiting area. The details of the floor and section plan of a counter, partitions, a speaker, and a listener are shown in Fig. 6.45.

The FDTD method is used for numerical simulation. Perfect Matched Layers (PMLs) are set to absorption boundary. Spatial intervals (grid size) $\Delta x = \Delta y = \Delta z$ are set to 0.02 m. Time interval is determined to 20 μs and the calculations were carried out 0.1 s. The initial sound pressure is set to sound pressure points whose values are weighted by the distance from the source point. After numerical simulation, the impulse responses obtained from each sound pressure point are modified as follows:

- To get rid of the color of direct sound, the impulse responses are convoluted the inverse filter made from the direct sound.
- To calculate 1/3 octave band levels, these responses are colored with the characteristics of—3 dB/oct.
- To demonstrate woman's voice, the characteristics of it are added to each band [26].
- To make a single number values, the A-weighted level is calculated from the 1/3 octave band levels.
- The sound pressure levels are normalized using the differential level at position P where the sound pressure level without partitions is equal to 60 dB.

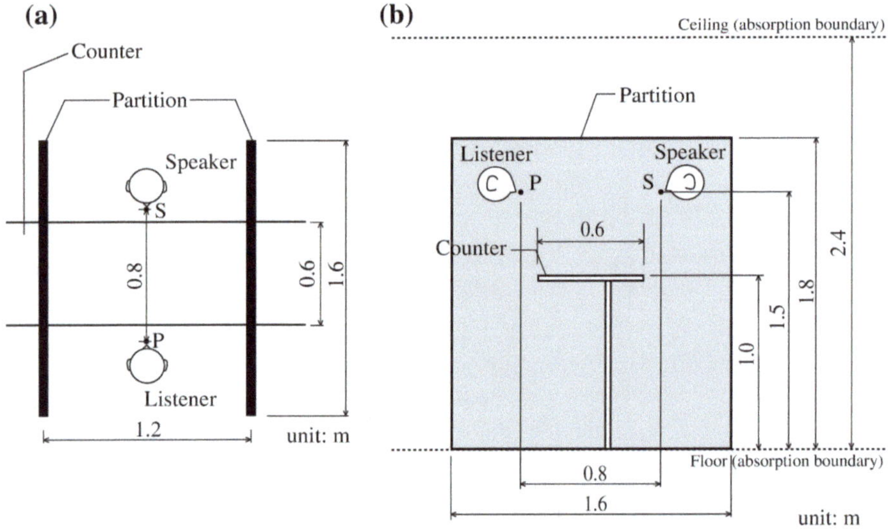

Fig. 6.45 Floor and section plan around speaker and listener. **a** Floor plan. **b** Section plan

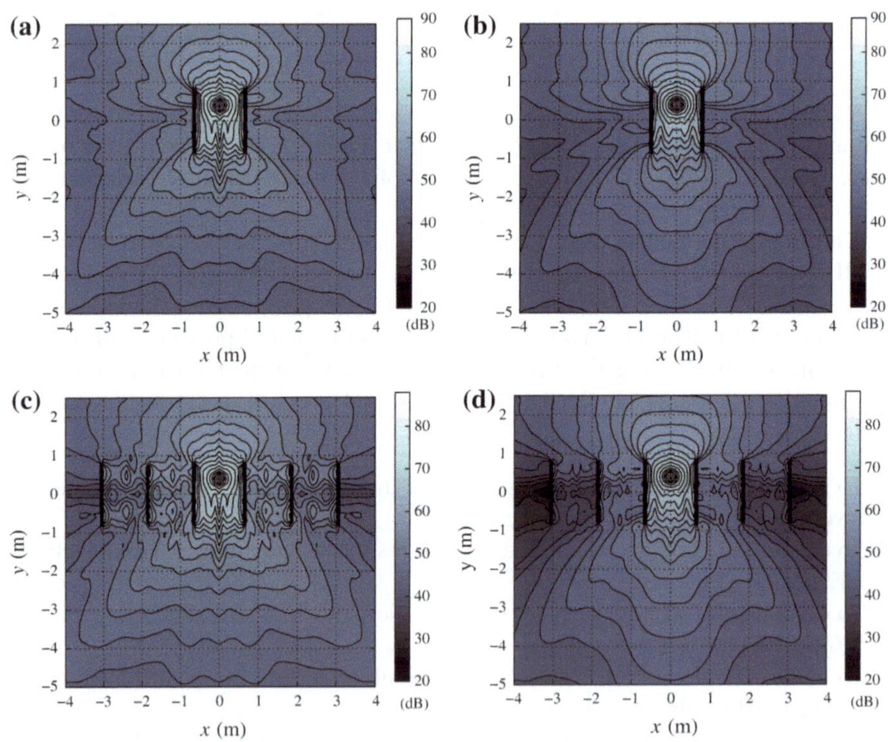

Fig. 6.46 Sound pressure level distribution. **a** b1α0. **b** b1α0.6. **c** b5α0. **d** b5α0.6

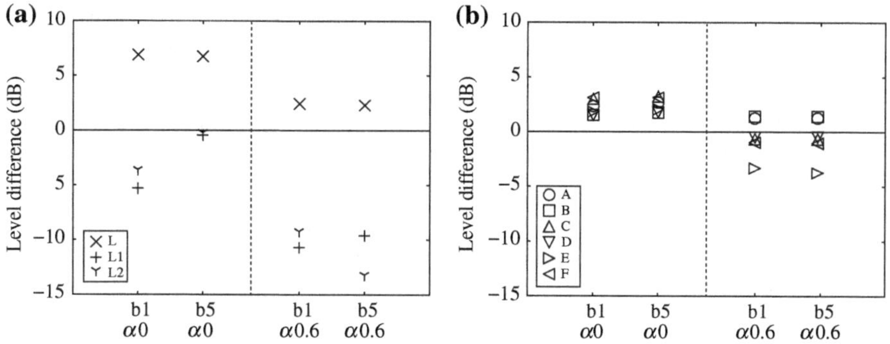

Fig. 6.47 Level differences due to inserting partitions. **a** Around conuter. **b** Waiting area

6.5.1.2 Results

Figure 6.46 shows the results of sound pressure level propagations in each condition. At first, focusing on Fig. 6.46a, it can seen that partitions insulate the sound for lateral direction, but the level of longitudinal direction seems to get worse, it looks partitions works such as a megaphone. What is worse is that the result of partitions set to one line shown in Fig 6.46c, the sound levels of other booths get worse. This result assumed to be caused by multiple reflections of the diffracted sounds in these booths. Fig. 6.46b and d show the results of absorptive partitions. These results expect a deterrent effect of these phenomena. Fig. 6.47a shows the level difference at the listening position in conversation booth and in the other booths. This result shows us type b5α0—reflective partitions set on a line—is the worst condition, b5α0.6—absorptive ones set on a line—is however the best one. Figure 6.47b shows the level difference due to partition installation at the waiting area in each condition. The increase of level can be appeared in types b1α0 and b5α0 , but cannot be appeared in types b1α0.6 and b5α0.6 . The results of this examination tell us absorption of partition is very important, especially when partitions are set on a line.

References

1. D.M. Sullivan, A frequency-dependent FDTD method using z transforms. IEEE Trans. Antennas propag. **40**, 1223–1230 (1992)
2. J. Escolano, F. Jacobsen, J. Lopez, An efficient realization of frequency dependent boundary conditions in an acoustic finite-difference time-domain model. J. Sound Vib. **316**, 234–247 (2008)
3. H. Suzuki, A. Omoto, K. Fujiwara, Treatment of boundary conditions by finite difference time domain method. Acoust. Sci. Tech. **28**, 16–26 (2007)
4. S. Sakamoto, H. Nagatomo, A. Ushiyama, H. Tachibana, Calculation of impulse responses and acoustic parameters in a hall by the finite-difference time-domain method. Acoust. Sci. Tech. **29**, 256–265 (2008)

5. Acoustics—measurement of room acoustic parameters—Part 1: performance rooms, annex a auditorium measures derived from impulse responses. ISO/DIS 3382–1 (2006)
6. Architectural Institute of Japan Sub. Comm. Computational Methods for Environmental Acoustics, Manag. Comm. Sound Environ. Benchmark platform on computational methods for architectural/environmental acoustics (http://gacoust.hwe.oita-u.ac.jp/aij-bpca/index.html)
7. N. Okamoto, R. Tomiku, T. Otsuru, Y. Yasuda, Numerical analysis of large-scale sound fields using iterative methods part II: application of Krylov subspace methods to finite element analysis. J. Comp. Acoust. **15**(4), 473–493 (2007)
8. Y. Yasuda, S. Sakamoto, Y. Kosaka, T. Sakuma, N. Okamoto, T. Oshima, Numerical analysis of large-scale sound fields using iterative methods part I: application of Krylov subspace methods to boundary element analysis. J. Comp. Acoust. **15**(4), 449–471 (2007)
9. G.M. Sessler, J.E. West, Sound transmission over theatre seats. J. Acoust. Soc. Am. **36**, 1725–1732 (1964)
10. T.J. Shultz, B.G. Watters, Propagation of sound across audience seating. J. Acoust. Soc. Am. **36**, 885–896 (1964)
11. Y. Kawai, T. Terai, Calculation of sound fields over audience seats by using integral equation method. Trans. ASME J. Vib. Acoust. **113**, 22–27 (1991)
12. D. Takahashi, Seat dip effect: the phenomena and the mechanism. J. Acoust. Soc. Am. **102**, 1326–1334 (1997)
13. T. Osa, K. Murakami, Y. Horinouchi, D. Takahashi, Applications of audience-seats characteristics to the sound field analysis for large enclosures. Appl. Acoust. **68**, 939–952 (2007)
14. T. Terai, Y. Kawai, in *BEM Applications in Architectural Acoustics*, ed. by R.D. Ciskowski, C.A. Brebbia, Boundary Element Methods in Acoustics, chapter 10 (Computational Mechanics, Southampton, 1991)
15. T. Hanyu, S. Kimura, O. Hashimoto, Y. Sakamoto, Analysis of diffusibility in sound field transient response. J. Archit. Plann. Environ. Eng. (in Japanese) **471**, 11–18 (1995)
16. M. Ikeda, Y. Furue, Y. Horinouchi, T. Terai, Calculation of sound intensity field using integral equation method and presentaiton of it. J. Archit. Plann. Environ. Eng. (in Japanese) **479**(1), 7–15 (1996)
17. T. Samejima, K. Hirate, M. Yasuoka, Optimum design of room shape and arrangement of absorptive patches by evaluating the distribution of poles of a transfer function in an acoustic system. J. Archit. Plann. Environ. Eng. (in Japanese) **511**, 9–14 (1999)
18. F. Jacobsen, T. Roisin, The coherence of reverberant sound fields. J. Acoust. Soc. Am. **97**(2), 1119–1123 (2000)
19. R.K. Cook, R.V. Waterhouse, R.D. Berendt, S. Edelman, J.M.C. Thompson. Measurement of correlation coefficients in reverberant sound fields. J. Acoust. Soc. Am. **27**(6), 1072–1077 (1955)
20. S.N. Yousri, F.J. Fahy, An analysis of the acoustic power radiated by a point dipole source into a rectangular reverberation chamber. J. Sound Vib. **316**, 234–247 (1972)
21. A. Craggs, Coupling of finite element acoustic absorption models. J. Sound Vib. **316**, 234–247 (1979)
22. S. Choi, H. Tachibana, Estimation of impulse response in a room by the finite element method. J. Acoust. Soc. Jpn. **21**(2), 87–95 (1993)
23. T. Sakuma, T. Iwase, M. Yasuoka, Prediction of sound fields in rooms with membrane materials -development of a limp membrane element in acoustical fem analysis and its application-. J. Archit. Plann. Environ. Eng. **505**, 1–8 (1998)
24. M. Koyasu, M. Yamashita, Evaluation of the degree of diffuseness in reverberation chambers by spatial correlation techniques. J. Acoust. Soc. Jpn. (in Japanese) **21**(2), 87–95 (1971)
25. H. Nelisse, J. Nicolas, Characterization of a diffuse field in a reverberant room. J. Acoust. Soc. Am. **97**(2), 1119–1123 (1997)
26. IEC-60268-16, Sound system equipment—part 16: objective rating of speech intelligibility by speech transmission index, edition 4.0. (2011)

Chapter 7
Noise Propagation Simulation

Masahiro Toyoda, Takuya Oshima, Takatoshi Yokota, Tomonao Okubo, Shinichi Sakamoto, Yosuke Yasuda, Takashi Ishizuka, Yasuhito Kawai and Takumi Asakura

Abstract This chapter shows examples of numerical analyses on various noise propagation problems including outdoor noise propagation, noise barriers, depressed roads, building façades, building windows, and floor impact sound. In each section,

M. Toyoda (✉) · Y. Kawai
Faculty of Environmental and Urban Engineering, Kansai University, 3-3-35 Yamate-cho, Suita-shi, Osaka 564-8680, Japan
e-mail: toyoda@kansai-u.ac.jp

Y. Kawai
e-mail: kawai@kansai-u.ac.jp

T. Oshima
Faculty of Engineering, Niigata University, 8050 Ikarashi-Ninocho, Nishi-ku, Niigata-shi, Niigata 950 2181, Japan
e-mail: oshima@eng.niigata-u.ac.jp

T. Yokota · T. Okubo
Kobayasi Institute of Physical Research, 3 20 41 Higashi-motomachi, Kokubunji-shi, Tokyo 185-0022, Japan
e-mail: t-yokota@kobayasi-riken.or.jp

T. Okubo
e-mail: okubo@kobayasi-riken.or.jp

S. Sakamoto
Institute of Industrial Science, The University of Tokyo, Tokyo, Japan
e-mail: sakamo@iis.u-tokyo.ac.jp

Y. Yasuda
Faculty of Engineering, Kanagawa University, 3-27-1 Rokkakubashi, Kanagawa-ku, Yokohama 221-8686, Japan
e-mail: yyasuda@kanagawa-u.ac.jp

T. Ishizuka · T. Asakura
Institute of Technology, Shimizu Corporation, 3-4-17 Etchujima, Koto-ku, Tokyo 135-8530, Japan
e-mail: ishiduka@shimz.co.jp

T. Asakura
e-mail: t_asakura@shimz.co.jp

T. Sakuma et al. (eds.), *Computational Simulation in Architectural and Environmental Acoustics*, DOI: 10.1007/978-4-431-54454-8_7,

considering the features of each problem, methodology of applying the simulation techniques presented in Part I to the practical problem is introduced and the calculated results are illustrated. Some of the calculated results are compared with measured ones and the applicability and efficiency of the analysis method are discussed.

Keywords Outdoor noise propagation · Noise barriers · Depressed roads · Building façades · Building windows · Floor impact sound

7.1 Outdoor Noise Propagation

In the field of noise control engineering, demand for prediction models that can precisely calculate outdoor propagation of sound, such as road traffic noise, railway noise and aircraft noise, is increasing every year. Parallel to this, outstanding progress in computer hardware has made it possible to apply wave-based numerical simulations to the problems of outdoor noise propagation.

When applying wave-based numerical simulation techniques to outdoor noise propagation problems, it is necessary to consider the following factors.

1. The sound field under consideration is very large;
2. The sound field has no boundaries at the end of the region (exterior problem);
3. The characteristics of sound propagation vary with the meteorological conditions such as wind speed and the vertical temperature gradient.

In this section, we will provide examples in which the characteristics of sound propagation vary due to the meteorological effects, and an effective technique for analyzing the long-range outdoor noise propagation will be introduced.

7.1.1 Influence of Wind

As an example of the effects of wind on outdoor noise propagation, we introduce some simulations that use the Parabolic Equation (PE) method. The PE method is widely used for analyzing the outdoor sound propagation by numerical simulation [1–4]. This method can take account of atmospheric refraction, which is caused predominantly by vertical gradients of the temperature and wind speed. We first outline the calculation method, and then we introduce simulation results for effects of the wind on variations in the spatial distribution of the sound pressure level.

7.1.1.1 Outline of the PE Method

In the PE calculation, a vertical profile of the effective sound speed is assumed based on a vertical distribution of the wind speed and temperature in the sound field. Here, we introduce the formulation of the PE method in a two-dimensional sound field (Detailed descriptions will be found in the Ref. [3]).

First, the Helmholtz equation described in Eq. (7.1) is divided into two components: an outward propagating part and an inward propagating part. By assuming an operator Q described in Eq. (7.2), these components can be represented as Eq. (7.3):

$$\left(\frac{\partial^2}{\partial x^2} + \frac{\partial^2}{\partial z^2} + k^2\right) p = 0, \tag{7.1}$$

$$\left(\frac{\partial^2}{\partial z^2} + k^2\right) p = Qp, \tag{7.2}$$

$$\left(\frac{\partial}{\partial x} + j\sqrt{Q}\right)\left(\frac{\partial}{\partial x} - j\sqrt{Q}\right) p = 0. \tag{7.3}$$

Considering only the outward propagating part of Eq. (7.3), the sound propagation from a source can be calculated as follows:

$$\frac{\partial p}{\partial x} = \left(j\sqrt{Q}\right) p, \tag{7.4}$$

where k is the wave number (ω/c), which is the variable that changes with the effective sound speed in the field, and p is the sound pressure.

Next, we define an operator q as in Eq. (7.5), and for convenience, the operator Q can be written as Eq. (7.6). Here, the right-hand side of Eq. (7.6) is approximately represented as Eq. (7.7) [5], and the basic formula of the PE method is represented as Eq. (7.8):

$$q = \frac{1}{{k_0}^2}\left(\frac{\partial^2}{\partial z^2} + k^2\right) - 1, \tag{7.5}$$

$$\sqrt{Q} = k_0\sqrt{\frac{1}{{k_0}^2}\left(\frac{\partial^2}{\partial z^2} + k^2\right) + 1 - 1} = k_0\sqrt{1+q}, \tag{7.6}$$

$$\sqrt{1+q} \simeq \frac{1+(3/4)q}{1+(1/4)q}, \tag{7.7}$$

$$\left(1 + \frac{1}{4}q\right)\frac{\partial p}{\partial x} = jk_0\left(1 + \frac{3}{4}q\right) p, \tag{7.8}$$

where k_0 is the reference value of k, and it is the value that corresponds to the temperature under no-wind condition.

Furthermore, if we select the solution of Eq. (7.1) as Eq. (7.9), then Eq. (7.8) can be rewritten as Eq. (7.10):

$$p(x, z) = \psi(x, z)\exp(jk_0 x), \tag{7.9}$$

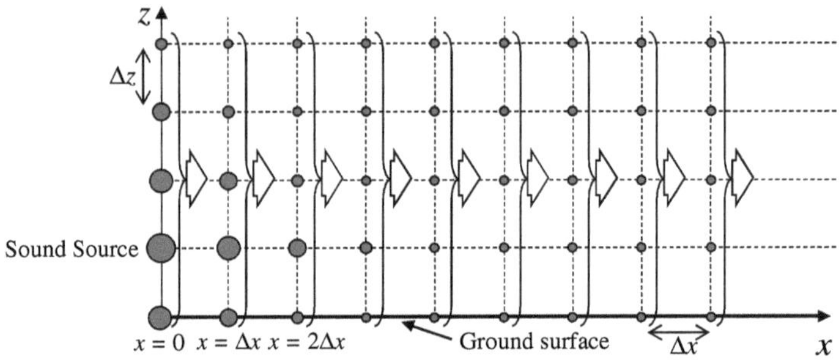

Step 1. As an initial condition, a spatial distribution of the complex sound pressure is set by the function $p\ (x = 0, z)$

Step 2. Sound pressure distribution $p\ (0 + \Delta x, z)$ is calculated based on the function $p\ (x = 0, z)$

Step n. [Spatial marching solution]
Sound pressure distribution $p\ ((n{+}1).\Delta x, z)$ is calculated based on $p\ (n{\cdot}\Delta x, z)$ as in Step 2

Fig. 7.1 Outline of the calculation method and the discretization of the sound field

$$\left(1 + \frac{1}{4}q\right)\frac{\partial \psi}{\partial x} = jk_0\frac{q}{2}\psi. \tag{7.10}$$

By dividing a sound field into a square grid as shown in Fig. 7.1, the partial differential terms for the x and z directions of Eq. (7.10) can be calculated by the central difference formula and by the Crank–Nicholson approximation, respectively. The vertical distribution of the sound pressure at the $(n+1)$th step $(p((n+1)\Delta x, z))$ can be calculated by using the vertical distribution of the sound pressure at the nth step $(p(n\Delta x, z))$. Consequently, if we set the vertical distribution of the sound pressure at $x = 0$ as an initial condition, we can perform a step-by-step calculation to obtain the distribution of the sound pressure from a source position to the receiving points.

7.1.1.2 Examples

In order to study the variation in the spatial distribution of the sound pressure level due to the effects of wind, sound propagation in hemi-free field with rigid boundary is calculated by using the PE method. Figure 7.2 shows the sound field and the vertical profile of the wind speed under consideration. In the calculations, the downwind (wind flows from the sound source to the receiver) and upwind (wind flows from the receiver to the sound source) propagation conditions are assumed. We assumed three different wind speeds at a height of 1.2 m: 2, 4, and 6 m/s.

Figure 7.3 shows the spatial distributions of the sound pressure level for the conditions of calm, downwind (+6 m/s), and upwind (−6 m/s). It can be seen that the sound pressure level in areas far from the sound source varies according to the wind speed. In particular, with different wind speeds, obvious differences can be seen

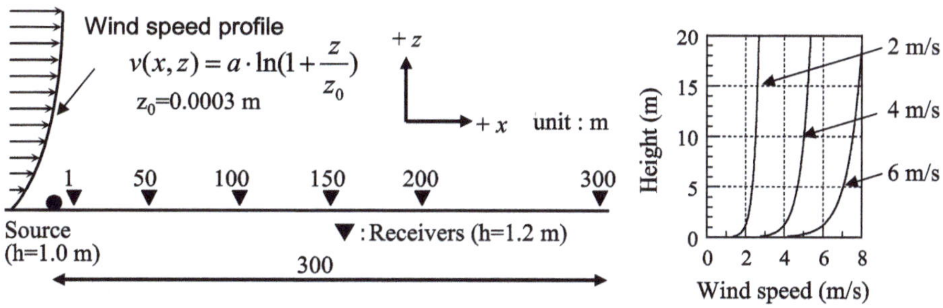

Fig. 7.2 Sound field and vertical profile of the wind

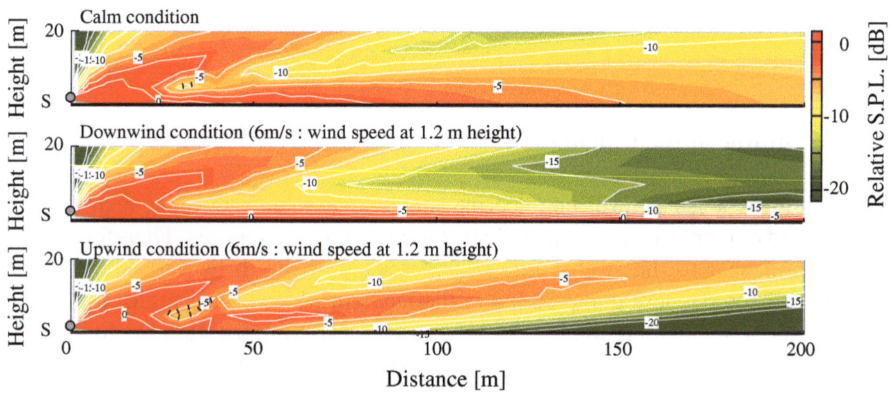

Fig. 7.3 Wind effect on outdoor sound propagation (500 Hz octave band)

near the ground, such that the sound pressure level increases in downwind areas and decreases in upwind areas.

Figure 7.4 shows the differences in sound pressure level at each receiving point for conditions with and without wind. It can be seen that variations in the sound pressure level increase with increasing frequency and propagation distance. We note that these results are in good agreement with those calculated by using the Linearized Euler Equation method introduced in Sect. 5.1 [6].

7.1.2 Efficient Calculation of Long-Range Propagation

When applying numerical simulations to long-range outdoor noise propagation predictions, the computational load becomes very large. In this section, we introduce an efficient calculation technique that combines two types of numerical simulations, the FDTD and the PE [7, 8].

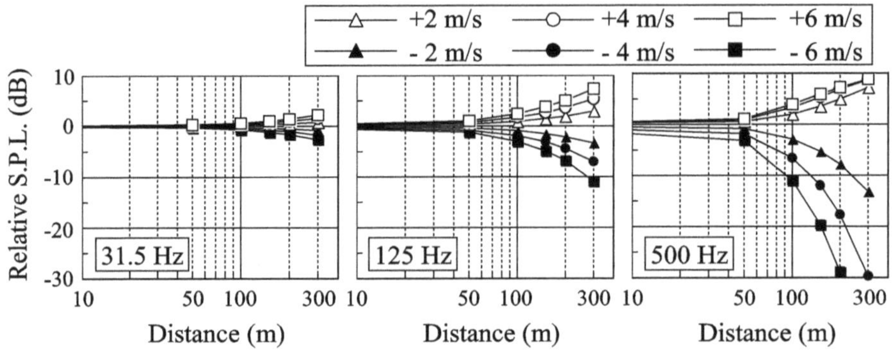

Fig. 7.4 Differences in the sound pressure level obtained at each receiving point for conditions with and without wind

7.1.2.1 Outline of the Calculation

In the field of noise control engineering, noise abatement facilities, such as noise barriers and embankments, are usually installed near noise sources. Therefore, sound propagation near to the source (the source region) tends to be very complicated, whereas it becomes less complicated in the far field (the receiver region). It is thus more efficient to calculate the long-range outdoor noise propagation by dividing the sound field under consideration into two regions and then applying a suitable numerical simulation method to each region, with consideration of the above-mentioned characteristics of each region. In the source region, such calculation methods as the BEM and the FDTD, which can take account of the reflections and diffractions caused by obstacles, are often applied. On the other hand, the ray model, the analytical solution, and the PE method, which can take account of ground reflections and the refraction due to meteorological effects, are often selected.

As an example of a coupled calculation technique, we here outline a coupling of the FDTD and PE methods:

1. Divide the sound field into two areas
   The sound field under consideration is divided into two areas at the position shown by a broken line in Fig. 7.5. Area 1 is the area near the sound source (the source region) where the sound propagation characteristics are very complicated due to the reflections and diffractions caused by noise abatement facilities. Area 2 is the area far from the noise source (the receiver region), where it can be assumed as a good approximation that the sound propagation is one way from the source to each receiver.
2. Perform a numerical simulation in the source region
   Sound propagation in Area 1 is calculated using the FDTD method, and the impulse responses at all grid points on the interface between Area 1 and Area 2 are obtained.

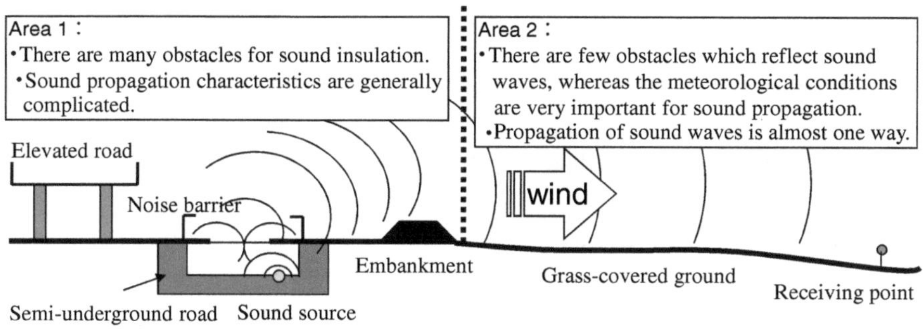

Fig. 7.5 Outline of the sound field

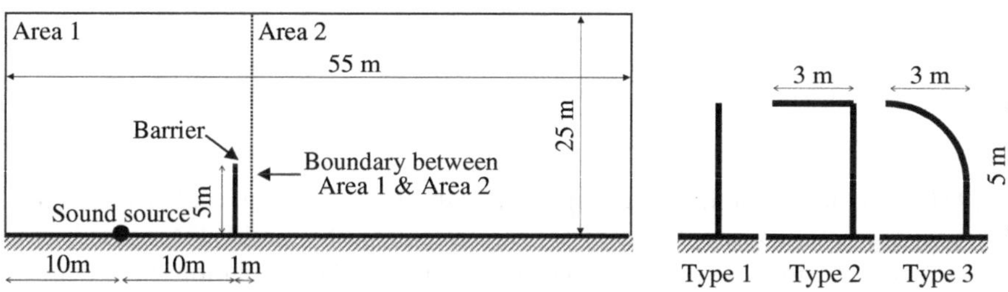

Fig. 7.6 Outline of the sound field under consideration

3. Prepare the spatial distribution of the sound pressure for the initial condition of the far-field simulation
   All the impulse responses at the grid points on the interface are converted to frequency response functions by using the FFT technique, and the spatial distribution of the complex sound pressure for a single frequency is prepared.
4. Perform a numerical simulation in the receiver region
   For each frequency, use the resultant spatial distribution of the complex sound pressure for that frequency as the initial condition, then conduct the PE calculation for Area 2. The frequency response function at the receiving points can be obtained by repeating the PE calculation for each frequency individually.

As described in the above four steps, the frequency response function at an arbitrary receiving point can be calculated and the impulse response can be obtained by applying the inverse FFT technique to the resultant frequency response function.

7.1.2.2 Example

As an example of a numerical simulation that applies the coupled FDTD-PE method, we show the results for the insertion loss due to a noise barrier [8]. Figure 7.6 shows a sound field and the type of noise barrier under consideration. We assume rigid boundary conditions at the ground and the barrier. Figure 7.7 shows the areas where

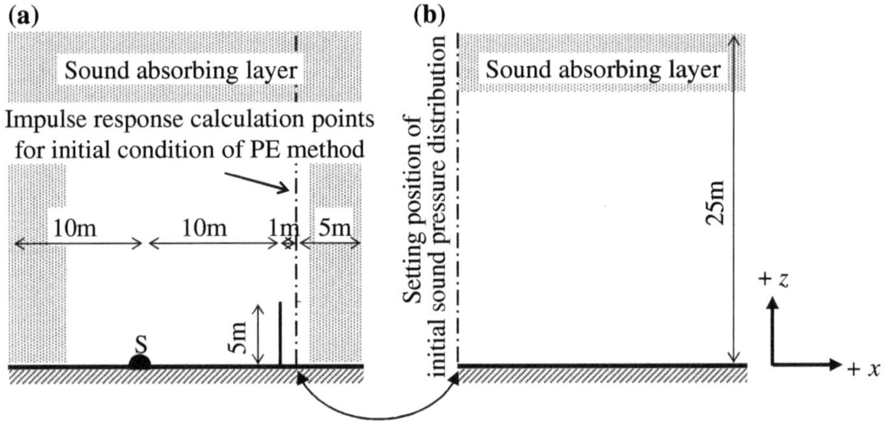

Fig. 7.7 Calculation area for each simulation method. **a** FDTD, **b** PE

the FDTD and the PE methods were applied. In this case, the interface between Area 1 and Area 2, where the impulse responses are calculated by the FDTD method, is set at a distance of 1 m from the noise barrier. In the FDTD calculation, the boundary of the computational area is set at a distance of 5 m from the interface and assumed to be the Perfectly Matched Layer (PML). In the PE calculation, it is not necessary to set boundary condition along the x-axis because of the assumption of one-way sound propagation. The size of spatial grids in the sound field was set to be 0.03 m in both methods. The time interval of 0.03 ms was assumed in the FDTD method and impulse responses of 245.76 ms duration (8,192 time-interval steps) were calculated. In the PE method, the calculation was repeated from 4.069 Hz to 10001.628 Hz at a frequency interval of 4.069 Hz, and the frequency response functions at the receiving points were obtained. We note that the frequency interval was determined according to the duration of the impulse responses that were obtained by the FDTD method.

7.1.2.3 Results

The insertion loss due to each noise barrier was calculated by two different methods, the coupled FDTD-PE method (described above) and the Full FDTD method (sound propagation was calculated only by the FDTD method). Figure 7.8 shows the spatial distributions of the insertion loss calculated by each method. It can be seen that the results are in good agreement, and that there is no obvious decrease in accuracy due to coupling of the two methods.

Table 7.1 shows the memory size and the computation time for each calculation by the coupled FDTD-PE method and by the Full FDTD method. Both the memory size and the calculation time decreased when using the coupled FDTD-PE method. Although the savings in the memory size were only some MB in this case study, the required memory size in the coupled FDTD-PE might not increase if the receiving

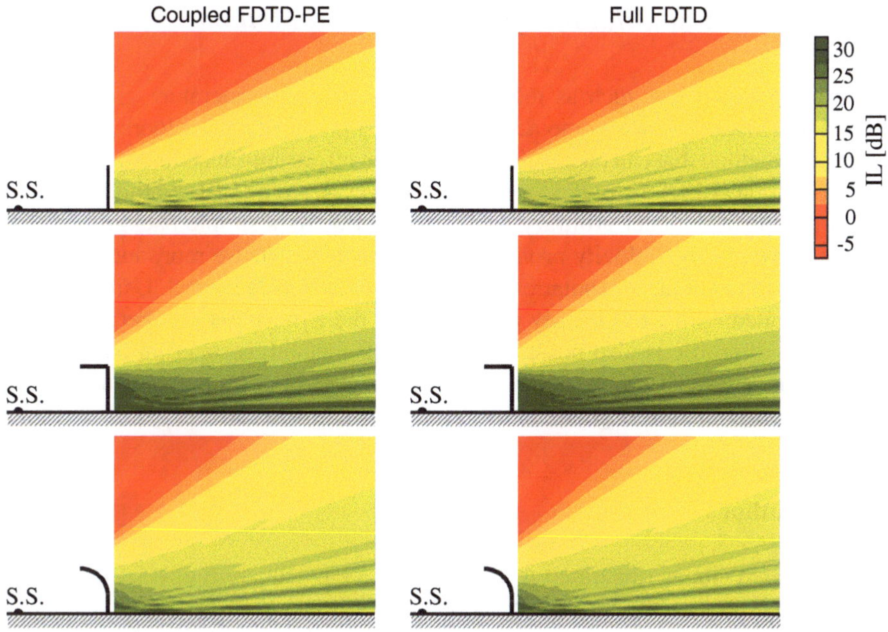

Fig. 7.8 Spatial distribution of insertion loss in each case (500 Hz octave band)

Table 7.1 Computation time and required memory size

	Coupled FDTD-PE	Full FDTD
CPU	Pentium 4 2.8 GHz	
Memory size	FDTD: About 19 MB	About 35 MB
	PE: About 0.06 MB	
Calculation time	Total: About 1 h 43 min	About 3 h 20 min
	FDTD: About 1 h 35 min	
	FFT: About 2 min	
	PE: About 6 min	

point were 1 km away from the sound source. The Full FDTD would require 650 MB of memory for the calculation of the long-range propagation. The amount of memory saved by coupling the two methods increases as the propagation distance increases.

As discussed above, long-range outdoor noise propagation can be calculated more efficiently by coupling suitable numerical simulation techniques with taking into consideration the characteristics of the sound field and the numerical simulation methods.

7.1.3 Chained Simulation of CFD and LEE Method

Meteorological effects such as the effects of wind and temperature gradients are factors that should be taken into account in sound propagation outdoors. As a computational method that can take the meteorological effects into account, the linearized Euler equation (LEE) method is introduced in Sect. 5.1. The LEE method can solve sound propagation under arbitrary background flow. As a simple background flow field, a vertical wind velocity profile is used. For more precise reproduction of the background flow field, computational fluid dynamics (CFD) is used. Using the flow field generated by a CFD is only possible with the LEE method, whereas using a vertical wind profile is also possible with the PE method as stated in the previous subsection. A case shown in this subsection is a conceptual example of using CFD for LEE background flow generation over a real-life urban area.

7.1.3.1 Outline

The simulated site as illustrated in Fig. 7.9 is a 1 km^2 square region of the central district of Nagaoka City, Japan, which includes Nagaoka railway station and a new city hall whose construction was scheduled to be finished in autumn of 2011.

Three-dimensional geometries of the buildings within the 1 km^2 region were generated by compositing 2-D digital geographic data in Shapefile format [9] (mentioned as GIS data hereafter) and aerial lidar scanned digital surface model [10] (mentioned as DSM data hereafter) except for the city hall whose geometry was created from a CAD model. The average resolution (the density of the point cloud) of the DSM data is around 2 m in horizontal direction. The standard error of the DSM data is stated to be 30 cm in horizontal direction and 15 cm in vertical direction [10]. For details about the geometry creation technique, see Oshima et al. [11].

7.1.3.2 Computational Fluid Dynamics

Mesh Generation

CFD and mesh generation for the CFD are performed using an open source CFD software OpenFOAM [12] which is based on unstructured finite volume methods. The origin and the center of the computational domain is taken at the city hall, and the x and y axes of the domain are taken so that they align with the façades of the city hall, with lengths of the domain being (1,024, 1,024, 512) m in (x, y, z) directions. The domain is meshed with hexahedral cells of edge lengths 8 m at high above the ground and increasingly refined by split-hex to cell sizes of 4 and 2 m as the mesh approaches ground. Furthermore, regions adjacent to the ground and to the GIS/DSM generated buildings are specifically refined to 1 m cell size. The ground is made flat since the difference of elevation in the district is small. The building façades are staircased following the geometries generated in Sect. 7.1.3.1. The mesh around the

Fig. 7.9 Extracted outline polygons of buildings. SP, RP1 and RP2 denote a source point and two receiving points

city hall is exceptional in that the cells adjacent to the city hall are refined to 25 cm and body-fitted since there was another purpose to the CFD of making an assessment of wind environment around the city hall [13]. The city hall, nearby buildings and ground portions of the resultant mesh is shown in Fig. 7.10. The number of cells in the mesh is about 15 million. It should be remarked that the edge display of the cells adjacent to the city hall in the figure is omitted due to too dense meshing.

Computational Setup

Since acoustic propagation in an urban spatial scale as dealt in the study is a phenomenon that has at most an order of several seconds in time, the unsteadiness of the wind velocity field may be considered sufficiently small to affect the sound propagation. Hence the steady-state semi-implicit method for pressure linked equation (SIMPLE) algorithm [14] was chosen as the solution technique. The turbulence model applied

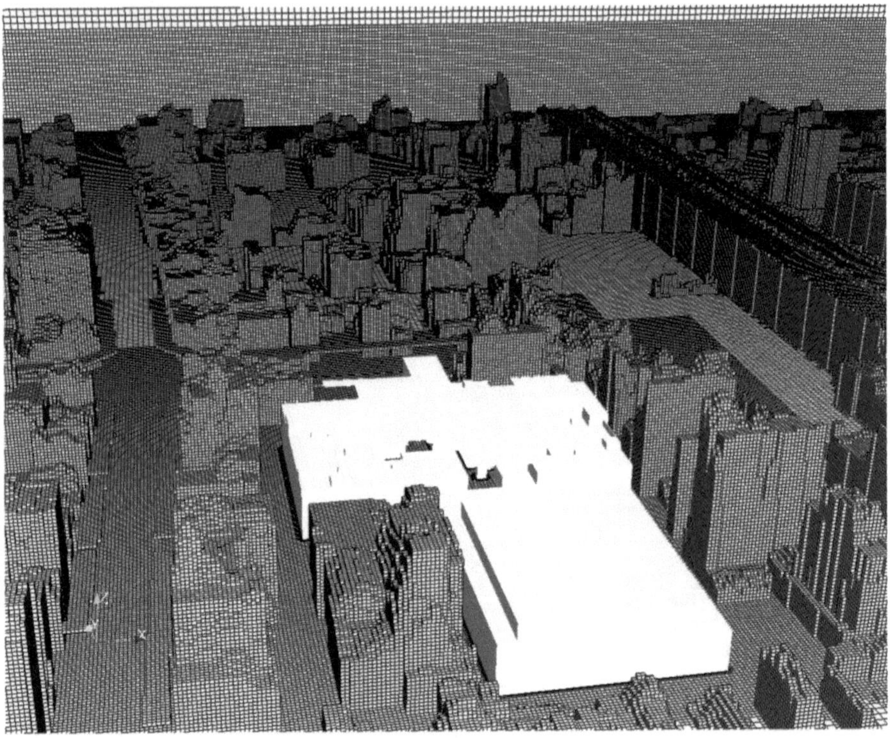

Fig. 7.10 Unstructured grid for CFD. Edge display of the cells adjacent to the city hall is omitted due to too dense meshing

in conjunction with the SIMPLE algorithm was the renormalization group (RNG) k-ε model [15], which turned out to have better overall accuracy and stability among the standard Reynolds averaged k-ε model and its variants [16] in prediction of urban wind environments.

The wind direction applied in the study is south-southwest (the y-direction of the computational domain) which is the main wind direction of 10 min averaged wind velocity during a decade of 1999 to 2008 at the Nagaoka automated meteorological data acquisition system (AMeDAS) station which situates at 23 m height above sea level. The initial profiles of wind velocity and turbulence kinetic energy were given using the 10 year average of the 10 min averaged wind speed $U_H = 2$ m/s, such that

$$U(z) = U_H \left(\frac{z}{H}\right)^{\alpha}, \tag{7.11}$$

where $\alpha = 0.27\,[-]$ and $H = 21.2\,[\mathrm{m}]$ (the roof height of the new city hall).

For more details about the CFD conditions, see [13].

7.1.3.3 Linearized Euler Equation Simulations

Mesh Generation and Flowfield Mapping

The sound propagation simulations by the LEE model require a uniform rectilinear grid with sufficiently smaller cell size than the characteristic acoustic wavelength in order to resolve acoustic waves throughout the computational domain, as opposed to the CFD where coarsened mesh may be applied at high above the ground. The requirement of the cell size leads to substantially larger number of cells than an unstructured solver can handle, and hence a necessity arises for a uniform structured mesh solver which is computationally more efficient than an unstructured solver. Thus the LEE simulations were conducted with a finite-difference structured algorithm under another uniform structured mesh than the mesh for CFD using the second-order scheme as stated in Sect. 5.1. The size of the computational domain without the absorbing boundary layers as described in the next subsection is (1,000, 1,000, 250) m which is slightly smaller than the CFD domain but identical to the original polygonal geometry size in x and y directions, and is halved in z direction than the CFD domain assuming that acoustic waves propagated to high above the ground no longer affects acoustic propagation near the ground. Due to constraints in computational resources, the cell size was set to 1 m which will allow simulations of spectral components up to around 35 Hz, which is barely within the lower frequency bound of human audible range. The number of cells was about 246 million. The generated mesh for the LEE simulations is shown in Fig. 7.11. Comparing to Fig. 7.10, one is able to see that the building shapes in the CFD mesh are accurately reproduced in the acoustic mesh as well.

Mapping of airflow velocity field obtained by the CFD to the acoustic mesh is done by assigning velocity at the corresponding cell in the CFD mesh to a cell in the acoustic mesh. Furthermore, in the study the velocity was scaled by a factor of 4 in order to clarify the wind effects.

For more detailed conditions, see [17]

Computational Setup

The boundary conditions for the buildings and the ground were all rigid. A perfectly matched layer (PML) [18] of 10-cell thickness was added to the sides and the top of the computational domain. A Gaussian acoustic pulse of wavelength 10 m in full width at half maximum is placed at $(x, y, z) = (-92.5, 80.5, 1.0)$ [m], which is the center of the crossing in front of the city hall marked as SP in Fig. 7.9. The Gaussian pulse contains spectral components up to around the frequency that corresponds to the wavelength equal to the full width at half maximum, which is around 35 Hz. The discrete time step was 1.6×10^{-3} s and the simulations was run up to 3.2 s in simulated time.

Besides the case with south-southwest wind, an additional LEE simulation was also run under without wind condition for reference. The LEE simulations were run on a single PC with dual quad-core 2.93 GHz processors and 16 GiBytes of RAM and took around 4 h per case.

Fig. 7.11 Uniform structured grid for LEE simulations

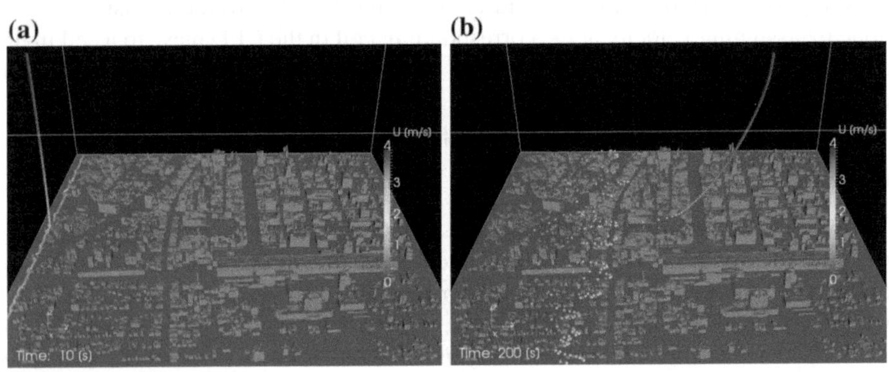

Fig. 7.12 Visualizations of CFD results by particle racing **a** Right after particle injection, $t = 10$ [s], **b** $t = 200$ [s]

7.1.3.4 Results

CFD

Figure 7.12 shows particle tracking of the computed flow field of the with-wind case where the wind blows from left to right in the figure. One is able to see that the

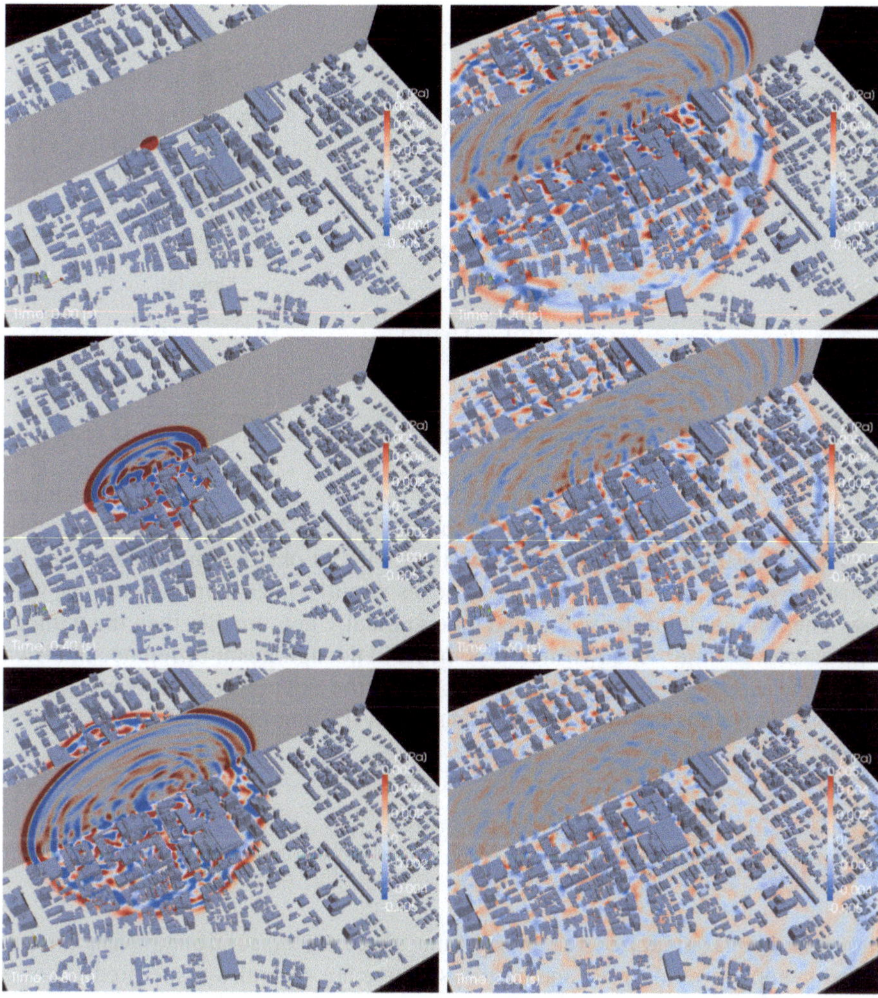

Fig. 7.13 Visualization of acoustic wave propagation. The horizontal contour surface is placed at 1.5 m height above the ground. The vertical surface is a surface for displaying contour

particles which entered the domain from near the ground of the inlet at left weaves slowly between the buildings, whereas those above the ground flows smoothly to the right along with the stratified wind velocity profile.

Acoustic Propagation

Figure 7.13 shows snapshots of instantaneous pressure contours taken in 0.4 s interval obtained by the LEE simulation for the with-wind case. Despite the low frequency components of the propagating wave up to around 35 Hz, one is able to see complex interference patterns created by reflections between the buildings inside the

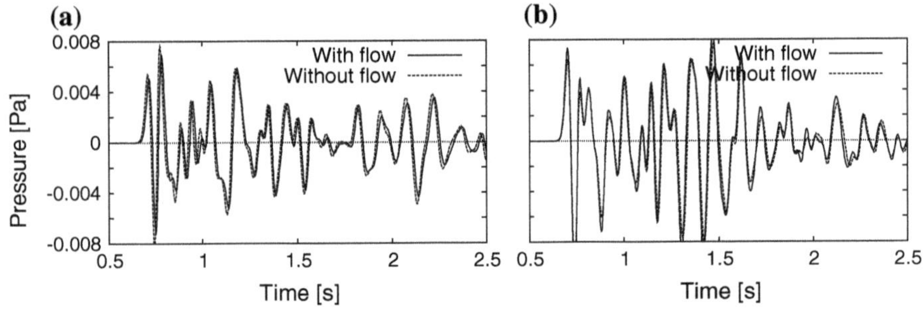

Fig. 7.14 Pressure waveforms at at **a** RP1 and **b** RP2

wavefront. From the snapshots, one is able to see that the simulation is producing reasonable results.

Figure 7.14 shows pressure waveforms of the with- and without-wind conditions at receiving points RP1 and RP2, shown in Fig. 7.9. At RP1 where the acoustic pulse propagates upstream from SP along the street without obstruction by buildings, the acoustic wave of the with-wind case reaches slightly later than the without-wind case. The amplitude of the with-wind case gradually gets smaller than the without-wind case over time. At downstream point RP2, no significant difference in arrival time of the acoustic waves is observed. However, the expansion of the acoustic wave amplitude is clearer for the with-wind case than the amplitude reduction observed for RP1.

7.2 Noise Barriers

Barriers have been applied for a long time to reduce noise around surface transportation such as road traffic and railways. There are several ways to predict noise-shielding efficiency of the barrier. The easiest way is the empirical engineering chart by Maekawa [19]: the insertion loss of a reflective barrier is calculated by using the only one parameter, the Fresnel number, defined as a ratio of path-length difference (between direct and diffracted sound rays) and a half wavelength. When absorption of barrier surface and/or ground surface is needed to be considered, wave-based analytical solutions should be applied [20]. In those analytical methods, shapes of the barrier and neighboring terrain are restricted to be simple shapes. For these decades, acoustic devices to reduce diffraction are sometimes installed on a top edge of the barrier [21]. The diffraction-reducing device usually has resonance cavities or absorbing materials, therefore bounary conditions on the barrier surface become complicated. When one develops barriers with such edge devices, efficiency prediction using numerical analyses should be useful.

In this section, numerical analyses on diffracted sound field behind noise barriers with arbitrary boundary conditions are described.

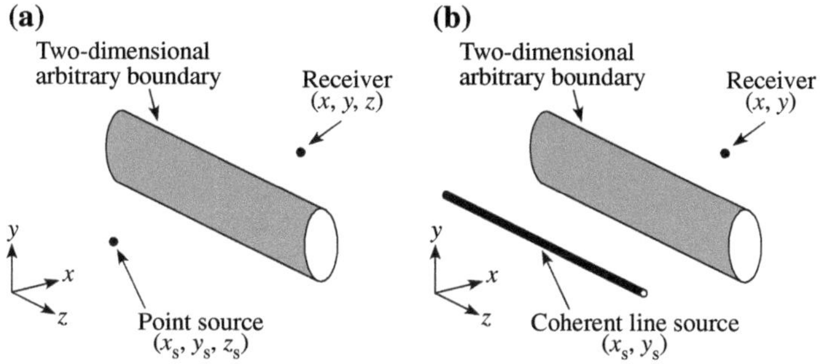

Fig. 7.15 **a** 2.5-D sound field to be analyzed, and **b** corresponding 2-D sound field

7.2.1 2.5-D Analysis Using BEM

As shown in Fig. 7.15a, let us consider a sound field with a point source and boundaries in a 2-D shape (i.e., the cross-sectional shape is constant infinitely along z axis). Such a sound field is sometimes referred to as a 2.5-D sound field. The boundary shape can include not only noise barriers but also slopes along embankment or depressed roads.

With a point source generating stationary pure tones, let us describe the way to calculate complex sound pressure $\Phi_{3D}(x, y, z, k)$ at a receiver. Time factor is defined as $e^{-i\omega t}$ with wavenumber $k = \omega/c$. When a point source in Fig. 7.15a is replaced by a coherent line source parallel to z axis in 2-D sound field as shown in Fig. 7.15b, complex sound pressure in the 2-D sound field, Φ_{2D}, and complex sound pressure in 2.5-D sound field, Φ_{3D}, are related by Fourier transform as [22]

$$\Phi_{3D}(x, y, z, k) = \frac{1}{2\pi} \int_{-\infty}^{+\infty} \Phi_{2D}\left(x, y, \sqrt{k^2 - k_z^2}\right) e^{ik_z(z-z_s)} dk_z. \tag{7.12}$$

The integral parameter k_z is the z-directional component of the wavenumber, and z and z_s is z coordinate for the receiver and point source, respectively. With wavenumber in 2-D field defined as $k_{2D} = \sqrt{k^2 - k_z^2}$, and considering symmetry of the integrand, Eq. (7.12) is rewritten as

$$\Phi_{3D}(x, y, z, k) = \frac{1}{\pi} \int_{0}^{+\infty} \Phi_{2D}(x, y, k_{2D}) \cos k_z (z - z_s) \, dk_z. \tag{7.13}$$

Calculate $\Phi_{2D}(x, y, k_{2D})$ for k_{2D} varying in a way $k_{2D} : k \to 0 \to j\cdot\infty$ as a function of the integral parameter k_z, and sum up all $\Phi_{2D}(x, y, k_{2D})$ using Eq. (7.13), then $\Phi_{3D}(x, y, z, k)$ for the 2.5-D field is obtained. In order to approximate integration by

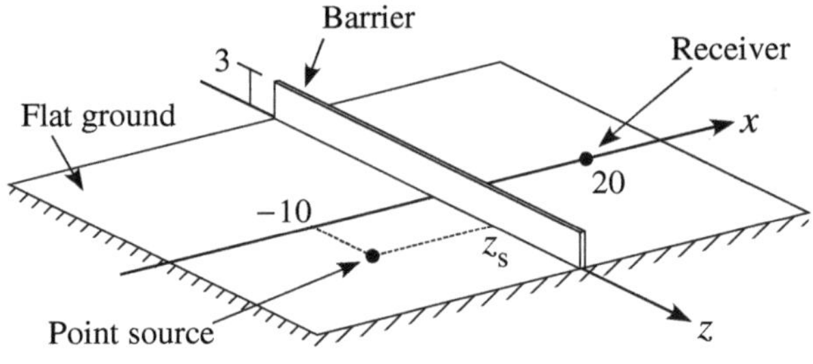

Fig. 7.16 Geometries of the ground, barrier, point source and receiver (Unit: m)

summation of $\boldsymbol{\Phi}_{2D}(x, y, k_{2D})$ at discrete wavenumbers k_{2D}, intervals of the discrete k_{2D} should be sufficiently small.

Arbitrary methods can be applied to calculate $\boldsymbol{\Phi}_{2D}(x, y, k_{2D})$: analytical analyses for simple boundaries or numerical analyses for complicated boundaries. Note that $\boldsymbol{\Phi}_{2D}(x, y, k_{2D})$ should be calculated for both real and imaginary numbers of k_{2D}. Boundary element method is introduced in examples in the following Sect. 7.2.2. Although arbitrary acoustic impedance can be set at boundary surfaces, computation time increases in case calculation of $\boldsymbol{\Phi}_{2D}(x, y, k_{2D})$ is needed for k_{2D} at a small intervals. In examples in Sect. 7.3.2, the finite difference method in time domain is applied. The latter is advantageous because the intervals of k_{2D} are arbitrarily configured in obtaining $\boldsymbol{\Phi}_{2D}(x, y, k_{2D})$ by Fourier transform of the calculate impulse response in time domain; in practice, the method is restricted to refletive boundaries because of difficulties in simulating actual absorbing boundaries.

The advantage of the method with Eq. (7.13) is that the memory size of the computer for 2-D numerical analyses is quite smaller compared to the direct 3-D computation for the sound field with infinitely long obstacles shown in Fig. 7.15a.

7.2.2 Numerical Examples

2.5-D analysis is carried out for the sound field shown in Fig. 7.16. A noise barrier infinite in length with a height of 3 m is standing between a point source and receiver on the ground surface. The point source simulates a vehicle running on a lane 10 m distant from the barrier; its position is indicated by z_s.

Noise-shielding efficiency against road traffic noise is investigated considering four kinds of barriers shown in Fig. 7.17 [23]. "Simple" stands for a thin simple barrier, "Thick" for a thick barrier, "Channels" for barriers with arrayed channels of 180 mm depth, and "Tubes" for barriers with arrayed rectangular tubes of 180 mm depth. All surfaces of the barriers are reflective. Because a cross-section of the boundary constant in the longitudinal direction in the 2.5-D analyses is assumed, shapes such as "Tubes" are out of scope of the 2.5-D analyses in nature. Therefore, as shown in Fig. 7.18, the theoretical acoustic impedance of the open tubes are applied as the

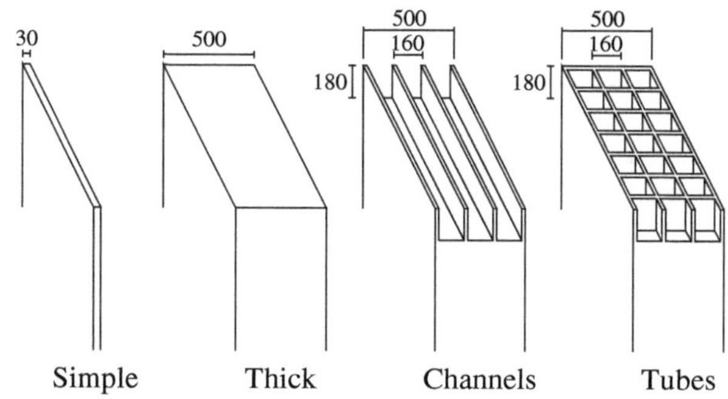

Fig. 7.17 Barrier shapes investigated by using 2.5-D analyses (Unit: mm)

Fig. 7.18 Approximation of boundary condition for the tubes barrier

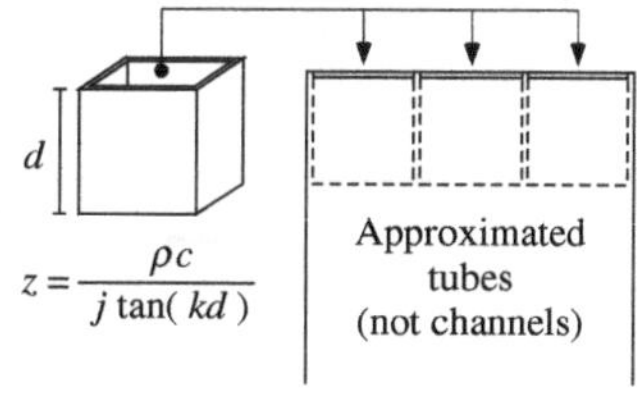

boundary conditions to approximate array of open tubes. Meanwhile, in analyses for the shape "Channels," rigid boundaries are set along the cross-sectional shape.

2.5-D analyses are carried out based on Eq. (7.13). 2-D analysis frequencies $f_{2D} = k_{2D} \cdot (c/2\pi)$ are set at intervals of 2 Hz or less. Because $\boldsymbol{\Phi}_{2D}\,(k_{2D})$ diverges at $f_{2D} = 0$ [24], f_{2D} around zero should be set such as $f_{2D} = \ldots, 6, 4, 2, 1, 0.5, (0.5)^2, \ldots, (0.5)^{10}, (0.5)^{10}j, \ldots, (0.5)^2 j, (0.5)j, 1j, 2j, \ldots$ [Hz], avoiding $f_{2D} = 0$. The 2-D analyses are done by the boundary element method, where the upper limit of the boundary element is $\lambda/8$. $\boldsymbol{\Phi}_{2D}\,(k_{2D})$ fluctuates as a function of k_{2D}, so the integration of Eq. (7.13) is done approximately by interpolation of $\boldsymbol{\Phi}_{2D}\,(k_{2D})$ at discrete sets of k_{2D}. The vehicle speed is sufficiently smaller than the sound speed, the point source is assumed to be resting in the 2.5-D analyses.

For accuracy validation, the numerical analyses are compared to scale-model experiments. Reduction in sound pressure level by the Channels and Tubes barriers relative to the Thick barrier are calculated to be compared to results of scale-model experiments. In the numerical analyses, calculation of Eq. (7.13) is carried out assuming pure tones at intervals of 1/15 octave, then results for five frequencies are averaged to approximate 1/3 octave band levels. In the scale-model experiments, 1/20-scale models of the barriers are made with chloroethene boards to measure impulse responses by using a spark-pulse sound source. Results are shown in Fig. 7.19. Generally, results of 2.5-D analyses and model experiments agree well. In results for the Tubes barrier, the analyses and experiments agree in the frequency range below 800 Hz. Difference between them above 1 kHz is error due to limitation of approximation shown in Fig. 7.18; the approximation is no longer accurate in the

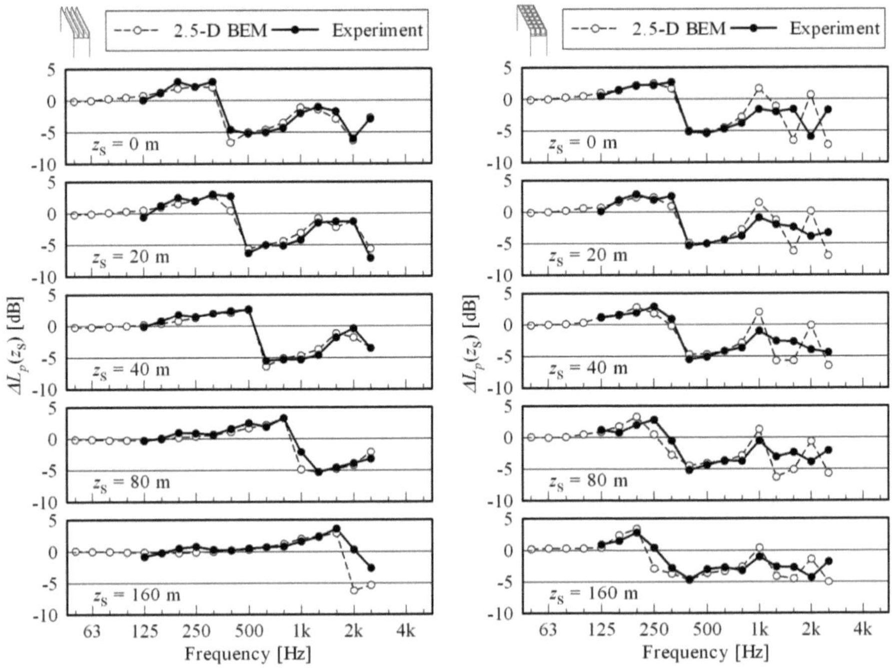

Fig. 7.19 Comparison between 2.5-D analyses and scale-model experiments: SPL reduction by channels and tubes barriers relative to thick barrier

frequency range above 1,060 Hz where the opening width 160 mm exceeds a half wavelength.

Figure 7.19 implies characteristics of diffraction reduction by the Channels and Tubes barriers. For the sound source set at $z_s = 0\,[\mathrm{m}]$, the diffraction-reducing efficiencies of both barriers show similar tendency. These efficiencies come maximized around 472 Hz where depths of channels and tubes correspond to a quarter wavelength. At lower frequencies, SPL increases due to antiresonance; that is, diffracted sound increases compared to that for the Thick barrier. For the sound source moving away the initial position $z_\mathrm{s} = 0\,[\mathrm{m}]$, the SPL reduction by the Tubes barrier is retained around 500 Hz, while the SPL reduction by the Channels barrier shifts toward high-frequency range. When the sound source reaches to $z_s = 160\,[\mathrm{m}]$, the SPL reduction disappears around 1 kHz where road traffic noise dominates.

Assuming A-weighted power spectrum to simulate vehicle noise for the point source, A-weighted sound pressure level $L_{p\mathrm{A}}$ [dB] is calculated. The power spectrum is set by definition described in ASJ RTN-Model 2008 [25], considering passenger vehicles running at constant speed 80 km/h on drainage pavement. In addition to the propagation characteristics computed by 2.5-D analyses, atmospheric absorption is also considered in homogeneous atmosphere with the temperature 15 °C, relative humidity 60 %, and pressure 1 atm.

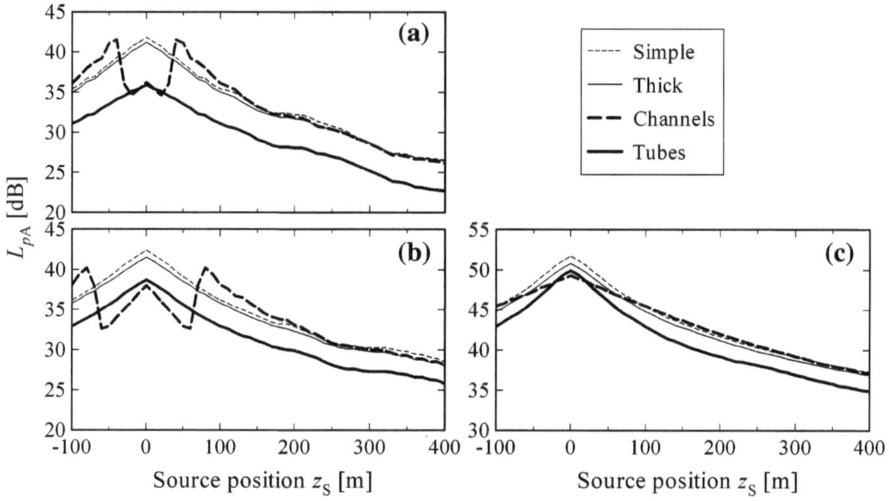

Fig. 7.20 Variation of A-weighted sound pressure level as a function of source position: **a** 500 Hz, **b** 800 Hz, **c** overall

Variation in L_{pA} [dB] as a function of the source position z_s [m] is shown in Fig. 7.20. The Fig. 7.20a is for 1/3 octave band level at 500 Hz. Curves for Simple, Thick, Tubes barriers decrease monotonically around their peak at $z_s = 0$ [m] closest to the receiver. Thickness effect makes the curve for the Thick barrier lower than that for the Simple barrier. The curve for the Tubes is 5 dB less than those two. Meanwhile, the curve for the Channels barrier is almost the same as the curve for the Tubes barrier for the source position $|z_s| \leq 30$ [m], although it exceeds the Simple for $40 \leq |z_s| \leq 130$ [m] and converges with the Simple and Thick barriers. This tendency is related to results in Fig. 7.19. In Fig. 7.20b, 1/3-ocvate band level at 800 Hz shown; the source position where the effect of the Channels barrier falls into negative is different from that for 500 Hz. As shown in Fig. 7.19, the effect in the low-frequency range disappears at the source positions far from the receiver. Overall (OA) level is shown in the Fig. 7.20c. Although effect of the Channels relative to the Simple appears only for the source positions $|z_s| \leq 80$ [m], the stable effect of the Tubes appears independently from the source position.

Figure 7.21 shows the equivalent continuous A-weighted sound pressure level L_{Aeq} for the traffic volume of 1,000 vehicles in an hour, integrating L_{pA} for the source position range $|z_s| \leq 1{,}000$ [m]. Difference between curves for the Simple, Thick, and Channels barriers are quite small; namely diffraction-reducing effect of Channels is quite small. In contrast, the Tubes barrier reduces L_{Aeq} in the frequency range from 315 to 800 Hz.

It is preferred that the integration range of z_s to calculate L_{Aeq} should be set where L_{pA} drops by 10 dB from its maximum. From results shown in Fig. 7.20, L_{pA} for the range of $|z_s| \leq 200$ [m] should be integrated at least. In addition, results in Fig. 7.21 indicate that calculation should be made up to 1.25 kHz or more. Whereas direct

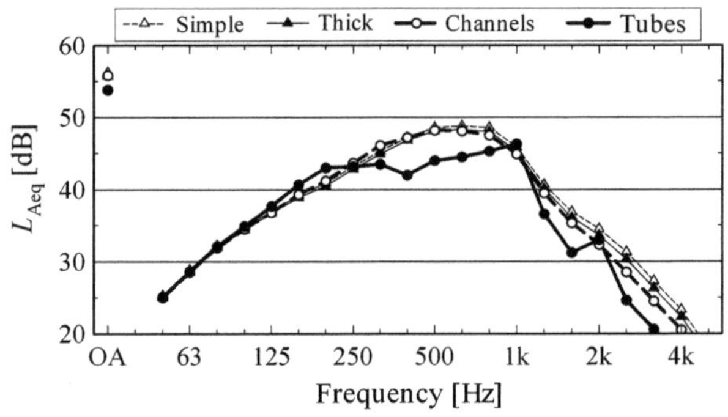

Fig. 7.21 Equivalent continuous A-weighted sound pressure level based on integration for the range $|z_s| \leq 1,000$ [m]

3-D computation requires much computational resources, the 2.5-D computation described in this section is executable even on general personal computers with less memory spaces.

7.2.3 Notes on Absorbing Boundaries

In the examples above, barriers without absorbing materials are considered. Practically, absorbing materials are sometimes installed on the top of the barrier. If impedance boundary (i.e., locally reacting boundary) is set for the absorbing material, diffraction-reducing effect will be overestimated; in other words, diffracted sound will be underestimated [26]. It is recommended to apply the analyses where sound fields inside and outside the absorbing material are coupled.

7.3 Depressed Roads

For suburban express ways with heavy traffic volume, depressed or semi-underground road structures are often used. In these road structures, multiple reflections and diffractions often occur and therefore sound propagation inside and outside the road structures become very complicated. On noise prediction around these road structures, enough accuracy cannot be secured by energy-base geometrical acoustics and the wave-based numerical analysis is needed in order to obtain reliable prediction results.

7.3.1 2-D Analysis Using FDTD Method

In order to make practical prediction of road traffic noise, noise propagation paths from sources to a receiver are traced in 3-D space in wide frequency range. To cal-

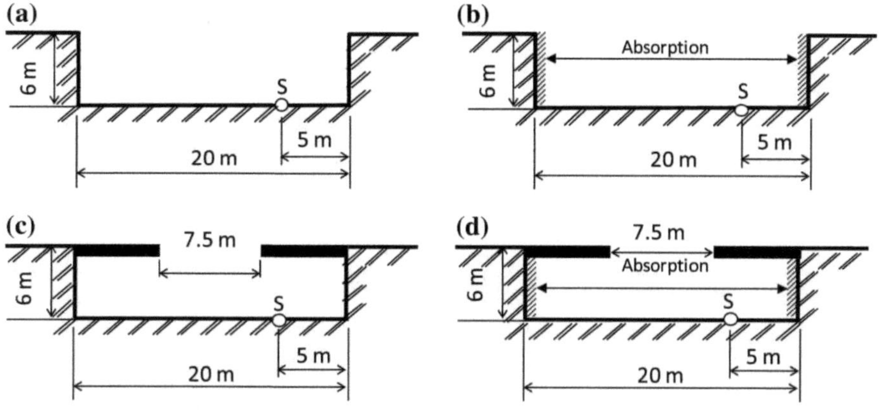

Fig. 7.22 A depressed and a semi-underground road structures under investigation. **a** Depressed structure without absorption. **b** Depressed structure with absorption. **c** Semi-underground structure without absorption. **d** Semi-underground structure with absorption

culate A-weighted sound pressure level, which is used for general road traffic noise assessment, sound propagation in a frequency range from 125 to 2kHz at least is needed. It seems to be very difficult to obtain such a sound propagation using a wave-based numerical analysis because huge computational resource is necessary in order to correspond a 3-D wide area and small wavelength. Considering many road traffic noises problems exist on almost straight roads, 2-D analysis for a sound field to which the cross-sectional shape of the road structure is modeled is one of efficient and smart solutions to simplify the problem. The 2-D model is approximately correspondent with calculation of equivalent continuous A-weighted sound pressure level by a steady flow of automobiles.

For a depressed road and a semi-underground road shown in Fig. 7.22, numerical analysis using the finite-difference time-domain method was performed. In the road structures under investigation, Fig. 7.22a, c have reflective surfaces and Fig. 7.22b, d have partially absorptive treatment on the wall surfaces. As an absorbing boundary condition, normal surface acoustic impedance correspondent with 0.8 absorption coefficient for all frequency range under investigation was given for the parts of the absorptive treatment.

As a result of the numerical analysis, snapshots of instantaneous sound pressure distribution every 50 ms after an impulsive source was generated at source point, S, as shown in Fig. 7.23. As for the depressed structure, in the case where the sidewalls are assumed reflective shown in the series of Fig. 7.23a, remarkable multiple reflections are seen between the sidewalls and strong sound waves caused by the multiple reflections propagate intermittently outside the road structure. On the other hand, in the case where the sidewalls are absorptive shown in the series of Fig. 7.23b, the multiple reflection is much reduced and consequently the subsequent reflections after the direct sound is much diminished compared to the former case. In the case of semi-underground structure without absorption treatment shown in the series

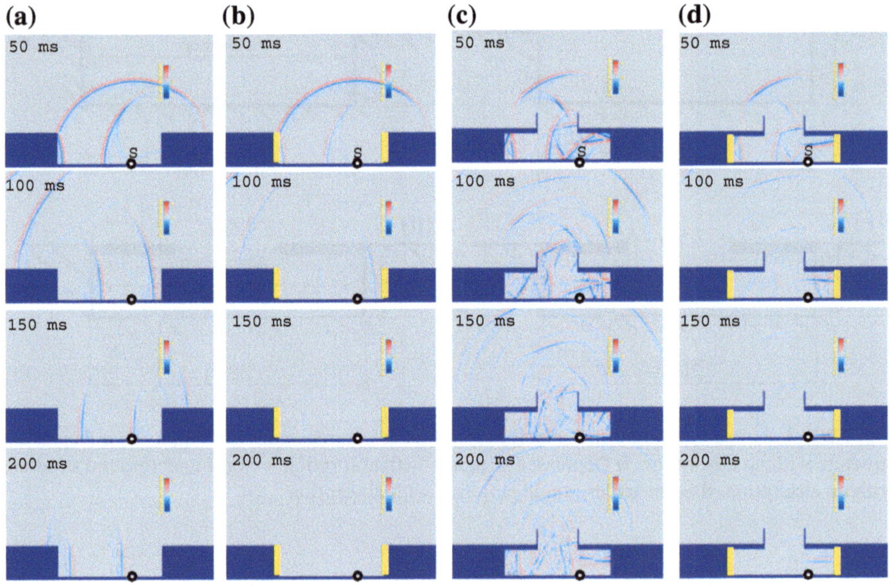

Fig. 7.23 Calculation results of sound propagation for a depressed and a semi-underground road structures road structure. **a** Depressed structure without absorption. **b** Depressed structure with absorption. **c** Semi-underground structure without absorption. **d** Semi-underground structure with absorption

of Fig. 7.23c, very complicated multiple reflections inside the structure and sound propagation through the opening are seen. In the case where the sidewalls are assumed absorptive as shown in the series of Fig. 7.23d, the multiple reflections and sound propagation outside are much diminished in the same manner as in the former case. In the series of Fig. 7.23d, it should be noted that multiple reflection still remains between the ceiling and the road surface.

General environmental noise is usually assessed by equivalent continuous sound pressure level. When assessing the road traffic noise, Since road traffic noise can be considered to be generated by series of passing automobiles with a certain average running speed, the sources can be modeled as a 2-D source. Here, it should be noticed that the source is an incoherent line source and it is different from a coherent cylindrical source which is assumed in 2-D numerical analysis. In case of application to road traffic noise problems, it has been reported by numerical and experimental studies that the difference between A-weighted sound pressure level from an incoherent line source and that from an coherent cylindrical line source is relatively small [27]. In the literature [25], a calculus which converts sound pressure level obtained in a 2-D sound field to equivalent continuous sound pressure level in the consistent 3-D sound field is shown. Here, a 2-D numerical analysis was made for a cross-sectional shape of a semi-underground road structure and the result was compared with a result of in-situ experiment [28]

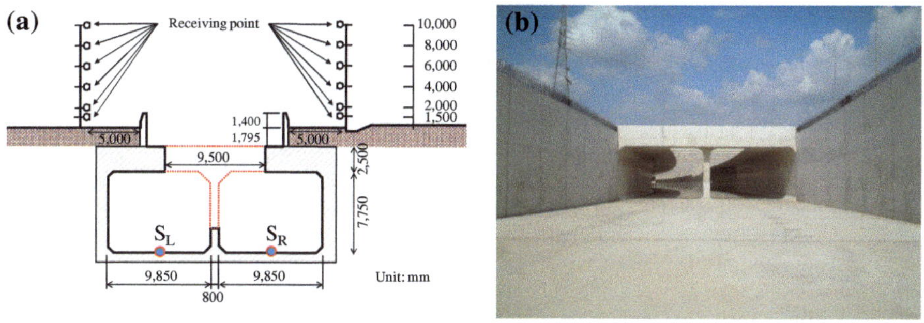

Fig. 7.24 A semi-underground road under investigation: **a** Sectional shape and positions of a source and receivers, **b** picture of experiment cite

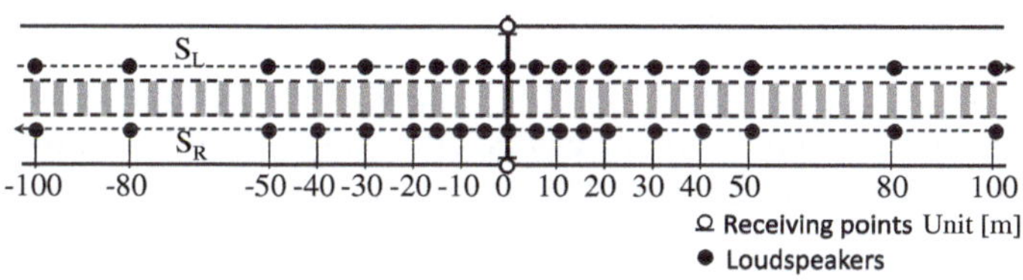

Fig. 7.25 Discrete points of sound sources

Figure 7.24 shows the cross-sectional shape of the semi-underground road under investigation. Two vertical receiving lines were set on both sides of the semi-underground road and six receiving points in each receiving line were distributed in the height range from 1.5 to 10 m above the ground as shown in Fig. 7.24a. In the experiment, discrete source points were set along the semi-underground road as shown in Fig. 7.25, and propagating sound energy from each source point was measured. All of the energies for all source positions were integrated in energy-base to simulate incoherent line source. In the calculation, a 2-D finite-difference time-domain analysis was made under the calculation condition of discrete spatial grid size of 0.04 m and of discrete time step of 0.05 ms. Comparisons between calculation and experiment are shown in Fig. 7.26. Sound pressure exposure levels for both of calculation and experimental results are corrected so that the overall sound energy level of the line source is 100 dB/m. The results are A-weighted sound pressure level calculated from band limited sound pressure between 125 Hz and 2 kHz in 1/1 octave band. The sound pressure exposure levels become higher as the position of receiving point becomes high. From the figures, it is clearly seen that the calculation results agree well with the experimental ones.

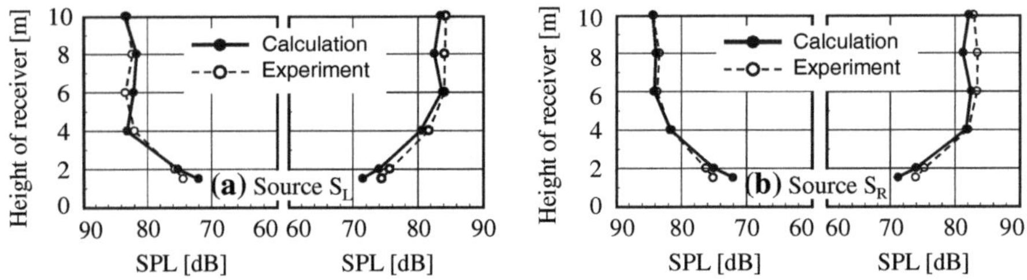

Fig. 7.26 Comparison between experiment (level of integrated sound pressure exposure of contributions from all sources) and calculation (2-D numerical analysis): **a** Source S_L, **b** Source S_R (Corrected results in which a sound power level is 100 dB/m)

7.3.2 2.5-D Analysis Using FDTD Method

As shown in Fig. 7.2, a 3-D sound field which has infinitely continuous homogeneous cross section and a spherical source can be efficiently solved by following Fourier integral of a solution with a 2-D sound field to which the cross section is modeled [22].

$$\boldsymbol{\Phi}_{3\mathrm{D}}(k) = \frac{1}{\pi}\int_{0}^{+\infty} \boldsymbol{\Phi}_{2\mathrm{D}}(k_{2\mathrm{D}}) \cos k_z z dk_z, \tag{7.14}$$

where, $\boldsymbol{\Phi}_{3\mathrm{D}}(k)$ is a solution with 3-D sound field with k as a wave number, $k_{2\mathrm{D}}$ is 2-D wave number and $k_{2\mathrm{D}} \equiv \sqrt{k^2 - k_z^2}$, $\boldsymbol{\Phi}_{2\mathrm{D}}(k')$ is a solution with 2-D sound field with k', z is relative coordinate in z direction of a receiver to a source. Considering that frequency and time correspond one to one by Fourier transform, Eq. (7.14) can be applied to solutions in time domain [29]. Let us express Eq. (7.14) by a sum of two integral equations as follows:

$$\boldsymbol{\Phi}_{3\mathrm{D}}(k) = \frac{1}{\pi}\int_{0}^{k} \boldsymbol{\Phi}_{2\mathrm{D}}\left(\sqrt{k^2 - k_z^2}\right) \cos k_z z dk_z + \frac{1}{\pi}\int_{k}^{\infty} \boldsymbol{\Phi}_{2\mathrm{D}}\left(j\sqrt{k_z^2 - k^2}\right) \cos k_z z dk_z, \tag{7.15}$$

then the first term on the right-hand side involves a real wave number $\sqrt{k^2 - k_z^2}$. It means that the integral term is calculated using frequency response of an obtained solution in time domain. On the other hand, the second term involves a imaginary wave number of $j\sqrt{k_z^2 - k^2}$ with j as an imaginary unit, and therefore an imaginary frequency domain should be treated with. Fourier transform of a 2-D solution in time domain, $\phi_{2\mathrm{D}}(t)$, is expressed as follows:

$$\Phi_{2D}(k) = \int_{-\infty}^{+\infty} \phi_{2D}(t) e^{j\omega t} dt. \tag{7.16}$$

Here, substitution of a real angular frequency by an imaginary angular frequency leads to

$$\Phi_{2D}(jk) = \int_{-\infty}^{+\infty} \phi_{2D}(t) e^{-\omega t} dt. \tag{7.17}$$

Equation (7.17) means that the imaginary frequency components of ϕ_{2D} is obtained by the Laplace transform of the transient solution. Therefore, the integration transformation described by Eq. (7.15) can be calculated using the results of the Fourier and Laplace transforms of a transient solution. In order to adapt the FFT results obtained from the transient solution, which is defined at discrete frequencies with constant intervals of $\Delta f = f_s/N$, f_s and N being the sampling frequency and the number of FFT points, respectively, to the integration procedure, Eq. (7.16) is transformed through a change of variables as follows.

In the case of $z \neq 0$,

$$\begin{aligned}\Phi_{3D}(k) &= \frac{1}{\pi} \sum_{i=0}^{M-1} \frac{\Phi_{2D}(\alpha[i]) + \Phi_{2D}(\alpha[i+1])}{2} \cdot \frac{\sin\sqrt{k^2-\alpha[i]^2}z - \sin\sqrt{k^2-\alpha[i+1]^2}z}{z} \\ &+ \frac{1}{\pi} \sum_{i=0}^{M-1} \frac{\Phi_{2D}(j\alpha[i]) + \Phi_{2D}(j\alpha[i+1])}{2} \cdot \frac{\sin\sqrt{k^2+\alpha[i]^2}z - \sin\sqrt{k^2+\alpha[i+1]^2}z}{z}.\end{aligned} \tag{7.18}$$

In the case of $z = 0$,

$$\begin{aligned}\Phi_{3D}(k) &= \frac{1}{\pi} \sum_{i=0}^{M-1} \frac{\Phi_{2D}(\alpha[i]) + \Phi_{2D}(\alpha[i+1])}{2} \cdot \left(\sqrt{k^2-\alpha[i]^2} - \sqrt{k^2-\alpha[i+1]^2}\right) \\ &+ \frac{1}{\pi} \sum_{i=0}^{M-1} \frac{\Phi_{2D}(j\alpha[i]) + \Phi_{2D}(j\alpha[i+1])}{2} \cdot \left(\sqrt{k^2+\alpha[i+1]^2} - \sqrt{k^2+\alpha[i]^2}\right).\end{aligned} \tag{7.19}$$

A solution of 2.5-D sound field in frequency domain is obtained by the calculation procedure mentioned above. After the frequency components at equally placed frequencies are obtained, also the 2.5-D sound field solution in the time domain can be obtained by the inverse Fourier transform process [29].

The calculation method mentioned above was applied to calculation of noise radiation from semi-underground road structures shown in Fig. 7.27 [30]. This calculation was aimed at the development of engineering prediction model of road traffic noise around semi-underground road structure. The width of the mouth part,

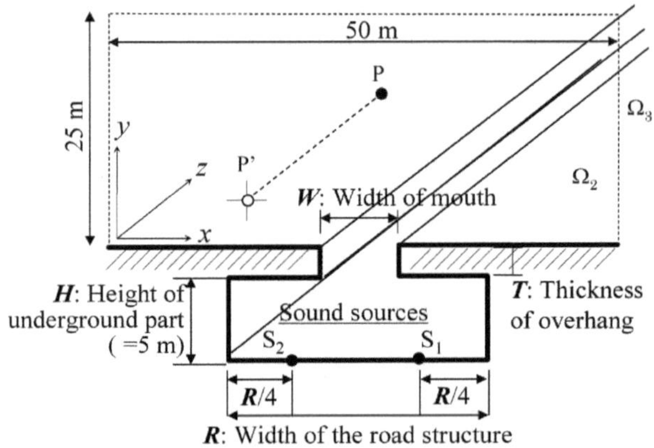

Fig. 7.27 Sectional shape of the semi-underground road structure under investigation

W, the width of road, R and the thickness of the overhangs, T, were varied as the parameters by which the semi-underground road structures are characterized, and the effects of the parameters to noise radiation characteristics were studied. In order to see a directivity of noise radiation, 120 receiving points in total were distributed on a hemi-spherical surface with radius of 20 m with its center point was positioned at the center of the mouth part just upper of the point source, as shown in Fig. 7.28. 2-D finite-difference time-domain analysis were performed for a cross section shown in Fig. 7.27. In the calculation, the spatial grid size, Δh, and discrete time interval, Δt were set as 0.0025 [m] and 0.01 [ms], respectively, and transient responses for 600,000 time steps (6 s duration) were calculated, since a long reverberation remained in the road structure caused by multiple reflections. As an example of calculation results, variation of sound radiation characteristics emitted from the mouth part of the semi-underground road structure by changing the width of the mouth part, W, from $W = 5$ m to $W = 15$ m are shown in Fig. 7.29. The sound pressure level in the figure was A-weighted, and was so corrected that the point source positioned on the road surface has spectral characteristics of vehicle noise of steady running condition with A-weighted sound power level of 100 dB. We can see that the radiation directivity varies with the sectional shape. In both cases, the directivity in the x–y plane is sharp in the upper and oblique directions, whereas that in the longitudinal section of the y–z plane is rather gentle. In order to validate the result of the numerical analysis, in this study, 1/20 scale model experiment using a spark discharge impulse source was performed. The experimental results are also shown in Fig. 7.29 and fair agreement between results of numerical analysis and of experiment can be seen.

7.3.3 3-D Analysis Using FMBEM

Numerical methods for 2.5-D problems are applicable to straight roads, including depressed ones, each of which has the same cross section along the road. These

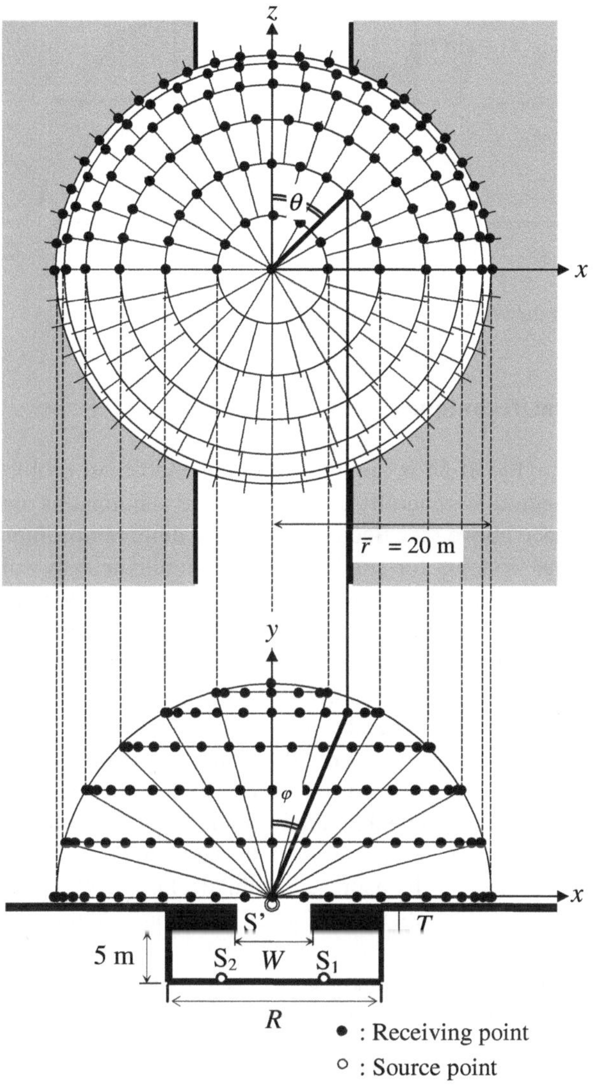

Fig. 7.28 Geometrical condition of sources and receivers in order to measure noise radiation from mouth part of the semi-underground road structure

methods, however, cannot deal with many depressed roads with columns and beams, as shown in Fig. 7.24. In order to know the reflection effect of such structures, numerical methods for 3-D problems are required. As mentioned above, it has been practically impossible to solve such large-scale 3-D problems using basic numerical methods such as the standard BEM. Here we briefly present an efficient numerical technique based on the fast multipole BEM (FMBEM) for analyzing 3-D depressed roads (see Sect. 4.3).

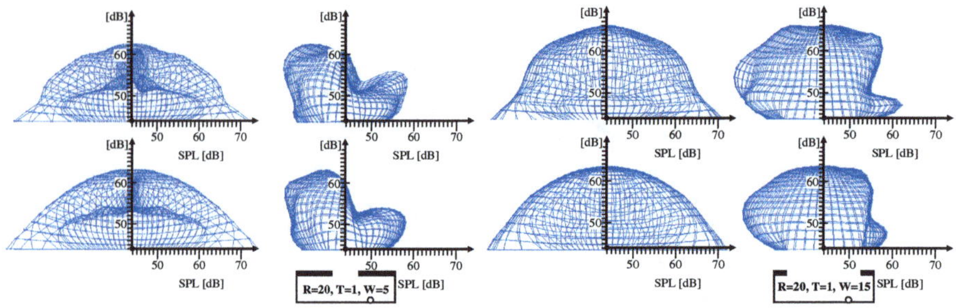

Fig. 7.29 Results of the analysis

7.3.3.1 Numerical Technique

When the BEM or FMBEM is applied to a depressed road problem, the domain decomposition method is generally required, where subdomains are the depressed space and half space above (see Sect. 4.2.2). In the domain decomposition method, two or more linear systems for subdomains are simultaneously solved, while the FMBEM is basically applicable only to single-domain problems, i.e., to the linear system for each subdomain in the domain decomposition method. There are some techniques to solve multi-domain problems based on separate calculation for each subdomain. One is the iterative domain coupling technique, where separate calculation for each subdomain and successive renewal of values on the interface boundaries between subdomains are performed until the whole domain converges [31]. Another is a technique to obtain the matrix-vector products for the linear system including all unknown vectors in the whole domain, by adding matrix-vector products for subdomains. The latter technique is presented in the following. It has been reported that a good convergence can be achieved in this technique by consideration of the ordering of unknowns and application of an effective preconditioner for the iterative method [32].

A linear system including all unknown vectors in the whole domain is expressed as Eq. (4.77). This equation can be transformed into the following expression:

$$\begin{bmatrix} \left[\boldsymbol{A}_{\mathrm{I}1}\ \boldsymbol{A}_{\mathrm{I}2}\right]\begin{Bmatrix}\boldsymbol{p}_{\mathrm{I,I}}\\ \boldsymbol{p}_{\mathrm{I,II}}\end{Bmatrix} - \left[\boldsymbol{G}_{\mathrm{I}1}\ \boldsymbol{G}_{\mathrm{I}2}\right]\begin{Bmatrix}\boldsymbol{0}\\ \boldsymbol{v}_{\mathrm{I,II}}\end{Bmatrix} \\ \boldsymbol{p}_{\mathrm{II,I}} - \boldsymbol{Z}_{\mathrm{II}}\boldsymbol{v}_{\mathrm{II,I}} \end{bmatrix} = \begin{bmatrix} \left[\boldsymbol{G}_{\mathrm{I}1}\ \boldsymbol{G}_{\mathrm{I}2}\right]\begin{Bmatrix}\boldsymbol{v}_{\mathrm{I,I}}\\ \boldsymbol{0}\end{Bmatrix} - \boldsymbol{p}_{\mathrm{d,I}} \\ \boldsymbol{0} \end{bmatrix}. \tag{7.20}$$

Each matrix-vector product in this expression can be efficiently calculated by applying the FMBEM to each subdomain. The total operation counts for the matrix-vector products are reduced to $O(N_{\mathrm{I}} \log N_{\mathrm{I}}) + O(N_{\mathrm{II}} \log N_{\mathrm{II}})$ ($\approx O(N_{\mathrm{a}})$, if N_{I} and N_{II} are large enough to be able to regard $\log N_{\mathrm{I}}$ and $\log N_{\mathrm{II}}$ almost constant), where N_{I}, N_{II},

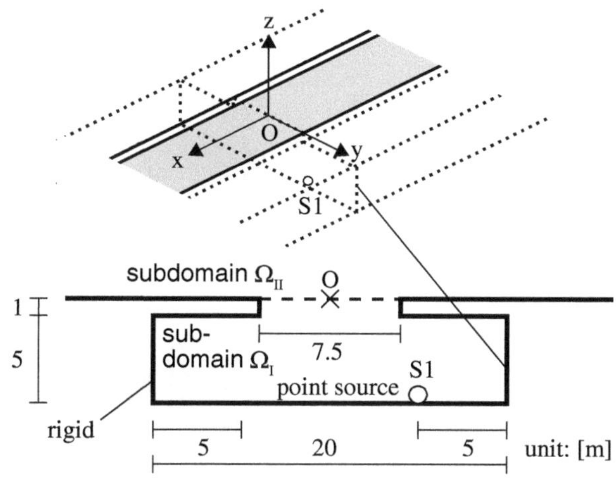

Fig. 7.30 Analysis model: depressed road 20 m *wide* and 100 m *long*

and N_a $(= N_I + N_{II})$ are the numbers of unknowns for subdomains Ω_I and Ω_{II}, and the whole domain, respectively. Highly efficient computation is achieved if the process of the iterative method for Eq. (4.77) rapidly converges. Since it is not necessary to calculate and store the matrices for subdomains, the memory requirement is similarly reduced.

7.3.3.2 Numerical Setup

Figure 7.30 shows an analysis model with a depressed road under an infinite rigid plane. The road is bounded in the length of 100 m, and one point source is located on the center cross section of the road. Only the real part of the surface impedance is assumed on both sides of the road corresponding to the normal incidence absorption coefficient $\alpha = 1$, and all the other surfaces are rigid. A result for a 1 octave band is obtained by energy summation of the results at nine single frequencies in the band. GPBiCG with ILUT(10^{-5}, 50) preconditioner is used as the iterative method, and its stopping criterion is $\varepsilon = 10^{-3}$.

7.3.3.3 Numerical Results

Figure 7.31 shows radiation directivities from the center point O of the opening. The directivities at single frequency and for 1 octave band become more complex with frequency.

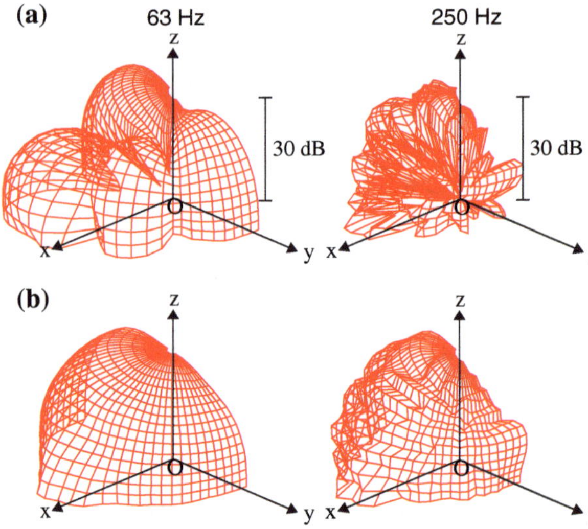

Fig. 7.31 Directivities from center point O of opening: **a** at single frequencies, and **b** for 1 octave bands

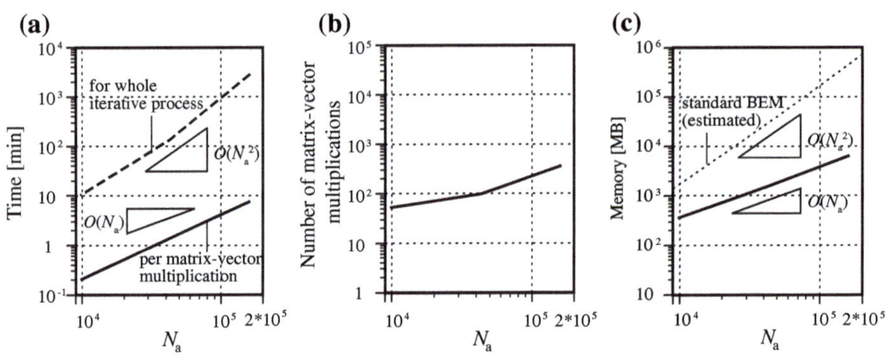

Fig. 7.32 Relations between number of unknowns for the whole domain N_a and computational efficiency of FMBEM: **a** computation time, **b** number of matrix-vector multiplications, and **c** memory requirement

7.3.3.4 Computational Efficiency

Figure 7.32 shows relations between the number of unknowns for the whole domain N_{a} and the computational efficiency of the FMBEM: (a) the computation time, (b) the number of matrix-vector multiplications, and (c) the memory requirement. The computation time per matrix-vector multiplication is about $O(N_{\mathrm{a}})$, whereas the time for the whole iterative process is about $O(N_{\mathrm{a}}^2)$, because the number of matrix-vector multiplications increases with N_{a}. Like this case, the convergence behavior of the iterative method spoils the rapid computation of the FMBEM; therefore, special consideration should be required on the improvement of convergence (see Sect. 4.3.4 for general consideration on convergence). In the domain decomposition method, the ordering of unknowns of the system matrix in Eq. (4.77) is important for good convergence. It has been reported that a nearly diagonal ordering (an ordering that diagonal entries of the system matrix are nearly dominant) is effective [32]. In the present case, all the numbers of iterations are less than $1/100$ of N_a, which are practically rapid convergence for each N_a. The memory requirement is $O(N_a)$ by the FMBEM, much more efficient than the standard BEM.

7.4 Building Façades

Form of building façade can affect noise propagating from outdoor into rooms, such as road traffic noise and railway noise. Noise-shielding efficiency of balconies on the building façade has been investigated by using field measurements, experimental models, and numerical calculations [33–38].

Sound field in the balcony is intricate due to diffraction, scattering, and multiple reflection in a semi-closed field. To predict effect of the balcony taking such complex factors into consideration, wave based analyses are effective. This section shows examples of a 2-D analysis using a hybrid method combining the BEM and the mode expansion method and a 3-D analysis using FMBEM.

7.4.1 2-D Analysis Using BEM

High-rise dwellings often have uniform balconies for intermediate floors, that is, the façade of the dwelling has a periodical sectional view in the vertical direction. In such case, a hybrid boundary element and mode expansion method [39] can efficiently analyze the sound field around the façade taking one floor of the dwelling as one period of a periodical structure [37]. Calculations in this method will be carried out for just one period of the strucure, i.e., one floor, by using periodicity. Modeling just one floor reduces the computer memory and computational time that a full model of a high-rise façade would require. This enables examinations of the acoustical performance of various balcony forms saving computational costs.

7.4.1.1 Formulation

Here, the façade of intermediate floors of a high-rise dwelling with uniform balconies is assumed to be an infinitely periodical structure as shown in Fig. 7.33. Taking one floor as one period of the structure, the sound field around the façade is expressed using the mode expansion formulation for the field far from the façade and the boundary element formulation in one period of the structure for the near field.

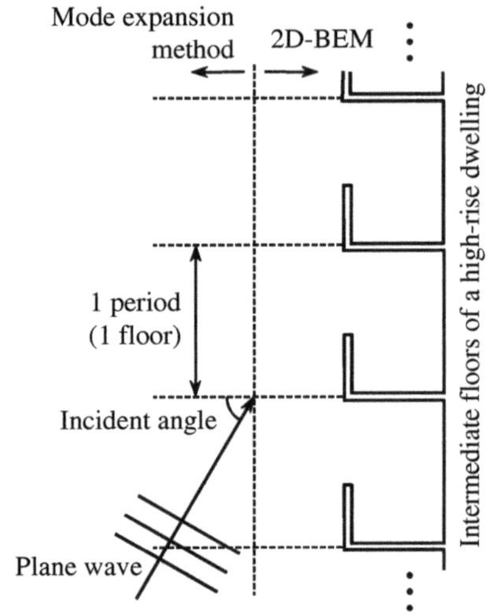

Fig. 7.33 2-D analysis model of a sound field around a high-rise dwelling façade with an infinitely periodical form

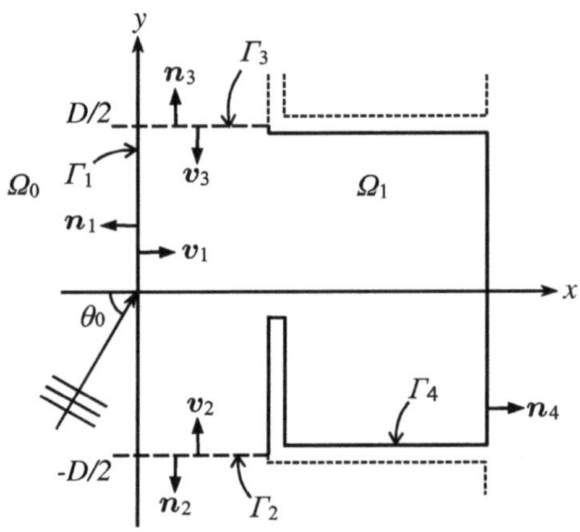

Fig. 7.34 Geometry of the 2-D analysis model for an infinitely periodical structure and difinitions of symbols

Figure 7.34 shows the geometry of the calculation model and the definitions of symbols used in the formulation. Ω_0 and Ω_1 denote the far ($x < 0$) and the near ($x > 0$) field, respectively. It is assumed that a plane wave is incident at an angle θ_0 upon the structure with periodical length of D. Boundaries Γ_1 to Γ_4 define the sound field in one period. Γ_1 on the y-axis ($-D/2 \leq y \leq D/2$) divides the field into the far (Ω_0) and the near fields (Ω_1). Γ_2 and Γ_3 on $y = \pm D/2$ ($x \geq 0$) are period boundaries, and Γ_4 represents a structure surface. Γ_i is divided into N_i constant elements, that is the sound pressure $p_{i,m}$ and particle velocity $v_{i,m}$ are constant on mth element ($i = 1$–4, $m = 1$–N_i). Positive directions of velocities ($\boldsymbol{v}_i$) as well as normal vectors ($\boldsymbol{n}_i$) are defined by the arrow shown in Fig. 7.34.

In the far field (Ω_0), assuming the incident wave to be a sine wave with the angular frequency of ω and the unit amplitude, the velocity potential is expressed as the sum of an the incident wave and scattered waves,

$$\varphi_0(x, y) = e^{j(\alpha_0 x+\beta_0 y)} + \sum_{s=-\infty}^{\infty} R_s e^{j(-\alpha_s x+\beta_s y)}, \tag{7.21}$$

where a time factor $e^{-j\omega t}$ is omitted. R_s is the complex reflection factor of the sth scattered wave and α_s and β_s are the directional componet of its wave number letting the wave number of incident wave be k,

$$\alpha_0 = k\cos\theta_0, \qquad \beta_0 = k\sin\theta_0, \tag{7.22}$$

$$\beta_s = k\sin\theta_0 + \frac{2\pi s}{D} \qquad (s = -\infty, \cdots, \infty), \tag{7.23}$$

$$\alpha_s = \begin{cases} \sqrt{k^2 - \beta_s^2} & (k^2 \geq \beta_s^2) \\ -\sqrt{\beta_s^2 - k^2} & (k^2 < \beta_s^2) \end{cases}. \tag{7.24}$$

Equations (7.23) and (7.24) imply that the directions of the scattered waves are discrete and depend on the periodicity of the structure. Equation (7.23) means a condition where the sth scattered waves from neighboring periods have the same phase in the far field. If the scattered waves have different phases, they would cancel each other and would not propagate in the far field.

The pressure and velocity on Γ_1 are derived from Eq. (7.21),

$$p_1(y) = -j\omega\rho\varphi_0(x, y)|_{x=0} = -j\omega\rho\left(e^{j\beta_0 y} + \sum_{s=-\infty}^{\infty} R_s e^{j\beta_s y}\right), \tag{7.25}$$

$$v_1(y) = -\left.\frac{\partial\varphi_0(x, y)}{\partial x}\right|_{x=0} = -j\alpha_0 e^{j\beta_0 y} + \sum_{s=-\infty}^{\infty} j\alpha_s R_s e^{j\beta_s y}, \tag{7.26}$$

where ρ denotes the air density. Integrating both sides of Eq. (7.26) on Γ_1 after multiplying them by $e^{-j\beta_\tau y}$ ($\tau = -\infty, \ldots, \infty$) gives

$$\int_{-D/2}^{D/2} v_1(y)e^{-j\beta_\tau y}dy = \begin{cases} -j\alpha_0 D(1 - R_0) & (\tau = 0) \\ j\alpha_\tau R_\tau D & (\tau \neq 0) \end{cases}. \tag{7.27}$$

The above equation is transformed to the following expression letting $\tau \to s$:

$$R_s = \mu_s + \frac{1}{j\alpha_s}\sum_{n=1}^{N_1} v_{1,n} w_{s,n}, \tag{7.28}$$

$$\mu_s = \begin{cases} 1 & (s = 0) \\ 0 & (s \neq 0) \end{cases}.$$

where $w_{s,n}$ is an integral on $\Gamma_{1,n}$ ($a_{n-1} \leq y \leq a_n$),

$$w_{s,n} = \frac{1}{D} \int_{a_{n-1}}^{a_n} e^{-j\beta_s y} dy. \tag{7.29}$$

It should be noted that $v_{1,n}$ expresses the velocity assumed to be constant on $\Gamma_{1,n}$. Substituting Eq. (7.28) into Eq. (7.25), the following linear equation related to the pressure on Γ_1 is obtained:

$$p_{1,m} = -2j\omega\rho e^{j\beta_0 y} - \sum_{n=1}^{N_1} v_{1,n} \sum_{s=-\infty}^{\infty} \frac{\omega\rho}{\alpha_s} w_{s,n} e^{j\beta_s y_m}, \tag{7.30}$$

where y_m is the middle point of $\Gamma_{1,m}$.

In the near field (Ω_1), applying the boundary element formulation, the sound pressure at $\boldsymbol{r}$ is expressed with an integral equation:

$$\begin{aligned} \varepsilon p(\boldsymbol{r}) = &- \sum_{i=1}^{3} \sum_{m=1}^{N_i} \left\{ p_{i,m} \int_{\Gamma_{i,m}} \frac{\partial}{\partial n} G(\boldsymbol{r}, \boldsymbol{r}') d\Gamma + j\omega\rho v_{i,m} \int_{\Gamma_{i,m}} G(\boldsymbol{r}, \boldsymbol{r}') d\Gamma \right\} \\ &- \sum_{m=1}^{N_4} p_{4,m} \int_{\Gamma_{4,m}} \left\{ \frac{\partial}{\partial n} G(\boldsymbol{r}, \boldsymbol{r}') - \frac{j\omega\rho}{Z_{4,m}} G(\boldsymbol{r}, \boldsymbol{r}') \right\} d\Gamma, \end{aligned} \tag{7.31}$$

where $Z_{4,m}$ denotes the surface impedance of $\Gamma_{4,m}$. $\varepsilon = 1$ when $\boldsymbol{r}$ lies in Ω_1 other than on $\Gamma_{i,m}$ and $\varepsilon = 1/2$ when $\boldsymbol{r}$ is on $\Gamma_{i,m}$. $\boldsymbol{r}'$ is the point on $\Gamma_{i,m}$ and $G(\boldsymbol{r}, \boldsymbol{r}')$ is the Green's function,

$$G(\boldsymbol{r}, \boldsymbol{r}') = \frac{j}{4} H_0^{(1)}(k|\boldsymbol{r}' - \boldsymbol{r}|), \tag{7.32}$$

where $H_0^{(1)}$ is the Hankel function of the first kind of order zero.

In the case where the boundary elements $\Gamma_{2,m}$ and $\Gamma_{3,m}$ are symmetric about the x-axis, the periodicity D of the structure gives the following equations:

$$p_{3,m} = p_{2,m} e^{j\beta_0 D}, \tag{7.33}$$

$$v_{3,m} = -v_{2,m} e^{j\beta_0 D}. \tag{7.34}$$

It should be noted that the positive direction of $v_{3,m}$ is opposite that of $v_{2,m}$.

A set of linear equations for unknown $p_{i,m}$ and $v_{i,m}$, created by setting $\boldsymbol{r}$ to the middle points of the boundary elements $\Gamma_{i,m}$ ($i = 1$–4, $m = 1$–N_i) in Eq. (7.31) and those of Eq. (7.30) for $m = 1$–N_1, is solved for the periodicity shown as Eqs. (7.33) and (7.34). After solving the linear equations, substituting the solutions into Eq. (7.31) enables to calculate the sound pressure at any $\boldsymbol{r}$ in the near field around the balcony.

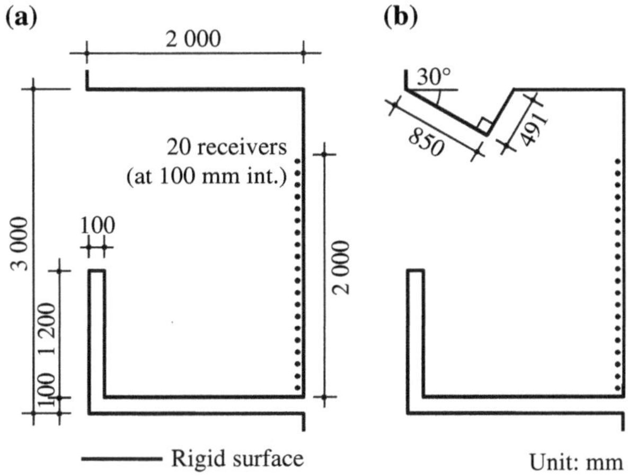

Fig. 7.35 Cross-sectional views of balconies calculated in 2-D analysis. **a** Flat ceiling. **b** Inclined ceiling

7.4.1.2 Analysis Model

It is assumed that balconies of the same form are installed on the façade of a dwelling for each floor and that the façade has periodicity equal to a floor height. Figure 7.35 shows cross-sectional views of calculated balconies based on a floor height of 3 m and a depth of 2 m. The one has a flat ceiling (Fig. 7.35a), and the other has partially inclined ceiling. We examine here the acoustical performance of the inclined ceiling.

Assuming that the height of the window opening is 2 m, 20 receivers are positioned at 10 mm from the window surface within a height range 50–1950 mm from the balcony floor. The calculations are carried out for pure tones at 1/75 octave intervals and a set of 25 results included in 1/3 octave band is incoherently summed to the approximate 1/3 octave-band sound pressure level (SPL). The mean SPL at the window surface is defined as an incoherent average of the reults for all of the receivers.

Balcony surfaces are assumed to be rigid with $Z_4 = \infty$. The order of scattered waves in Eq. (7.30) is limited as $-80 \le s \le 80$.

7.4.1.3 Numerical Results

Figure 7.36 shows the spectra for the reduction in the mean SPL at the window surface for the balcony with the inclined ceiling compared to the balcony with the flat ceiling, at incident angles of noise of 30°, 45°, 60°. The spectra illustrate the noise-shielding effect of the inclined ceiling. The results reveal that the effect of the inclined ceiling at a larger incident angle is greater than that at a smaller incidnet angle. In the results at an incident angle of 60°, the effect increases as the frequency increases. It is considered that the diffraction effect of the balcony parapet result in these trends as

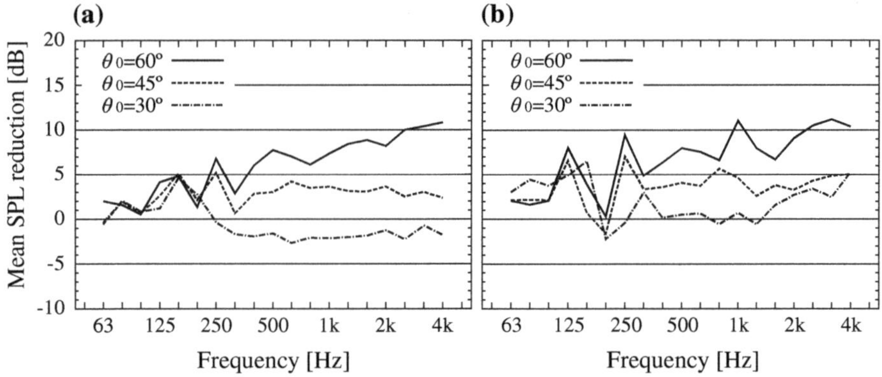

Fig. 7.36 Spectra of noise-shielding effect of an inclined balcony ceiling compared with a flat ceiling. **a** Calculated results. **b** Experimental results

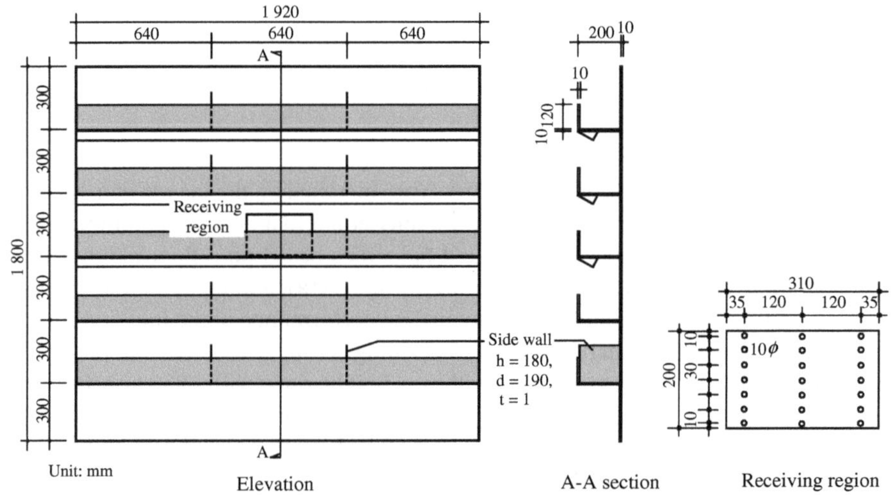

Fig. 7.37 A 1/10-scale experimental model of a high-rise dwelling façade consisting of 6 intermediate floors × 3 dwelling unit

most of the noise directly incident on the ceiling surface are reflected outside and the main component of noise incident on the window are waves diffracted at the parapet.

Figure 7.36 also shows result of a 3-D scale-model experiment [37]. A one-tenth-scale acrylic façade model, shown in Fig. 7.37, consists of six floors × three dwelling unit that are assumed to be the intermediate floors of a high-rise dwelling. A location corresponding to the window surface at the center dwelling unit was designated as the receiving region. SPLs were measured at 21 positions in the receiving region and were incoherently averaged as the mean SPL at the window surface. A point source was located at the front of the model. In spite of differences between the

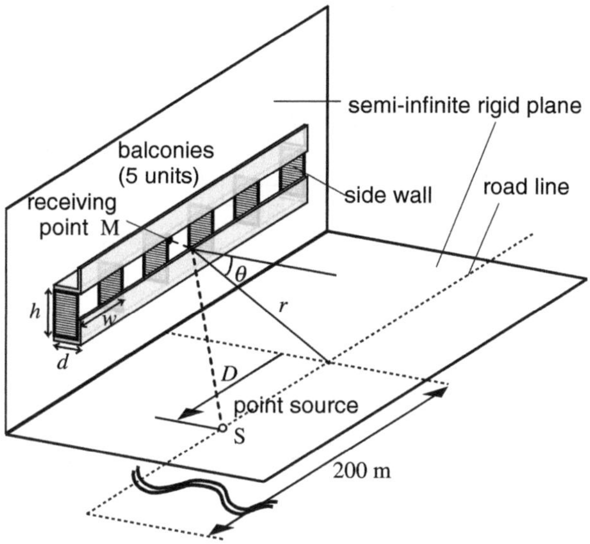

Fig. 7.38 3-D analysis model for building façade with balconies

2-D numerical analysis and the 3-D experiment, calculated and experimental results correspond approximately to each other in terms of the effect of the inclined ceiling.

7.4.2 3-D Analysis Using FMBEM

A 2-D sound field analysis in front of a façade with balconies using its cross section model is presented in Sect. 7.4.1. Here we present a 3-D analysis [36] using the FMBEM (see Sect. 4.3) for plane-symmetric problems [40, 41]. This enables one to obtain the time history of the sound energy propagated from moving sources such as vehicles along a street.

7.4.2.1 Analysis Model

Figure 7.38 shows an analysis model, in which a building façade with balconies and a straight road are assumed in a 1/4 free space with two semi-infinite rigid planes, a ground and wall of the building. Five balcony units are settled on the wall of the building. A receiving point, M, is located on the wall inside the center balcony unit, 1.5 m height from the floor of the balcony and at the center of the balcony in the horizontal direction. A point source is located on the ground and its position is changed along the road at intervals of 0.5 m, to obtain a "unit pattern", which is the time history of sound energy from a single moving source.

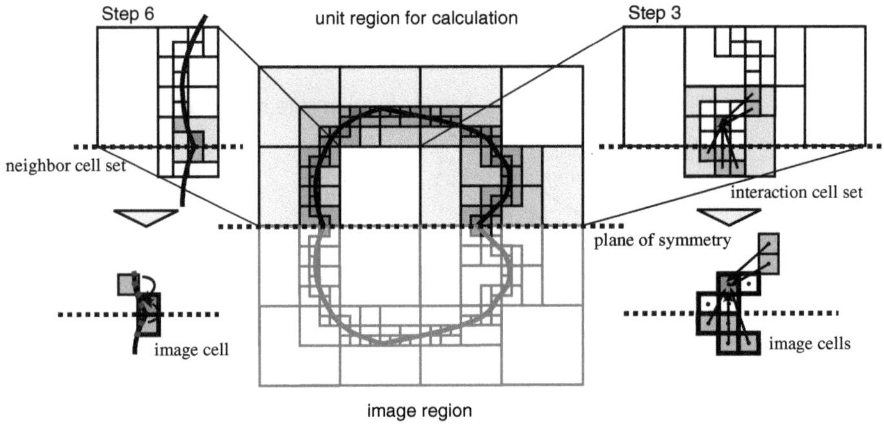

Fig. 7.39 Plane-symmetric boundary and hierarchical cell structure of FMBEM

7.4.2.2 Numerical Techniques

FMBEM for Plane-symmetric Problems

Since the analysis field is plane-symmetric with respect to three planes (including the two infinite rigid ones) that are orthogonal to one another, one can efficiently calculate this field considering only $1/8$ free space. An efficient technique in the FMBEM for such plane-symmetric problems presented here is based on a symmetrical relation among multipole expansion coefficients (LF-FMBEM) [41] or far-field signature functions (HF-FMBEM) [40], and it reduces both the operation count and required memory to about $1/2^{n_{\mathrm{sym}}}$ of those using the standard LF/HF-FMBEM, where n_{sym} is the number of planes of symmetry.

Figure 7.39 shows a plane-symmetric boundary and hierarchical cell structure of the FMBEM. The hierarchical cell structure is arranged symmetrically corresponding to the symmetry of the boundary. Here we define a unit part of the symmetrical shape as a unit region for calculation (the upper half region in Fig. 7.39), and the other parts as image regions. Calculation only in the unit region is sufficient in Steps 1 and 2 in the FMBEM (see Sect. 4.3.3.2), where coefficients/functions for elements or cells are translated to those of their parent cells, and in Steps 4 and 5, where coefficients/functions for cells are translated to those of their child cells or elements. In Steps 3 and 6, contributions from cells in the image regions have to be counted, only when a cell in the unit region is located near the plane of symmetry, as shown in Fig. 7.39.

Reciprocity Theorem

The reciprocity theorem allows the exchange of positions of the point source and receiving point; therefore, many problems to be individually solved, each of which has a point source on the road line and the receiving point M, can be replaced with one problem with a point source at M and many receiving points on the road line.

Fig. 7.40 Chest used for scale-model experiment ($1 \times 0.45 \times 1.35\,\mathrm{m}^3$)

This enables one to obtain an entire unit pattern at an analysis frequency through only one calculation.

7.4.2.3 Numerical Results

Figure 7.41 shows the effect of absorption parts on a unit pattern for 500-Hz 1 octave band at the receiving point M, obtained by (a) the scale-model experiment and (b) calculation using the HF-FMBEM [40] with the Burton–Miller formulation [42]. The values for 1 octave bands are obtained by summation of sound energies calculated at $1/9$ octave band center frequencies. A wooden chest of drawers shown in Fig. 7.40 was used as a $1/16$ scale model of the building. The experimental and numerical results show the similar tendency. The effect of the ceiling absorption is great against the sound energy from near sources (where D is small). The side wall absorption is less effective against near sources than the ceiling absorption, whereas it is greatly effective against far sources (about $D > 5$).

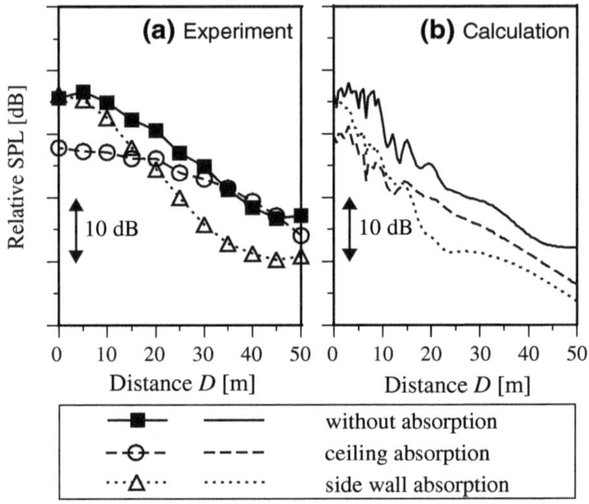

Fig. 7.41 Comparison between experimental and numerical unit patterns at receiving point M (500-Hz 1 octave band, $d = 1.6$, $h = 2.9$, $w = 5$, $r = 10.8$ [m], $\theta = 42$ [deg]): **a** experiment, **b** calculation

7.5 Building Windows

7.5.1 Flanking Propagation Between Apertures

Sound transmission between adjoining rooms is often influenced by flanking sound passing through open windows in the shared exterior wall of the rooms. It is thought that such flanking sound might be more influential than sound transmitted through the partition wall, especially in the summer when the windows are frequently open. Some experimental studies have been made with respect to the problems of how sound emitted from the open window of the source room is transmitted to the open window of the other room [43, 44]. In this paragraph, an example for predicting flanking sound obtained from 3-D analysis using boundary integral equations is presented.

Let a source room Ω_1, which includes an omnidirectional point sound source emitting a sinusoidal wave with wavenumber k, be adjacent to a receiving room Ω_2 through the partition wall Γ_t, as shown in Fig. 7.42. Each room has an aperture, A_1 or A_2, in its exterior wall, Γ_1 or Γ_2. The Mflanking sound' means here the sound which is emitted from the aperture A_1 propagated in the exterior space Ω_o and entering through the aperture A_2 into the receiving room. For simplicity's sake, in this analysis, the source room and the receiving room are assumed to be semi-infinite. Though the two rooms are illustrated as quarter-infinite spaces in Fig. 7.42, this assumption means that the surfaces of the partition wall between them are both perfectly absorbent. This assumption also presumes a one-directional sound incidence on the aperture A_1

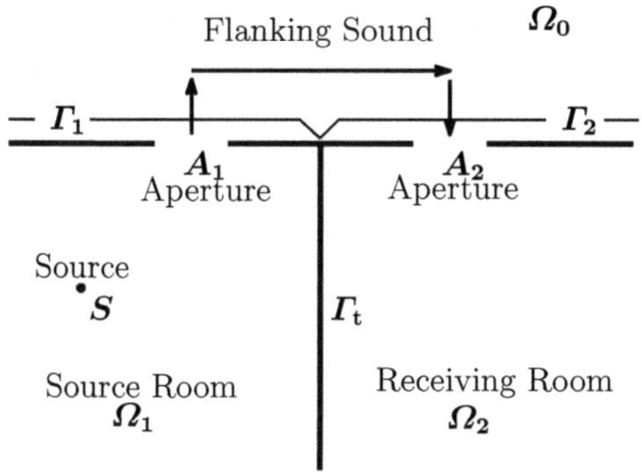

Fig. 7.42 Flanking sound passing through open windows in the shared exterior wall of the rooms

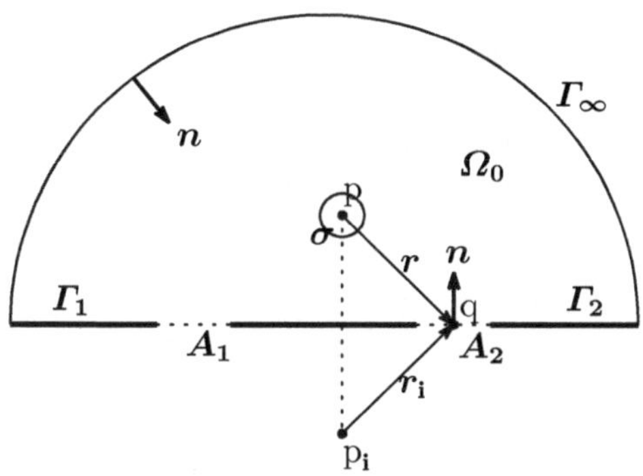

Fig. 7.43 Derivation of the boundary integral equation for the semi-infinite region Ω_0: Ω_0 is bounded by infinite sphere Γ_∞ of center P, plane boundaries Γ_1, Γ_2 and apertures A_1, A_2; $\boldsymbol{n}$ denotes inward normal; P and P_i denote a receiving point and the image point of P with respect to the plane Γ_1, Γ_2, A_1, A_2, respectively; Q is a point inside Ω_0 or on Γ_1, Γ_2, A_1, A_2; σ denotes small sphere of center P

and no reradiation from the aperture A_2. With this assumption, the flanking sound, which is derived from the energy emitted from aperture A_1 and that entering through the aperture A_2, can be precisely estimated. Here, thc thickncss of the exterior wall is assumed to be infinitely thin.

Let us consider an infinitely large rigid flat surface, in two parts of which are apertures A_1 and A_2. Also, let a semi-infinite space Ω_0 bounded by the infinitely large semi-sphere Γ_∞, the rigid surfaces Γ_1 and Γ_2, and the apertures A_1 and A_2 include a receiving point P, and let P_i be the image point of P with respect to the surfaces Γ_1, Γ_2, A_1 and A_2 (see Fig. 7.43).

In order to derive an integral formula with respect to the space Ω_0, we will use

$$G(\boldsymbol{r}_P, \boldsymbol{r}_Q) = \frac{e^{jkr}}{4\pi r} + \frac{e^{jkr_i}}{4\pi r_i}, \tag{7.35}$$

as a fundamental solution, in which the image point P_i is taken into account (see Eq. (4.73)). Here, $r = |\boldsymbol{r}| = |\boldsymbol{r}_Q - \boldsymbol{r}_P|$, $r_i = |\boldsymbol{r}_i| = |\boldsymbol{r}_Q - \boldsymbol{r}_{Pi}|$. We apply Green's theorem to the space $\Omega_0 - \sigma$, where σ is a small sphere of center P with radius ϵ. Taking into consideration (1) the normal component of particle velocity vanishes throughout Γ_1 and Γ_2, (2) $\frac{\partial G}{\partial n} = 0$ throughout Γ_1, Γ_2, A_1 and A_2, and (3) Sommerfeld's radiation condition [45], we can obtain

$$\boldsymbol{\Phi}(\boldsymbol{r}_P) = -\frac{1}{2\pi} \int_{A_1+A_2} \frac{\partial \boldsymbol{\Phi}(\boldsymbol{r}_Q)}{\partial n_q} \frac{e^{jkr}}{r} d\Gamma, \quad (P \in \Omega_0, A_1, A_2, \Gamma_1, \Gamma_2), \tag{7.36}$$

where $\boldsymbol{\Phi}(\boldsymbol{r}_P)$ denotes velocity potential at P and $\boldsymbol{n}$ the inward drawn normal. Equation (7.36) is valid when P is located on Γ_1, Γ_2, A_1 or A_2 (i.e., $P = P_i$), since $\lim_{\epsilon \to 0} \int_{\partial\sigma} d\Gamma = -\boldsymbol{\Phi}(\boldsymbol{r}_P)$ $(P \in A_1, A_2, \Gamma_1, \Gamma_2)$ [46].

As for the space Ω_1, considering the point source is located within it and Eq. (7.35) being used as the fundamental solution, we have

$$\boldsymbol{\Phi}(\boldsymbol{r}_P) = \boldsymbol{\Phi}_D(\boldsymbol{r}_P) + \boldsymbol{\Phi}_D(\boldsymbol{r}_{Pi}) + \frac{1}{2\pi} \int_{A_1} \frac{\partial \boldsymbol{\Phi}(\boldsymbol{r}_Q)}{\partial n_q} \frac{e^{jkr}}{r} d\Gamma, \quad (P \in \Omega_1, A_1, \Gamma_1), \tag{7.37}$$

where $\boldsymbol{\Phi}_D$ denotes the direct sound and $\boldsymbol{n}$ the outward drawn normal. The point P_i also denotes the image point of $P \in \Omega_1$ with respect to Γ_1 and A_1.

Considering that no source is located, we can also obtain an integral formula for the space Ω_2 in the same way: i. e.,

$$\boldsymbol{\Phi}(\boldsymbol{r}_P) = \frac{1}{2\pi} \int_{A_2} \frac{\partial \boldsymbol{\Phi}(\boldsymbol{r}_Q)}{\partial n_q} \frac{e^{jkr}}{r} d\Gamma, \quad (P \in \Omega_2, A_2, \Gamma_2), \tag{7.38}$$

When P is located on A_1 or A_2, then Eqs. (7.36), (7.37), and (7.38) are boundary integral equations with an unknown function $\frac{\partial \boldsymbol{\Phi}}{\partial n}$ on A_1 and A_2. When P is located on A_1, subtracting the difference between Eqs. (7.36) and (7.37) yields

$$\frac{1}{\pi} \int_{A_1} \frac{\partial \boldsymbol{\Phi}(\boldsymbol{r}_Q)}{\partial n_q} \frac{e^{jkr}}{r} d\Gamma + \frac{1}{2\pi} \int_{A_2} \frac{\partial \boldsymbol{\Phi}(\boldsymbol{r}_Q)}{\partial n_q} \frac{e^{jkr}}{r} d\Gamma = -2\boldsymbol{\Phi}_D(\boldsymbol{r}_P), \quad (P \in A_1). \tag{7.39}$$

Also, when P is located on A_2, subtracting the difference between Eqs. (7.36) and (7.38) yields

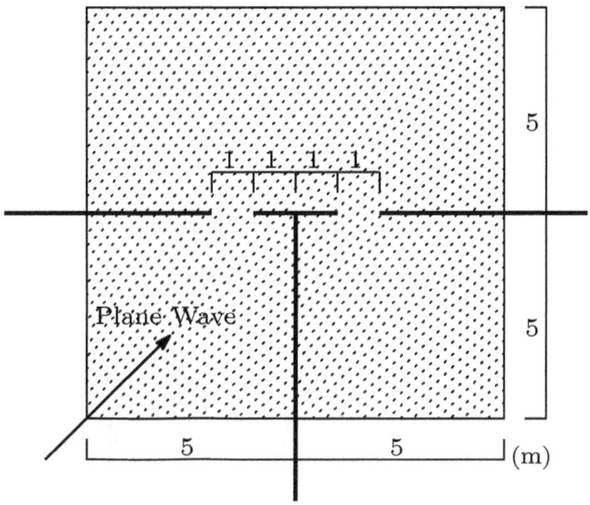

Fig. 7.44 A numerical example for a franking sound propagation

$$\frac{1}{2\pi}\int_{A_1}\frac{\partial\boldsymbol{\Phi}(\boldsymbol{r}_Q)}{\partial n_q}\frac{e^{jkr}}{r}d\Gamma+\frac{1}{\pi}\int_{A_2}\frac{\partial\boldsymbol{\Phi}(\boldsymbol{r}_Q)}{\partial n_q}\frac{e^{jkr}}{r}d\Gamma=0,\quad (\mathrm{P}\in A_2). \tag{7.40}$$

Solving the simultaneous integral equations (7.39) and (7.40), we can obtain $\frac{\partial\boldsymbol{\Phi}}{\partial n}$ on A_1 and A_2, which gives the velocity potential $\boldsymbol{\Phi}$ in Ω_0, Ω_1 and Ω_2 by substituting it into Eqs. (7.36), (7.37) and (7.38), respectively.

Under the conditions shown in Figs. 7.44 and 7.45 shows the results of calculation of the sound pressure distribution near the apertures of the receiving room and the source room. Figure 7.44 shows a cross section at the center of the apertures. The sound pressure distribution of the Fig. 7.45 shows a state that the sound emitted from the aperture A_1 is transmitted through the external space Ω_0 and enters the receiving room.

The energy I_1 emitted from the aperture A_1 and the energy I_2 entering through the aperture A_2 can be obtained by the velocity potential $\boldsymbol{\Phi}$ and $\frac{\partial\boldsymbol{\Phi}}{\partial n}$ on A_1 and A_2 using the following equations:

$$p=-i\omega\rho\boldsymbol{\Phi},\qquad v=-\frac{\partial\boldsymbol{\Phi}}{\partial n_q}, \tag{7.41}$$

and

$$I_1=\int_{A_1}\frac{1}{4}(pv^*+p^*v)d\Gamma,\qquad I_2=\int_{A_2}\frac{1}{4}(pv^*+p^*v)d\Gamma, \tag{7.42}$$

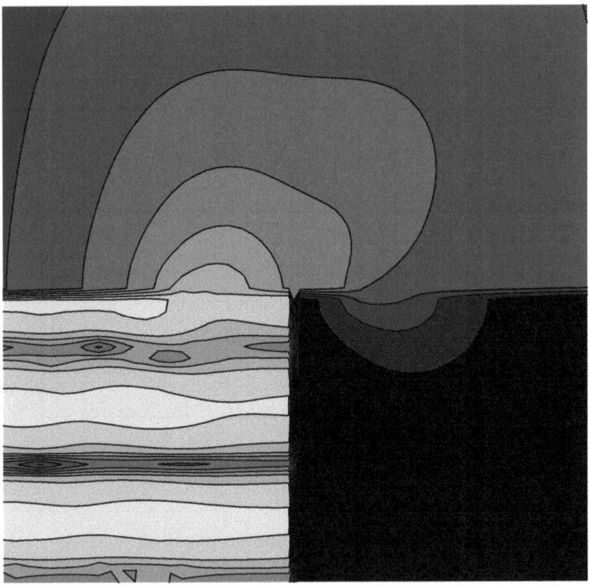

Fig. 7.45 The contour representation of sound pressure amplitude [dB], 125 Hz

where p^* and v^* are complex conjugates of sound pressure P and particle velocity v, respectively [47]. The transmission rate (in decibels) of the flanking sound is obtained by $10 \log_{10} \frac{I_2}{I_1}$.

For purposes of noise control, it is useful to know the transmission rate of flanking sound under the condition of 1 octave band random incidence noise. In order to simulate the above condition, numerical calculations are carried out for 825 plane waves that are incident on the aperture A_1 from all directions at regular solid angle intervals and for six frequencies taken in the octave band. The energies of both the emitted wave from A_1 and of the incoming one are calculated for each condition and summed up separately. The transmission rate under conditions of random incidence can thus be obtained using the resultant values.

Figure 7.46 shows a chart of the distance attenuation of flanking sound calculated using the above-mentioned method when the dimension of both apertures is $0.5 \times 0.5\,\text{m}^2$. The abscissa in this chart denotes the distance between the midpoints of the apertures. We can see that the transmission rate reduces by approximately 6 dB for every doubling of distance.

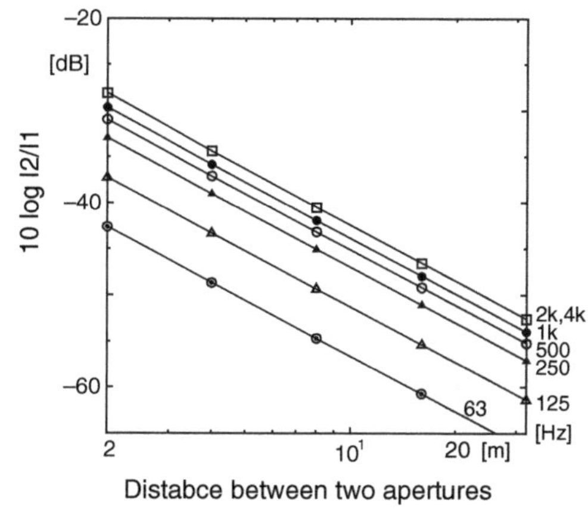

Fig. 7.46 Attenuation characteristics of the franking sound with the distance between the midpoints of the apertures

7.5.2 Propagation Through Facing Apertures

One sometimes encounters the sound propagation problem in the summer season such that the sound emitted from the open window of one of the room placed in apartment and condominium complexes enters the room placed in the opposite building through the open window. The sound propagation between rooms that are located on both sides of the middle corridor of such a museum or a hospital ward is also a similar problem. Though this problem can be considered as a double diffraction, it must also take into account multiple reflections at the parallel walls, including the respective aperture. In the following, a method of combining the boundary integral equations applied to appropriately divided regions is proposed. The effective image method also introduced which restricts unknown functions to be solved to particle velocity over the apertures and reduces the computation time.

Let us consider a sound source room Ω_1 and a receiving room Ω_2 each aperture is facing as shown in Fig. 7.47. For simplicity, both walls Γ_1 and Γ_2 are assumed to spread infinitely, rigid and thinner than the wavelength. In addition, it is assumed that the diffuse sound field is satisfied in the sound source room and that the plane wave is incident uniformly from all directions to the sound source room aperture. Also, the receiving room and the sound source room are supposed to be semi-infinite. Though the sound source S is displayed as a point sound source in Fig. 7.47, it is assumed to be far enough and is treated as a plane wave incidence in the analysis.

We apply Green's formula to the region Ω between the parallel walls where inward normal is taken. Further, as a fundamental solution, we use Green's function considering the mirror images $\mathrm{P}_m (m = 1, \ldots, \infty)$ of the receiving point P between parallel walls as shown in Fig. 7.48: i.e.,

$$G'(\boldsymbol{r}_\mathrm{P}, \boldsymbol{r}_\mathrm{Q}) = \frac{\mathrm{e}^{jkr}}{4\pi r} + \sum_{m-1}^{\infty} \frac{\mathrm{e}^{jkr_m}}{4\pi r_m}, \tag{7.43}$$

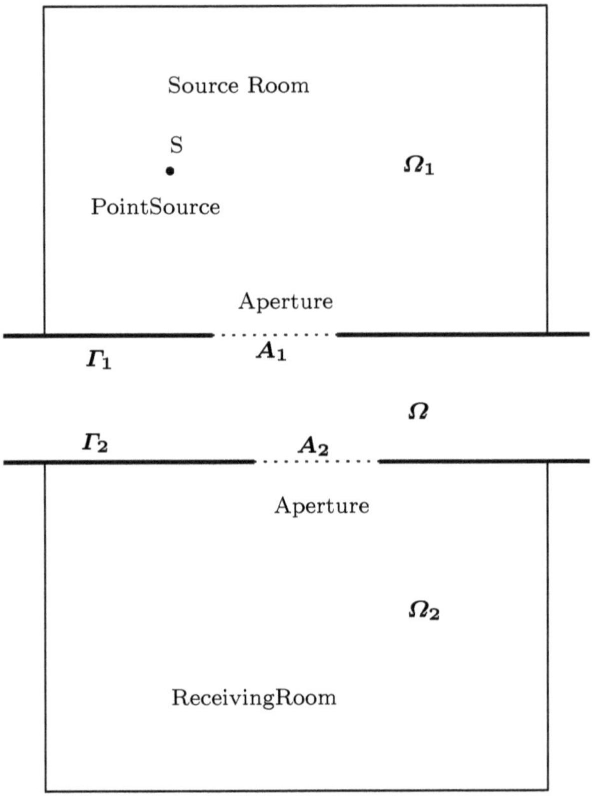

Fig. 7.47 Sound propagation between the apertures taken on the wall surface facing

where, $r = |\boldsymbol{r}| = |\boldsymbol{r}_Q - \boldsymbol{r}_P|$, $r_m = |\boldsymbol{r}_m| = |\boldsymbol{r}_Q - \boldsymbol{r}_{Pm}|(m = 1, \ldots, \infty)$. Influence of the potential $\boldsymbol{\Phi}$ on the boundary Γ_1 and Γ_2 being canceled by the summation of the integration for the mirror image points, one can obtain the following equation in which the unknown function to obtain is only the particle velocity $\dfrac{\partial \boldsymbol{\Phi}}{\partial n_q}$ over the apertures:

$$\boldsymbol{\Phi}(\boldsymbol{r}_P) = -\int_{A_1} \frac{\partial \boldsymbol{\Phi}(\boldsymbol{r}_Q)}{\partial n_q} G'(\boldsymbol{r}_P, \boldsymbol{r}_Q) d\Gamma - \int_{A_2} \frac{\partial \boldsymbol{\Phi}(\boldsymbol{r}_Q)}{\partial n_q} G'(\boldsymbol{r}_P, \boldsymbol{r}_Q) d\Gamma,$$
$$(\mathrm{P} \in \Omega, A_1, A_2). \qquad (7.44)$$

Equation (7.44) holds in the region Ω and the boundary A_1, A_2 [48].

On the other hand, taking into account that the sound source exists and outward normal is taken as shown in Fig. 7.43, we have the following boundary integral equation with respect to the semi-infinite region Ω_1 in consideration of the mirror image $\mathrm{P_i}$ of the point P to the boundary A_1, Γ_1:

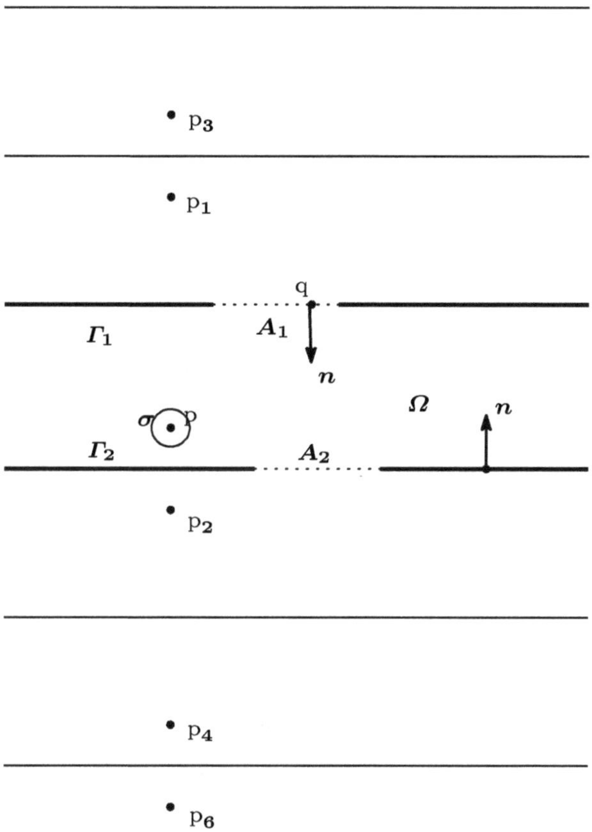

Fig. 7.48 Mirror images of the receiving point P between parallel walls

$$\boldsymbol{\Phi}(\boldsymbol{r}_{\mathrm{P}}) = \boldsymbol{\Phi}_{\mathrm{D}}(\boldsymbol{r}_{\mathrm{P}}) + \boldsymbol{\Phi}_{\mathrm{D}}(\boldsymbol{r}_{\mathrm{Pi}}) + \int_{A_1} \frac{\partial \boldsymbol{\Phi}(\boldsymbol{r}_{\mathrm{Q}})}{\partial n_{\mathrm{q}}} G(\boldsymbol{r}_{\mathrm{P}}, \boldsymbol{r}_{\mathrm{Q}}) d\Gamma, \quad (\mathrm{P} \in \Omega_1, A_1), \tag{7.45}$$

where $\boldsymbol{\Phi}_{\mathrm{D}}(\boldsymbol{r}_{\mathrm{P}})$ denotes the direct wave at P and Green's function Eq. (7.35) is used in the region Ω_1.

For the region Ω_2, taking into account that no sound source exists, we can obtain the following boundary integral equation in the similar way:

$$\boldsymbol{\Phi}(\boldsymbol{r}_{\mathrm{P}}) = \int_{A_2} \frac{\partial \boldsymbol{\Phi}(\boldsymbol{r}_{\mathrm{Q}})}{\partial n_{\mathrm{q}}} G(\boldsymbol{r}_{\mathrm{P}}, \boldsymbol{r}_{\mathrm{Q}}) d\Gamma, \quad (\mathrm{P} \in \Omega_2, A_2). \tag{7.46}$$

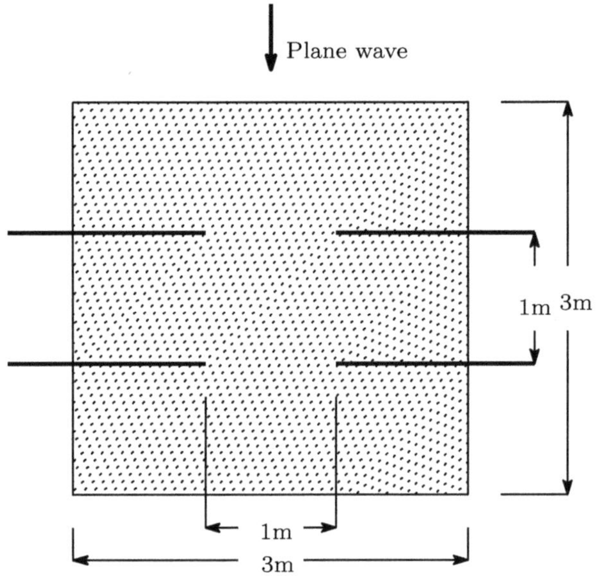

Fig. 7.49 A numerical example

When P $\in A_1$, the difference between Eqs. (7.45) and (7.44) yields

$$\int_{A_1} \frac{\partial \boldsymbol{\Phi}}{\partial n_{\mathrm{q}}}\{G(\boldsymbol{r}_{\mathrm{P}}, \boldsymbol{r}_{\mathrm{Q}}) + G'(\boldsymbol{r}_{\mathrm{P}}, \boldsymbol{r}_{\mathrm{Q}})\}d\Gamma + \int_{A_2} \frac{\partial \boldsymbol{\Phi}}{\partial n_{\mathrm{q}}} G'(\boldsymbol{r}_{\mathrm{P}}, \boldsymbol{r}_{\mathrm{Q}})d\Gamma = 2\boldsymbol{\Phi}_{\mathrm{D}}(\boldsymbol{r}_{\mathrm{P}}),$$

$$(\mathrm{P} \in A_1). \qquad (7.47)$$

Also, when P $\in A_2$, the difference between Eqs. (7.46) and (7.44) yields

$$\int_{A_1} \frac{\partial \boldsymbol{\Phi}}{\partial n_{\mathrm{q}}} G'(\boldsymbol{r}_{\mathrm{P}}, \boldsymbol{r}_{\mathrm{Q}})d\Gamma + \int_{A_2} \frac{\partial \boldsymbol{\Phi}}{\partial n_{\mathrm{q}}}\{G(\boldsymbol{r}_{\mathrm{P}}, \boldsymbol{r}_{\mathrm{Q}}) + G'(\boldsymbol{r}_{\mathrm{P}}, \boldsymbol{r}_{\mathrm{Q}})\}d\Gamma = 0,$$

$$(\mathrm{P} \in A_2). \qquad (7.48)$$

By solving boundary integral equations (7.47) and (7.48) simultaneously, we can obtain $\frac{\partial \boldsymbol{\Phi}}{\partial n_{\mathrm{q}}}$ over the boundaries A_1, A_2. The velocity potential in each region can be obtained from $\frac{\partial \boldsymbol{\Phi}}{\partial n}$ solved above using Eqs. (7.44), (7.45), and (7.46).

The incident energy I_{i} and the transmitted energy I_{t} is obtained as in the case of Eq. (7.42) [49]. The ratio $\frac{I_{\mathrm{t}}}{I_{\mathrm{i}}}$ yields the transmission rate.

As a numerical example of the method proposed here, the distribution of the sound pressure amplitude in the shaded region shown in Fig. 7.49 is calculated. In this

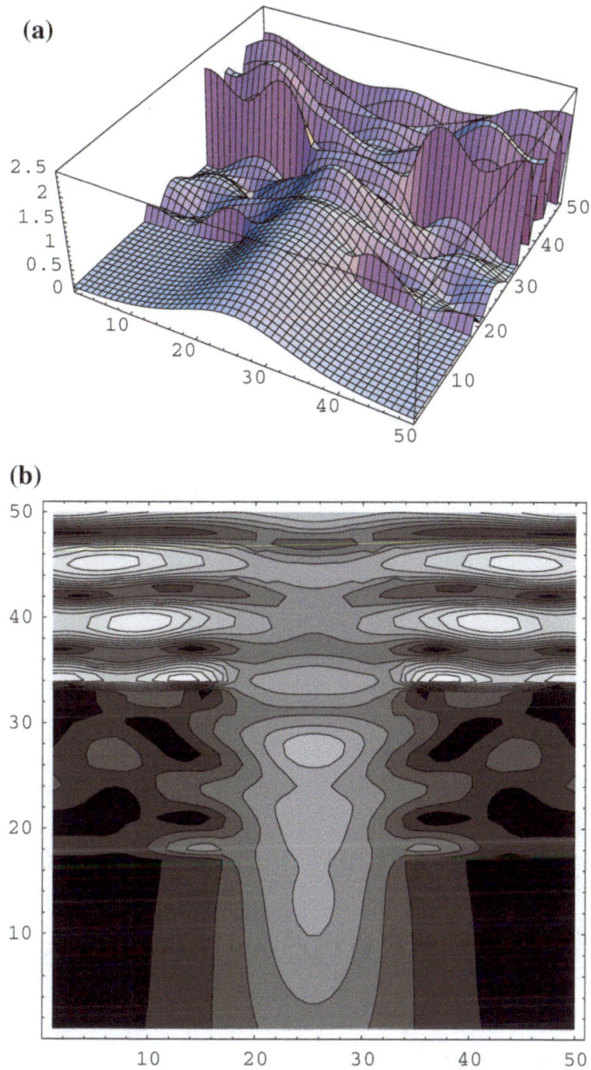

Fig. 7.50 The sound pressure amplitude in the shaded region shown in Fig. 7.49: **a** bird's eye view, **b** contour representation; the number of images 20; 500 Hz

calculation, the distance between two parallel walls is taken 1 m and the dimensions of two apertures $1 \times 1\,\text{m}^2$ where the central axis of both is the same. Figure 7.50 shows the result at frequency of 500 Hz for plane wave incidence. One finds that the state in which the sound wave propagates two apertures are captured.

Figure 7.51a, b shows the result of transmission losses (TL) in 1 octave band noise for rondom incidence at 500 and 1 kHz in which they are plotted at the aperture center-to-center distance. Results of the calculations for a number of conditions, including a shift of the axis of the two apertures, it has been found that TL is determined

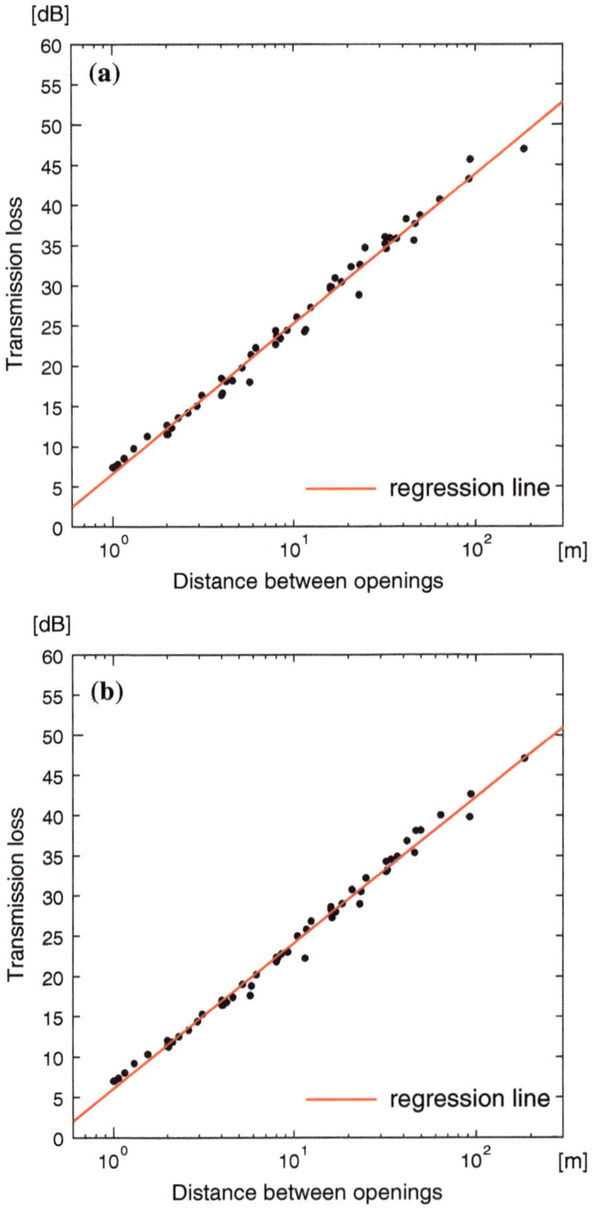

Fig. 7.51 The transmission loss (TL) in 1 octave band noise for rondom incidence: **a** 500 Hz, **b** 1 kHz; dimensions of two apertures are $1 \times 1\,m^2$

approximately by the distance between the centers of the aperture. Also, it appears regression lines by the least squares method to fit calculated values for various conditions [50].

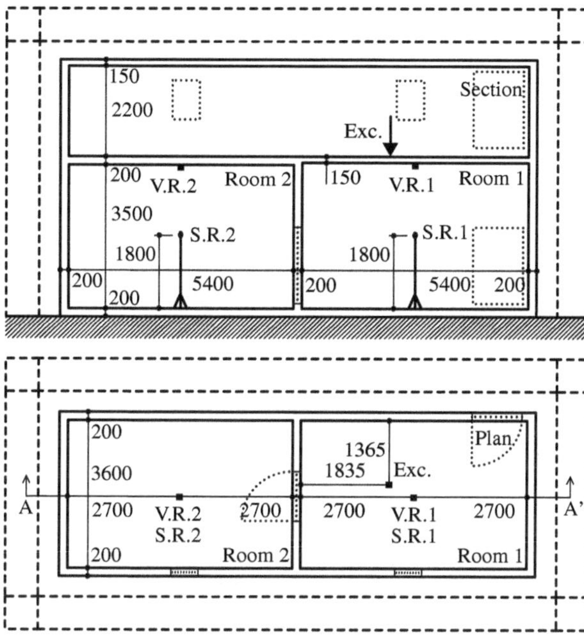

Fig. 7.52 Section and plan of the two-story concrete building

7.6 Floor Impact Sound

Sound insulation performance is one of the important factors that dominate building quality. Among various sound insulation problems, heavyweight floor impact sound caused by soft and heavy impacts such as a child's jumping is recognized as a serious problem because it is strongly affected by a building structure and the insulation performance is difficult to be improved after construction. Therefore precise prediction for heavyweight floor impact sound at the planning phase is necessary. The impedance method [51] has been widely used as a practical prediction method, while the finite element method and the finite-difference time-domain method are sometimes employed for more accurate prediction. In numerical calculations of floor impact sound, wave propagation between building components should be taken into account because the excited components vibrate other walls, floors, and ceilings, which cause sound to radiate. Vibration-mode transformations can occur in these propagations, thus it seems that considering only the flexural wave is insufficient for these analyses. In this section, the predicted results for heavyweight floor impact sound with the vibroacoustic finite-difference time-domain method as shown in Sects. 2.3.1 and 2.3.2 are introduced and discussed [52, 53].

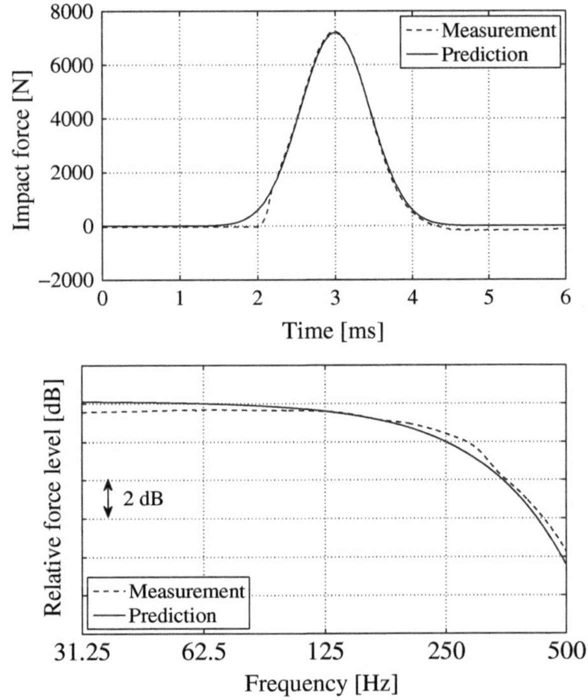

Fig. 7.53 Impact forces for the prediction and the measurement

7.6.1 Solid Modeling

7.6.1.1 Configurations

Figure 7.52 shows the cross section and the plan of the two-story concrete building to be calculated. In the figure, areas enclosed with broken lines are assumed to be Perfectly Matched Layers (PMLs) and the ground surface is assumed to be a fixed boundary. In the actual building, there are three doors and two windows, which are drawn with dotted lines in the figure. In addition, a few plywood boards are supported by wooden gratings on the walls in the lower story. However, for simplicity, these doors, windows, plywood boards, and wooden gratings are neglected, and all walls and floors are assumed to be flat concrete (density 2,400 kg/m^3, Young's modulus 2.4×10^{10} N/m^2, Poisson's ratio 0.2) in the calculations. All rooms as well as outdoors are assumed to be filled with air (density 1.2 kg/m^3, bulk modulus 1.4×10^5 N/m^2). The maximum target frequency in the calculations is set to 500 Hz. Allowing for this, spatial intervals $\Delta x = \Delta y = \Delta z$ are set to 0.05 m. Friction, second viscosity, and shear viscosity coefficients are respectively set as 5 Ns/m^4, 0 Ns/m^2, 1.8×10^{-5} Ns/m^2 for air and 25,000 Ns/m^4, 1.658×10^5 Ns/m^2, 2.487×10^5 Ns/m^2 for concrete. All initial values of velocities and stresses are set to zero. Considering the stability conditions, Δt is determined to 0.001465 ms and the calculations were carried out until 3 s.

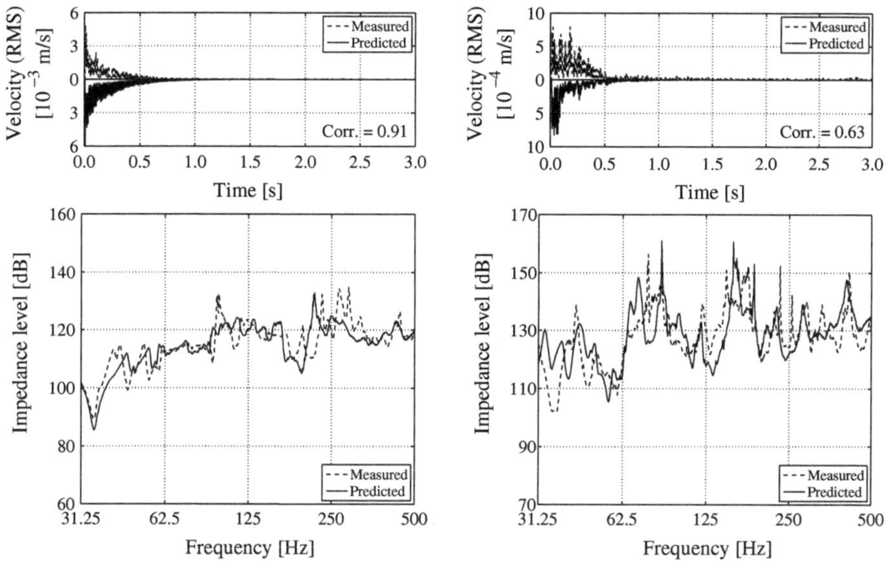

Fig. 7.54 Echo diagrams and impedance levels at vibration receiver points V. R. 1 (*left*) and V. R. 2 (*right*)

In the measurements, acceleration pickups and microphones were located at two vibration receivers (V. R. 1 and 2) and two sound receivers (S. R. 1 and 2). Velocities in the normal direction were calculated by integrating measured acceleration values and sound pressures were measured at a sampling frequency of 44,100 Hz. The excitation point (Exc.) was hit by an impulse hummer and the force history between the floor and the hummer was measured. The force history used in the calculations is determined so that the waveforms are similar to the measured data as shown in Fig. 7.53.

7.6.1.2 Results

Figures 7.54 and 7.55 show echo diagrams at all four receivers, the driving-point impedance levels at the vibration receiver (V. R. 1), the transfer impedance levels at the vibration receiver (V. R. 2), and the transfer functions by a unit excitation at the sound receivers (S. R. 1 and 2). The echo diagrams are obtained by passing the signals through a low-pass filter with the cutoff frequency of 500 Hz and a numerical RMS detector with a time constant of 1 ms. In these figures, the maximum value of the cross-correlation-coefficient vector between the echo diagrams of the predicted and measured data calculated in a three second range are also shown. The correlation value at V. R. 2 is lower than that at V. R. 1 and some disagreements between predicted and measured impedance at V. R. 2 can also be observed. As for transfer function, relatively significant disagreements both at S. R. 1 and 2 can be seen at low frequencies. The disagreement between these figures should be due to the following:

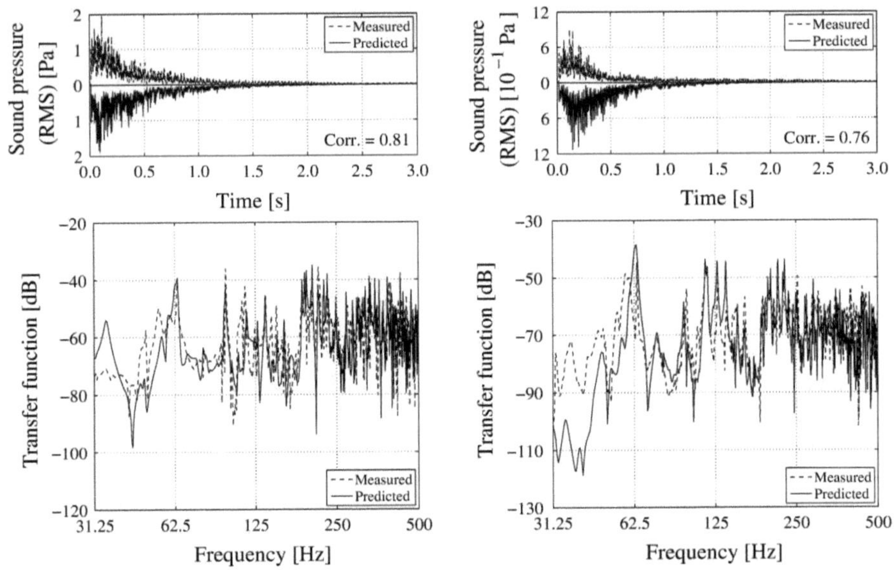

Fig. 7.55 Echo diagrams and transfer functions at sound receiver points S. R. 1 (*left*) and S. R. 2 (*right*)

1. Accuracy of shape modeling
2. Using common material constants
3. Accuracy of damping modeling
4. Numerical dispersion of the finite difference method

Items 1 and 2 can be overcome by modeling the object more accurately and using measured material constants. For example, in this case, the door between rooms in the lower story is neglected, which is why the velocity at vibration receiver 2 could not be predicted precisely. Item 3 can be possibly improved by adding more degrees of freedom to frictional and viscous terms. That however will require more parameters which are difficult to be measured and make the physical meanings of the terms more complicated. As for item 4, the prediction accuracy can be typically improved by discretizing with smaller spatial intervals and employing a higher order difference scheme. Although some dicrepancies can be seen as discussed above, the correlation values of echo diagrams at V. R. 1, S. R. 1 and 2 are about 0.8 and therefore it can be said that the energy decay in the time domain is predicted well. Especially at V. R. 1, predicted results of driving impedance are in good agreement with measured ones where the first natural frequency and the absolute values can be predicted precisely.

In the FDTD method, responses of stress and velocity in time domain at all reference points must be calculated. This means visualization and auralization are easy by saving the step-by-step results in a target volume, on a surfaces, or a point. Figure 7.56 shows the visualized fields of displacement and sound pressure until about 15 ms at a section AA' shown in Fig. 7.52. Displacements are obtained by

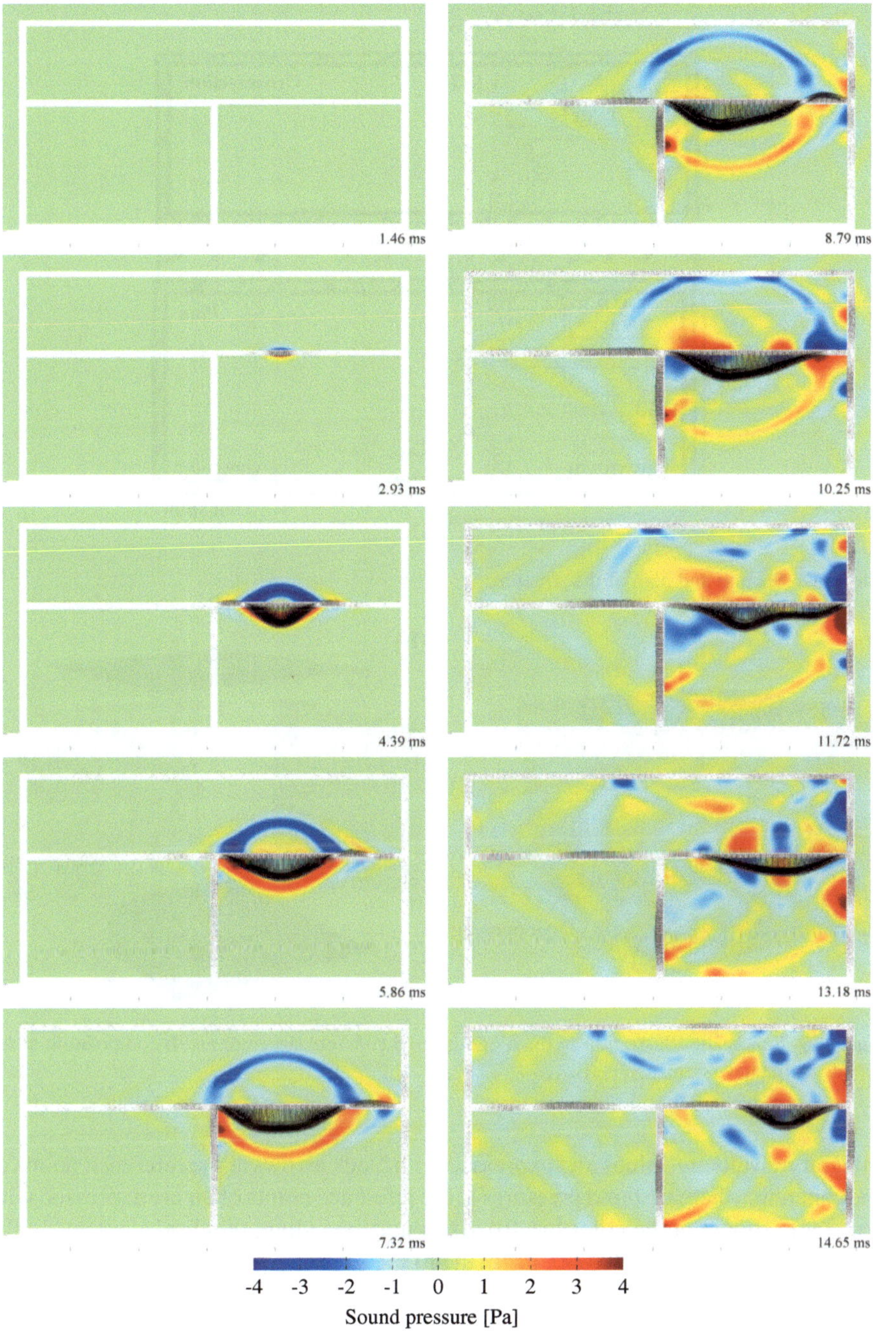

Fig. 7.56 Distributions of displacements and sound pressures at section AA' until 15 ms

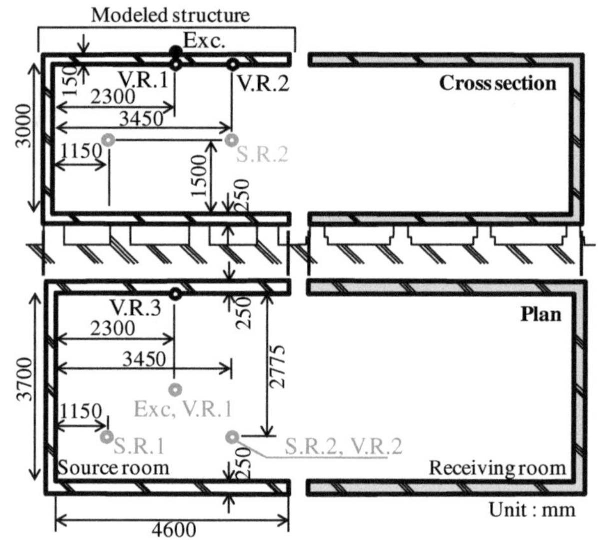

Fig. 7.57 Section and plan of the structure

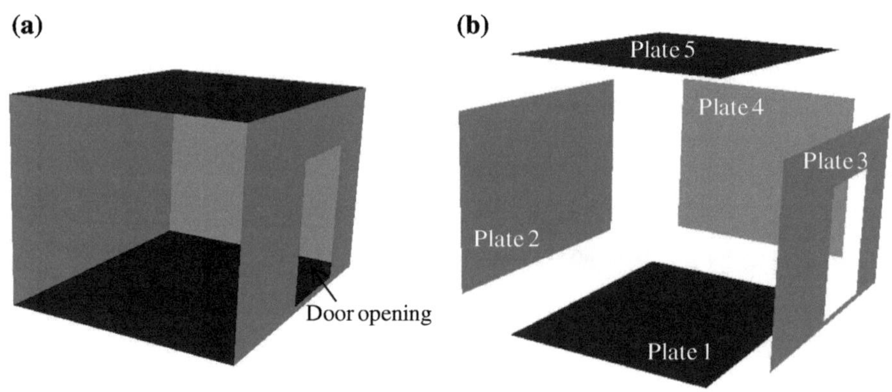

Fig. 7.58 Schematic illustration of the **a** one-layered concrete structure and **b** composition of the 5 plate elements

integrating velocity values and expressed with black arrows at the reference points of the concrete, whereas sound pressures at the reference points of air are expressed with the colors shown in the color bar. To aid in understanding, the displacements in the figure are multiplied by 40,000. From the figures, curved-surface sound radiation by fast longitudinal-wave propagation can be seen on the left-hand side in the upper room and in the lower left room, and then, spherical sound radiation by strong and slow shear-wave propagation can be seen on the right-hand side in the upper room and in the lower right room. For auralization, downsampling process is necessary in the case where the time interval of calculation is very small as shown here. Low-pass-filtering

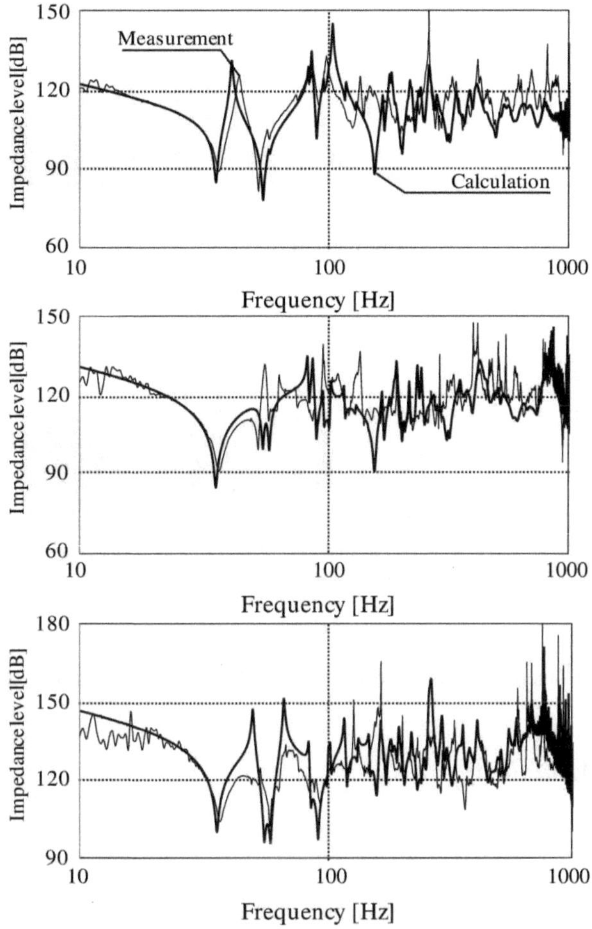

Fig. 7.59 Impedance levels at vibration receiver points V. R. 1 (*upper*), V. R. 2 (*middle*, and V. R. 3 (*lower*)

process would be sometimes desirable because reliability of the calculated data is low below the maximum target frequency. Although auralization from frequency domain data needs a careful treatment in causality, auralization from time domain data does not need such a careful treatment when the data seems to be stable because the results will diverge if there is a problem in causality. Such visualization and auralization would give instinctive understandings of complicated wave propagation and sound radiation phenomena and also make sound insulation planning more effective and convincing.

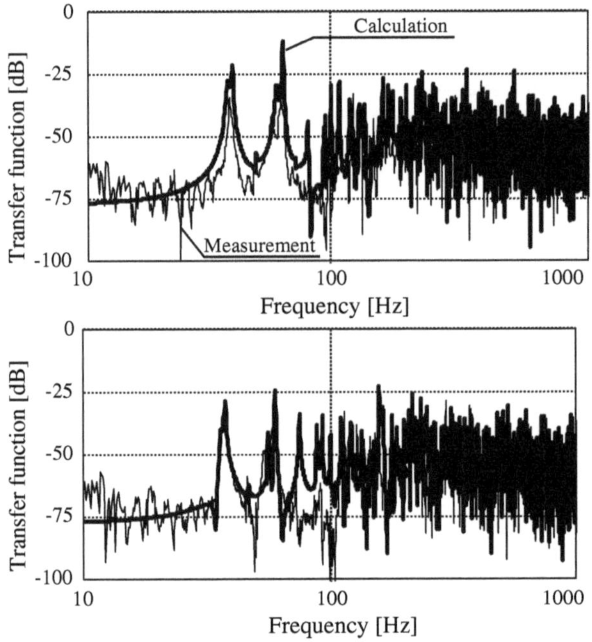

Fig. 7.60 Transfer functions at sound receiver points S. R. 1 (*upper*) and S. R. 2 (*lower*)

7.6.2 Plate Modeling

7.6.2.1 Configuration

Figure 7.57 shows the cross section and plan of a one-story concrete structure. In the measurement, the driving point of Exc. on the ceiling slab is hit by the impulse hammer, and the vibration velocities at V. R. 1, V. R. 2, and V. R. 3, and the sound pressures at S. R. 1 and S. R. 2 are measured. In this study, only the left-side structure shown as "modeled structure" in the figure is aimed.

Figure 7.58 shows the modeling scheme. A box-type concrete structure, which is composed of five concrete slabs as shown in Fig. 7.58a, is simulated as a composition of five plate elements like Fig. 7.58b. As for the basic equation, the bending equation based on the Mindlin's thick plate theory was applied on behalf of Eq. (2.106) shown in Sect. 2.3.2. Excitation by an impulse hammer is modeled in the FDTD by giving the time-transient force with a Gaussian profile in time domain by following the same procedure as described in the previous section. The following physical properties are given for simulating the concrete slab: Young's modulus of 2.4×10^{10} N/m^2, density of 2,500 kg/m^3 and Poisson's ratio of 0.2. In addition, the spatial intervals of $\Delta x = \Delta y = 25$ mm, and the time sampling of 48 kHz are adopted.

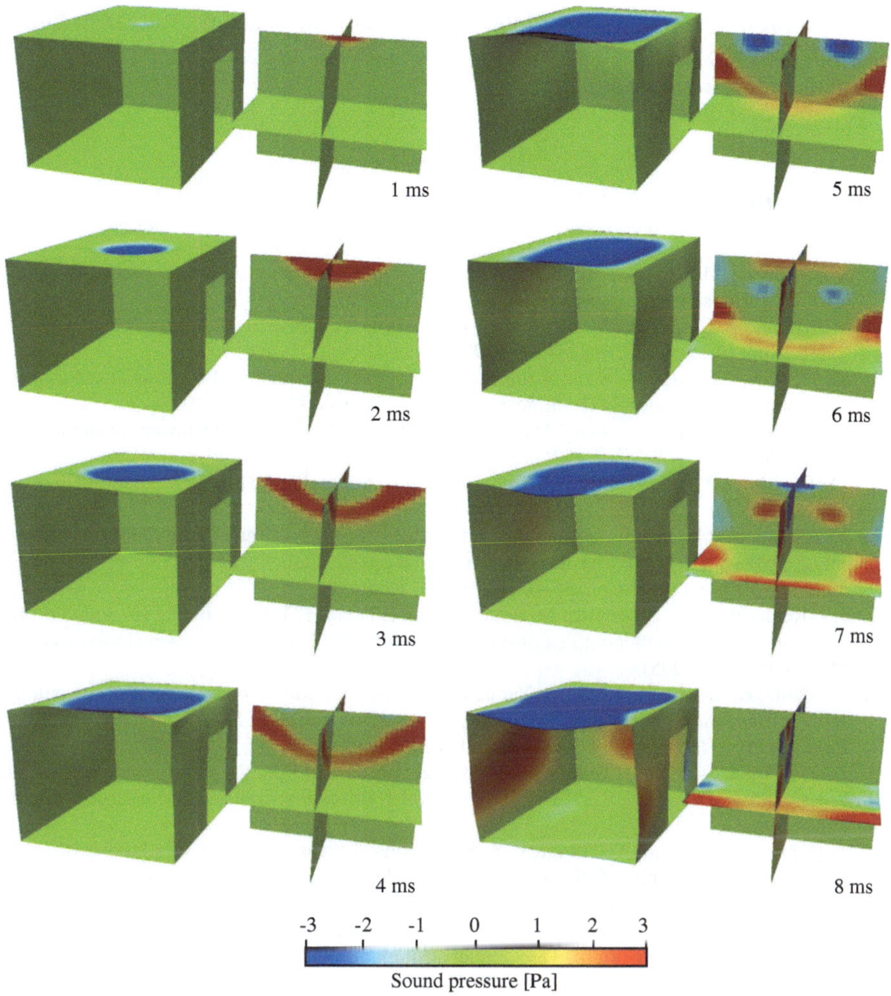

Fig. 7.61 Deformation of the structure and sound pressure distribution inside

7.6.2.2 Results

Figure 7.59 shows the driving-point and transfer impedance levels at each vibration receiver obtained from the calculation and measurement. The results show good agreement, especially from the viewpoint of the frequency characteristics of the peaks and dips. Figure 7.60 shows the transfer functions at each sound receiver. In this result, the calculated results also accurately describe the peak-dip characteristics in the lower frequency range, but the correspondence in the middle and high frequency ranges is difficult to evaluate. Figure 7.61 visualizes the time-transient characteristics of the out-of-plane deformation of the structure, and the sound pressure distribution inside. It provides easy understanding representation of the excited vibration and the sound radiation from the structure.

References

1. K.E. Gilbert, M.J. White, Application of the parabolic equation to sound propagation in a refracting atmosphere. J. Acoust. Soc. Am. **85**, 630–637 (1989)
2. M.J. White, K.E. Gilbert, Application of the parabolic equation to the outdoor propagation of sound. Appl. Acoust. **27**(3), 227–238 (1989)
3. M. West, K. Gilbert, R.A. Sack, A tutorial on the parabolic equation (PE) model used for long range sound propagation in the atmosphere. Appl. Acoust. **37**, 31–49 (1992)
4. E.M. Salomons, *Computational Atmospheric Acoustics*, Chapter 4.5. (Kluwer Academic Publishers, 2001)
5. J.F. Claerbout, *Fundamentals of Geophysical Data Processing*. (Blackwell Scientific Publications, 1985)
6. T. Yokota, K. Makino, K. Yamamoto, Finite difference time domain simulation of outdoor sound propagation under the influences of wind speed gradient, in *Proceedings of INCE/J Autumn Meeting*, Niigata, pp. 59–62, (2010) (in Japanese)
7. T.V. Renterghem, E. Salomons, D. Botteldooren, Efficient FDTD-PE model for sound propagation in situations with complex obstacles and wind profiles. Acta Acustica United Acustica **91**(4), 671–679 (2005)
8. T. Yokota, K. Makino, Y. Hirao, K. Yamamoto, Numerical simulation on long-distance outdoor sound propagation over complex-shaped boundaries by coupling FDTD and PE methods, in *Proceedings of Inter-Noise 2006*, Honolulu, 632 (2006)
9. Zenrin Co., Ltd, Zmap-TOWNII. http://www.zenrin.co.jp/product/gis/zmap/zmaptown.html
10. Kokusai Kogyo Co., Ltd, RAMS-e Surface Level 1. http://www.ramse3d.com/laser/lib/index.html
11. T. Oshima, M. Imano, Y. Hiraguri, Y. Kamoshida, Linearized Euler simulations of sound propagation with wind effects over a reconstructed urban terrain using digital geographic information. Appl. Acoust. **74**(12), 1354–1366 (2013)
12. H.G. Weller, G. Tabor, H. Jasak, C. Fureby, A tensorial approach to computational continuum mechanics using object-oriented techniques. Comput. Phys. **12**(6), 620–631 (1998)
13. M. Imano, A. Nishimura, K. Ogino, Y. Sakamoto, Assessment of wind environment in an open-air space surrounded by large-scale low-rise buildinds and a large roof, in *Proceedings of 23rd CFD Symposium*, Miyagi, G4-3 (2009) (in Japanese)
14. S.V. Patankar, D.B. Spalding, A calculation procedure for heat, mass and momentum transfer in three-dimensional parabolic flows. Int. J. Heat Mass Transf. **15**(10), 1787–1806 (1972)
15. V. Yakhot, S.A. Orszag, Renormalization group analysis of turbulence. I. Basic theory. J. Sci. Comput. **1**(1), 3–51 (1986)
16. M. Imano, H. Hoshino, Y. Akamine, M. Kamata, A. Nishimura, Y. Sakamoto. Validation of CFD analysis of cross-ventilation in detached house, in *Proceedings of 11th International Conference on Air Distribution in Rooms (ROOMVENT)*, (2009). (In CD-ROM)
17. T. Oshima, M. Imano, Unified automatic mesh generation of urban terrain for linearized euler simulations using digital geographic information, in *Proceedings of Inter-Noise 2011*, Osaka, 429662 (2011) (In CD-ROM)
18. Q. Qui, T.L. Geers, Evaluation of the perfectly matched layer for computational acoustics. J. Comput. Phys. **139**, 166–183 (1998)
19. Z. Maekawa, Noise reduction by screens. Appl. Acoust. **1**, 157–173 (1968)
20. T. Isei, T.F.W. Embleton, J.E. Piercy, Noise reduction by barriers on finite impedance ground. J. Acoust. Soc. Am. **67**, 46–58 (1980)

21. T. Okubo, T. Matsumoto, K. Yamamoto, O. Funahashi, K. Nakasaki, Efficiency of edge-modified noise barriers: Intrinsic efficiency determination of practical products and prediction of the diffracted sound field. Acoust. Sci. Tech. **31**(1), 56–67 (2010)
22. D. Duhamel, Efficient calculation of the three-dimensional sound pressure field around a noise barrier. J. Sound Vib. **197**, 547–571 (1996)
23. T. Okubo, K. Yamamoto, O. Funahashi, M. Yamamoto, Barriers against road traffic noise: relationship between L_{Aeq} reduction at roadside and L_pA reduction determined in normal-incidence alignment, in *Proceedings of Inter-Noise 2008*, Shanghai, (2008)
24. H. Nakajima, H. Tsuru, S. Ogata, Analysis and error investigation using fourier transform of sound field produced by moving source. J. Acoust. Soc. Jpn **60**, 717–726 (2004). (in Japanese)
25. K. Yamamoto, Road traffic noise prediction model "ASJ RTN-Model 2008": Report of the Research Committee on Road Traffic Noise. Acoust. Sci. Tech. **31**(1), 2–55 (2010)
26. T. Ishizuka, T. Okubo, Inaccuracy due to local-reaction model in numerical analyses of noise-shielding efficiency of barriers with absorbing edge. Acoust. Sci. Tech. **30**, 372–375 (2009)
27. A. Fukushima, K. Fujiwara, Numerical study on the insertion loss of noise barrier for coherent/incoherent line source. J. Acoust. Soc. Jpn. **58**, 10–19 (2002). (in Japanese)
28. S. Sakamoto, J. Hirano, F. Satoh, H. Tachibana, O. Funahashi, T. Mori, Experiment and calculation of sound radiation characteristics from semi-underground road, in *Proceedings of Inter-Noise 2005*, Rio de Janeiro, 1815 (2005) (In CD-ROM)
29. S. Sakamoto, Calculation of sound propagation in three-dimensional field with constant cross section by duhamel's efficient method using transient solutions obtained by finite-difference time-domain method. Acoust. Sci. Tech. **30**(2), 72–82 (2009)
30. S. Sakamoto, Development of energy-based calculation method of noise radiation from semi-underground road using a numerical analysis. Acoust. Sci. Tech. **31**(1), 75–86 (2010)
31. N. Kamiya, H. Iwase, E. Kita, Parallel implementation of boundary element method with domain decomposition. Eng. Anal. Bound. Elem. **18**, 209–216 (1996)
32. Y. Yasuda, S. Sakamoto, T. Sakuma, Application of the fast multipole boundary element method to sound field analysis using a domain decomposition approach, in *Proceedings of Inter-Noise 2006*, Honolulu, 624 (2006)
33. D. May, Freeway noise and high-rise balconies. J. Acoust. Soc. Am. **65**(3), 699–704 (1979)
34. D.C. Hothersall, K.V. Horoshenkov, S.E. Mercy, Numerical modeling of the sound field near a tall building with balconies near a road. J. Sound Vib. **148**(4), 507–515 (1996)
35. S.K. Tang, Noise screening effects of balconies on a building façade. J. Acoust. Soc. Am. **118**(1), 213–221 (2005)
36. Y. Yasuda, S. Gohara, T. Sakuma, S. Sakamoto, Numerical analysis of outdoor noise reduction due to building balconies, in *Proceedings of 19th International Congress of Acoustics*, Madrid, RBA-03-005 (2007)
37. T. Ishizuka, K. Fujiwara, Traffic noise reduction at balconies on a high-rise building façade. J. Acoust. Soc. Am. **131**(3), 2110–2117 (2012)
38. T. Ishizuka, K. Fujiwara, Full-scale test of reflective noise-shielding devices for balconies on high-rise buildings, J. Acoust. Soc. Am. **134**(2), EL185–EL190 (2013)
39. T. Fujimoto, K. Fujiwara, Analysis of sound absorption by periodic structures using a hybrid-boundary element/mode expantion method. Appl. Acoust. **64**, 525–532 (2003)
40. Y. Yasuda, T. Sakuma, A technique for plane-symmetric sound field analysis in the fast multipole boundary element method. J. Comput. Acoust. **13**(1), 71–85 (2005)
41. Y. Yasuda, K. Higuchi, T. Oshima, T. Sakuma, Efficient technique in low-frequency fast multipole boundary element method for plane-symmetric acoustic problems. Eng. Anal. Bound. Elem. **36**(10), 1493–1501 (2012)
42. A.J. Burton, G.F. Miller, The application of integral equation methods to the numerical solution of some exterior boundary value problems. Roy. Soc. Lond. Ser. A **323**, 201–210 (1971)
43. S. Kimura, H. Kozakai, Influence of flanking transmission through windows on sound insulation between rooms. Arch. Acoust. Noise Control **17**(1), 39–46 (1988). (in Japanese)
44. Y. Tobimatsu, S. Kimura, The effect of flanking transmission for sound insulation efficiency. Trans. Tech. Comm. Archi. Acoust. Acoust. Soc. Jpn., AA97-2 (1997). (in Japanese)

45. A. Sommerfeld. Die Greensche Funktion der Schwingungs-gleichung. J. -Ber. Deutsch Math.-Verein **21**, 309–353 (1912)
46. Y. Kawai, Prediction of noise propagation from a depressed road by using boundary integral equations. J. Acoust. Soc. Jpn. **56**(3), 143–147 (2000). (in Japanese)
47. T. Terai, Y. Furue, Y. Horinouchi, M. Ikeda, Calculation and representation of intensity field, in *Proceedings of AIJ Kinki Chapter Research Meeting (Planning)*, vol. 33, pp. 173–176 (1993) (in Japanese)
48. Y. Kawai, On flanking sound propagation passing through open windows placed in the exterior wall of neighboring rooms. J. Acoust. Soc. Jpn. **57**(2), 139–143 (2001). (in Japanese)
49. P.M. Morse, K.U. Ingard, *Theoretical Acoustics*. (McGraw-Hill, New York, 1968)
50. Y. Kawai, Sound propagation through openings placed in parallel walls. J. Acoust. Soc. Jpn. **63**(6), 143–147 (2007). (in Japanese)
51. S. Kimura, K. Inoue, Practical calculation of floor impact sound by impedance method. Appl. Acoust. **26**, 263–292 (1989)
52. M. Toyoda, D. Takahashi, Prediction for architectural structure-borne sound by the finite-difference time-domain method. Acoust. Sci. Tech. **30**, 265–276 (2009)
53. T. Asakura, T. Ishizuka, T. Miyajima, M. Toyoda, S. Sakamoto, Finite-difference time-domain analysis of structure-borne sound using a plate model. Acoust. Sci. Tech. **34**, 48–51 (2013)

Chapter 8
Acoustic Property Simulation for Building Components

Takumi Asakura, Yasuhito Kawai, Hisaharu Suzuki, Naohisa Inoue, Tetsuya Sakuma, Hirofumi Onitsuka and Takayuki Masumoto

Abstract This chapter shows practical examples of numerical simulation results for acoustical characteristics of building elements, such as the sound absorption, sound-scattering, and sound insulation performance. Additionally, radiation characteristics of speaker systems are also treated. In each section, methodologies and numerical modeling schemes of the simulation, and the calculated results for practical cases

T. Asakura (✉)
Institute of Technology, Shimizu Corporation, 3-4-17 Etchujima, Koto-ku, Tokyo 135-8530, Japan
e-mail: t_asakura@shimz.co.jp

Y. Kawai
Faculty of Environmental and Urban Engineering, Kansai University, 3-3-35 Yamate-cho, Suita-shi, Osaka 564-8680, Japan
e-mail: kawai@kansai-u.ac.jp

H. Suzuki
Evixar Japan, Inc., 1-8-6 Shinkawa, Chuou-ku, Tokyo 104-0033, Japan
e-mail: h.suzuki@evixar.com

N. Inoue · T. Sakuma
Graduate School of Frontier Sciences, The University of Tokyo, 5-1-5 Kashiwanoha, Kashiwa-shi, Chiba 277-8563, Japan
e-mail: 127628a@sbk.k.u-tokyo.ac.jp

T. Sakuma
e-mail: sakuma@k.u-tokyo.ac.jp

H. Onitsuka
Yamaha Corporation Research and Development Division, 203 Matsunokijima, Iwata-shi, Shizuoka 438-0192, Japan
e-mail: hirofumi.onitsuka@music.yamaha.com

T. Masumoto
Cybernet Systems Co., Ltd., Fujisoft Bldg, 3 Kanda-neribeicho, Chiyoda-ku, Tokyo 101-0022, Japan
e-mail: masumoto@cybernet.co.jp

T. Sakuma et al. (eds.), *Computational Simulation in Architectural, and Environmental Acoustics*, DOI: 10.1007/978-4-431-54454-8_8, are illustrated. The results are validated through comparison with the measurement results, and the applicability of the numerical methods is discussed.

Keywords Sound absorption coefficient · Random-incidence diffusion coefficient · Sound reduction index · Poler pattern

8.1 Absorbers

There are various absorbing materials used for noise control and room acoustics design. In the numerical calculation, two methods can be applied in order to take dissipative effects of absorbing materials into account. One is that frequency-dependent surface impedance is given to the boundary. This method is often used with a rough approximation of locally reacting boundary. Sound wave is assumed to go through the material, perpendicularly to the locally reacting surface regardless of incidence angle. The following sections present the effects of the approximation on numerical results. The other is that wave propagation within materials is also considered. For porous materials, wave propagation can be simulated based on the Rayleigh model or the Biot model. The frame of the porous material is assumed to be rigid in the Rayleigh model, whereas its elasticity is incorporated in the Biot model. The third subsection presents a finite-difference time-domain method based on the Rayleigh model, and the last subsection presents a finite element modeling method of layered materials based on the Biot theory.

8.1.1 Assumption of Locally Reacting Condition

Let us consider the meaning of the assumption of locally reacting condition for the case of a semi-infinite sound absorber by using boundary integral equations. Let semi-infinite sound absorber Ω_{a} and air Ω be in contact with boundary Γ as shown in Fig. 8.1. In the following, the speed of sound, effective density, and wave number of the sound absorber are denoted by c_a, ρ_a, and $k_a = \omega/c_a$, respectively. Also, the velocity potential at P in the sound absorber is denoted by $\Psi(\boldsymbol{r}_{\mathrm{P}})$.

Using the fundamental solution,

$$G_a(\boldsymbol{r}_{\mathrm{P}}, \boldsymbol{r}_{\mathrm{Q}}) = \frac{e^{jk_a r}}{4\pi r}, \tag{8.1}$$

if P is located on the boundary Γ, we obtain the bounary integral equation

$$-\int_{\Gamma} \left[\Psi(\boldsymbol{r}_{\mathrm{Q}}) \frac{\partial G_a(\boldsymbol{r}_{\mathrm{P}}, \boldsymbol{r}_{\mathrm{Q}})}{\partial n_{\mathrm{q}}} - \frac{\partial \Psi(\boldsymbol{r}_{\mathrm{Q}})}{\partial n_{\mathrm{q}}} G_a(\boldsymbol{r}_{\mathrm{P}}, \boldsymbol{r}_{\mathrm{Q}}) \right] d\Gamma = \frac{1}{2} \Psi(\boldsymbol{r}_{\mathrm{P}}), \quad (\mathrm{P} \in \Gamma), \tag{8.2}$$

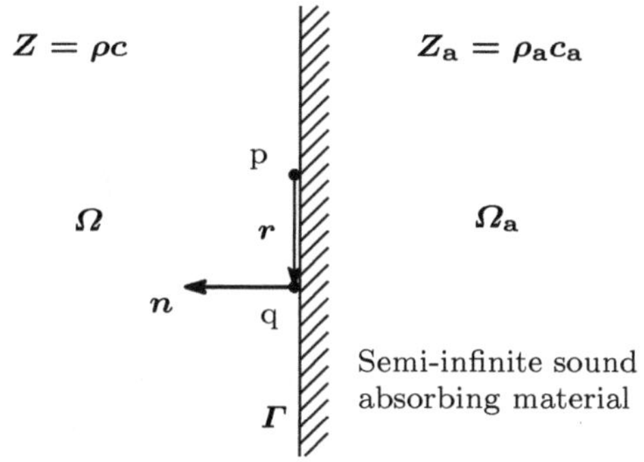

Fig. 8.1 Semi-infinite sound absorber: Ω the air side semi-infinite region, Ω_a sound-absorbing side semi-infinite region, $Z = \rho c$ characteristic impedance of air, $Z_a = \rho_a c_a$ characteristic impedance of sound absorber

where Q is a point on Γ, $r = |\boldsymbol{r}| = |\boldsymbol{r}_{\mathrm{Q}} - \boldsymbol{r}_{\mathrm{P}}|$, $\boldsymbol{r}_{\mathrm{P}}$ and $\boldsymbol{r}_{\mathrm{Q}}$ are position vectors of P and Q, respectively. The time factor $\exp(-j\omega t)$ is omitted here.

Since normal component of particle velocity and sound pressure of both the medium are equal at the boundary surface, i.e.,

$$\Psi = \frac{\rho}{\rho_a}\Phi, \qquad \frac{\partial \Phi}{\partial n} = \frac{\partial \Psi}{\partial n}, \tag{8.3}$$

where Φ denotes velocity potential in air. Therefore, substituting Eq. (8.3) into Eq. (8.2), we have

$$-\int_{\Gamma} \left[\frac{\rho}{\rho_a}\Phi(\boldsymbol{r}_{\mathrm{Q}})\frac{\partial G_a(\boldsymbol{r}_{\mathrm{P}}, \boldsymbol{r}_{\mathrm{Q}})}{\partial n_{\mathrm{q}}} - \frac{\partial \Phi(\boldsymbol{r}_{\mathrm{Q}})}{\partial n_{\mathrm{q}}} G_a(\boldsymbol{r}_{\mathrm{P}}, \boldsymbol{r}_{\mathrm{Q}}) \right] d\Gamma = \frac{\rho}{2\rho_a}\Phi(\boldsymbol{r}_{\mathrm{P}}), \quad (\mathrm{P} \in \Gamma). \tag{8.4}$$

On the other hand, we also have the normal derivative form of Eq. (8.4):

$$-\int_{S} \left[\frac{\rho}{\rho_a}\Phi(\boldsymbol{r}_{\mathrm{Q}})\frac{\partial^2 G_a(\boldsymbol{r}_{\mathrm{P}}, \boldsymbol{r}_{\mathrm{Q}})}{\partial n_{\mathrm{p}}\partial n_{\mathrm{q}}} - \frac{\partial \Phi(\boldsymbol{r}_{\mathrm{Q}})}{\partial n_{\mathrm{q}}} \frac{\partial G_a(\boldsymbol{r}_{\mathrm{P}}, \boldsymbol{r}_{\mathrm{Q}})}{\partial n_{\mathrm{p}}} \right] d\Gamma = \frac{1}{2}\frac{\partial \Phi(\boldsymbol{r}_{\mathrm{P}})}{\partial n_{\mathrm{p}}}, \quad (\mathrm{P} \in \Gamma). \tag{8.5}$$

The boundary integaral equation of the normal derivative type is reffered to as NDF (Normal Derivative Form) and that of the ordinary type as BF(Basic Form) in the following.

Taking into account that the boundary Γ is an infinte plane and the attenuation of the sound waves propagating through sound absorber, we obtain

$$\int_{\Gamma} G_a(\boldsymbol{r}_{\mathrm{P}}, \boldsymbol{r}_{\mathrm{Q}}) d\Gamma = \int_{\Gamma} \frac{e^{jk_a r}}{4\pi r} d\Gamma = \frac{1}{4\pi} \left(\oint_{\partial\Gamma} e^{jk_a r} d\phi - \frac{2\pi}{jk_a} \right) = -\frac{1}{2jk_a}. \tag{8.6}$$

Also, since $\dfrac{\partial G_a}{\partial n} = 0$ from $\cos(\boldsymbol{r}, \boldsymbol{n}) = 0$, we have

$$\frac{\partial \Phi(\boldsymbol{r}_{\mathrm{P}})}{\partial n} = -jk_a \frac{\rho}{\rho_a} \Phi(\boldsymbol{r}_{\mathrm{P}}) + \underbrace{2jk_a \int_{\Gamma} \frac{\partial[\Phi(\boldsymbol{r}_{\mathrm{Q}}) - \Phi(\boldsymbol{r}_{\mathrm{P}})]}{\partial n_{\mathrm{q}}} G_a(\boldsymbol{r}_{\mathrm{P}}, \boldsymbol{r}_{\mathrm{Q}}) d\Gamma}_{I_{\mathrm{B0}}} \quad [\mathbf{BF}]. \tag{8.7}$$

If it is possible to ignore the second term of the right-hand side I_{B0}, we can obtain the relational expression of locally reacting condition:

$$\frac{\partial \Phi(\boldsymbol{r}_{\mathrm{P}})}{\partial n_{\mathrm{p}}} = -jk \frac{\rho c}{Z_a} \Phi(\boldsymbol{r}_{\mathrm{P}}). \tag{8.8}$$

If we use Eq. (8.5), since the hyper singular integral becomes [1]

$$\int_{\Gamma} \frac{\partial^2 G_a(\boldsymbol{r}_{\mathrm{P}}, \boldsymbol{r}_{\mathrm{Q}})}{\partial n_{\mathrm{p}} \partial n_{\mathrm{q}}} d\Gamma = \frac{1}{4\pi} \left(2\pi jk_a - \oint_{\partial\Gamma} \frac{e^{jk_a r}}{r} d\phi \right) = \frac{jk_a}{2}, \tag{8.9}$$

then Eq. (8.5) reduces

$$\frac{\partial \Phi(\boldsymbol{r}_{\mathrm{P}})}{\partial n_{\mathrm{p}}} = -jk_a \frac{\rho}{\rho_a} \Phi(\boldsymbol{r}_{\mathrm{P}}) \underbrace{-2\frac{\rho}{\rho_a} \int_{\Gamma} [\Phi(\boldsymbol{r}_{\mathrm{Q}}) - \Phi(\boldsymbol{r}_{\mathrm{P}})] \frac{\partial^2 G_a(\boldsymbol{r}_{\mathrm{P}}, \boldsymbol{r}_{\mathrm{Q}})}{\partial n_{\mathrm{p}} \partial n_{\mathrm{q}}} d\Gamma}_{I_{\mathrm{N0}}} \quad [\mathbf{NDF}]. \tag{8.10}$$

If the term I_{N0} is negligibly small, we can obtain the relational expression of the locally reacting condition Eq. (8.8).

As a result of some numerical analyses for the case of plane wave incidence, the value of I_{N0} depends on the angle of incidence and become smaller according as the frequency decreases and as the flow resistance increases.

8.1.2 Area Effect

Assumption of diffuse sound field is often used in carrying out the acoustic design. Since sound waves incident from all directions to the peripheral wall under such a condition, the sound absorption coefficient of the material surface by the random incidence conditions is required. A reverberation room is a space for obtaining approximately a difuse sound field and the random-incidence absorption coefficient of various wall components is measured by using it. Sound absorption coefficient obtained there is called reverberation (or reverberation room method) absorbption coefficient, and is widely used conventionally.

The reverberation absorption coefficient of the material used for the wall is calculated by the following equation which was derived using Sabine's formula from the difference between the measured values of the reverberation time with and without the specimen set on the floor of the reverberation room.

$$A = 55.3\frac{V}{c}\left(\frac{1}{T_2} - \frac{1}{T_1}\right), \tag{8.11}$$

$$\alpha = \frac{A}{S}. \tag{8.12}$$

In this equation, A denotes equivalent absorption area, V room volume, c sound speed in air, S area of specimen, α reverberation absorbing coefficient, T_1 and T_2 the reverberation times with and without the specimen, respectively. It is noted that the method of measuring reverberation absorption coefficient is standardized and more detailed information is described in ISO 354, JIS A 1409: 1998 and the literature [2].

By the way, it is known that the random-incidence absorption coefficient increases when the dimensions of a sound-absorbing surface become as small as the dimension of wavelength. This phenomenon is usually referred to as area effect, and it becomes evident when the absorption coefficient is large. Therefore, the random-incidence absorption coefficient as the value of the material itself is that of the material with infinite area does not cause the area effect.

In predicting the reverberation time of an auditorium, the total sound absorption is necessary, and it is calculated as the sum of the product of the area of each peripheral wall portion and the sound absorption coefficient of each wall material obtained by the reverberation room method described above. Though it seems that such a calculation method is reasonable at first glance, area effect is included much on the measurement of reverberation absorption coefficient. Since area effect varies depending on the surrounding circumstances, the estimation of the total sound absorption may contain significant errors if we evaluate by 'Msound absorption = reverberation absorption coefficient × area." In the following, some examples for the analysis of the area effect by introducing the boundary integral equation method are shown.

Consider the absorbing energy when the plane wave is incident on the sound-absorbing surface of finite size from a certain direction [3–5]. Let semi-infinite space Ω_0 bounded by the infinitely large semi-sphere Γ_∞, the rigid surface Γ and the sound-absorbing surface A include a point source S at a position of $\boldsymbol{r}_\mathrm{S}$ and a receiving point P at a position of $\boldsymbol{r}_\mathrm{P}$, and P_i be the image point of P with respect to the surfaces Γ and A as shown in Fig. 8.2. Also, Q is a point on Γ or A at a position of $\boldsymbol{r}_\mathrm{Q}$.

In order to derive an integral formula, let us use

$$G(\boldsymbol{r}_\mathrm{P}, \boldsymbol{r}_\mathrm{Q}) = \frac{e^{jkr}}{4\pi r} + \frac{e^{jkr_i}}{4\pi r_i}, \tag{8.13}$$

as a fundamental solution in which the image point P_i is taken into account. Here, $r = |\boldsymbol{r}| = |\boldsymbol{r}_\mathrm{Q} - \boldsymbol{r}_\mathrm{P}|$, $r_i = |\boldsymbol{r}_i| = |\boldsymbol{r}_\mathrm{Q} - \boldsymbol{r}_{\mathrm{P}i}|$. We apply here Green's theorem or

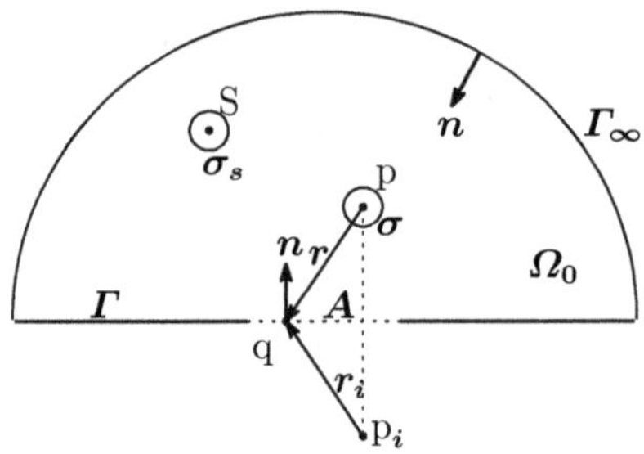

Fig. 8.2 Definition of symbols: Ω_0 semi-infinite space bounded by rigid surface Γ, sound-absorbing surface A, and infinitely large semi-sphere Γ_∞; n normal vector; S point source; P receiving point; P_i the image of P with respect to Γ and A; σ_s and σ small spheres of centers S and P respectively

integration by parts to the space $\Omega_0 - \sigma_s - \sigma$ where σ_s and σ are small spheres of centers S and P, respectively, with radius ϵ. Taking into consideration (1) the normal component of particle velocity vanishes on Γ, (2) $\dfrac{\partial G}{\partial n} = 0$ on Γ and A, and (3) Sommerfeld's radiation condition [6], we can obtain

$$\Phi(\boldsymbol{r}_\mathrm{P}) = \Phi_D(\boldsymbol{r}_\mathrm{P}) + \Phi_D(\boldsymbol{r}_\mathrm{Pi}) - \frac{1}{2\pi}\int_A \frac{\partial \Phi(\boldsymbol{r}_\mathrm{Q})}{\partial n_\mathrm{q}} \frac{e^{jkr}}{r} d\Gamma, \quad (\mathrm{P} \in \Omega_0, \Gamma, A), \tag{8.14}$$

where $\Phi(\boldsymbol{r}_\mathrm{P})$ denotes velocity potential at P and Φ_D the direct wave. Equation (8.14) is valid when P is located on Γ or A (i.e., $\boldsymbol{r}_\mathrm{P} = \boldsymbol{r}_\mathrm{Pi}$), since $\lim_{\epsilon \to 0} \iint_{\partial\sigma/2} d\Gamma = -\Phi(\boldsymbol{r}_\mathrm{P})$, $\quad (\mathrm{P} \in \Gamma, A)$.

If the sound-absorbing surface A is assumed to be locally reacting surface, using Eq. (8.8), we have

$$\Phi(\boldsymbol{r}_\mathrm{P}) = \Phi_D(\boldsymbol{r}_\mathrm{P}) + \Phi_D(\boldsymbol{r}_\mathrm{Pi}) + \frac{jk\rho c}{2\pi Z_a}\int_A \Phi(\boldsymbol{r}_\mathrm{Q}) \frac{e^{jkr}}{r} d\Gamma, \quad (\boldsymbol{r}_\mathrm{P} \in \Omega_0, \Gamma, A) \tag{8.15}$$

where Z_a denotes the specific impedance.[1] Morse and Ingard [7] and Thomasson [8] have also introduced equations similar above.[2] When P is located on A, Eq. (8.15) is a boundary integral equation and Φ is an unknown surface function. The sound field in the space Ω_0 can be calculated by substituting Φ obtained by solving Eq. (8.15) into the same equation. Sound-absorbing energy and incident energy per unit time is calculated by the following equations:

---

[1] Since the direction n is different from that in Eq. (8.8), the integral term has opposite sign.

[2] The former deals with the two-dimensional problem.

$$I_{\mathrm{i}} = \int_A \frac{1}{4}(p_{\mathrm{i}} v_{\mathrm{i}}^* + p_i^* v_{\mathrm{i}}) d\Gamma, \tag{8.16}$$

$$I_a = \int_A \frac{1}{4}(p_a v_a^* + p_a^* v_a) d\Gamma, \tag{8.17}$$

where p_{i} and v_{i} are sound pressure and the normal component of particle velocity of incident sound, and p_a and v_a are those on A, respectively. The superscript $*$ denotes complex conjugate.

In order to determine the flow of sound energy (intensity) within the space Ω_0, each direction component of the particle velocity is required. By differentiating Eq. (8.15) with respect to $\boldsymbol{r}_{\mathrm{P}}$ in $\boldsymbol{n}_p$ direction, we have

$$\begin{aligned} \frac{\partial \Phi(\boldsymbol{r}_{\mathrm{P}})}{\partial n_{\mathrm{p}}} &= \frac{\partial \Phi_D(\boldsymbol{r}_{\mathrm{P}})}{\partial n_{\mathrm{p}}} + \frac{\partial \Phi_D(\boldsymbol{r}_{\mathrm{Pi}})}{\partial n_{\mathrm{pi}}} \\ &\quad + \frac{jk\rho c}{4\pi Z_a} \int_A \Phi(\boldsymbol{r}_{\mathrm{Q}}) \left(\frac{\partial}{\partial n_{\mathrm{p}}} \frac{e^{jkr}}{r} + \frac{\partial}{\partial n_{\mathrm{pi}}} \frac{e^{jkr_i}}{r_{\mathrm{i}}} \right) d\Gamma. \end{aligned} \tag{8.18}$$

It should be noted that the direction of the differentiation with respect to the mirror image point P_i is taken in the $\boldsymbol{n}_{\mathrm{pi}}$ which is the the mirror image of $\boldsymbol{n}_{\mathrm{p}}$. Since each direction component of the particle velocity can be obtained from Eq. (8.18), we can calculated the time-average flow of sound energy at a point P in the same way as Eqs. (8.16) and (8.17).

The flow of sound energy (intensity) near the edge of the sound-absorbing surface calculated from the method described above is shown in Fig. 8.3. In the calculation, the size of the sound-absorbing surface is $2 \times 2\,\mathrm{m}$, sound pressure reflection coefficient $r_{\mathrm{p}} = 0.2 \exp(i\pi/12)$ and the plane wave of 63 Hz is incident vertically on it. In Fig. 8.3, in the vicinity of the absorber's edge sound energy flow from adjacent space to the absorber can be seen. This is because the sound pressure at the rigid surface Γ increases but decreases the sound-absorbing surface area A. Particle velocity $\boldsymbol{v} = -\mathrm{grad}\,\Phi$ is determined by the sound pressure gradient, we can therefore understand that the energy flows into the sound-absorbing surface from the adjacent area outside the edge. Accordingly, it is the factor of area effect that the energy flowing into absorbing surface from the adjacent area is added to the absorbing energy of the incident sound. In addition, it can be considered that, if other sound-absorbing area exists near sound-absorbing surface under consideration, flow energy is distributed to the other and the area effect will decrease. The term 'Minterference" or "interaction" is appropriate for this phenomenon.

To simulate random incidence noise, the numerical calculations were carried out for m plane waves that are incident on the sound-absorbing surface from all directions at regular solid angle ($m = 213$ in this calculation). The energy of the incident wave and of the absorbed one are calculated for each condition and summed up separately. The random-incidence absorption coefficient α can be obtained by:

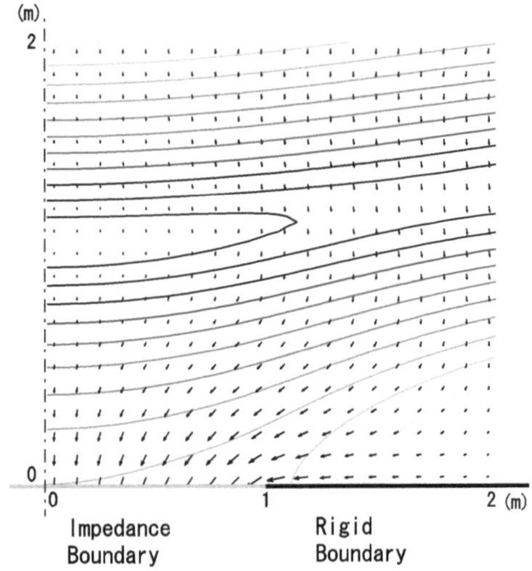

Fig. 8.3 Sound intensity near a locally reacting surface (2 × 2 m) in case of normal incidence of a plane wave, 63 Hz

$$\alpha = \sum_{l=1}^{m} (I_a)_l \;/\; \sum_{l=1}^{m} (I_{\mathrm{i}})_l. \tag{8.19}$$

For value of a frequency band would be calculated from the average of the values of several frequenies within the band.

In order to verify the effectiveness of the method introduced above, reverberation absorption coefficients of glass wool patches (25 mm thickness, 32 kg/m^3, no air space) are measured in the reverberation room in the General Building Research Corporation (GBRC). The reverberation room used for the measurements has a volume of 317.4 m^3. Six curved polyvinyl chloride panels with dimensions of 915×1830 mm^2 are hung in the room in order to diffuse sound. Since the thickness of the absorbing layer is relatively thin, it will be possible to assume the locally reacting condition in the frequency band considered.

Figure 8.4 shows the results of the measurements for 25 mm thick glass wool with dimensions of 0.5 × 0.5 m and 3.0 × 4.0 m, respectively,[3,4]. In these figures dashed dotted lines denote experimental results for one-third octave band noise, dotted lines the values obtained from the numerical calculations of the boundary integral equation, dashed lines normal incidence absorption coefficients α_0 and solid lines statistical random-incidence absorption coefficients α_s for infinitely large absorbing surface. Normal impedances used for obtaining the above-calculated absorption coefficients were measured by using the tube method (B & K 4206 Two-microphone impedance

[3] Since the specimen with dimensions of 0.5 × 0.5 m has little absorption, measurement error becomes large. Therefore, four specimens were set sufficiently apart from each other in the reverberation room in the measurement.

[4] Since the averaged absorption coefficients within one-third octave band have approximately same values as those for the center frequencies of the bands, the latter values are used in Fig. 8.4.

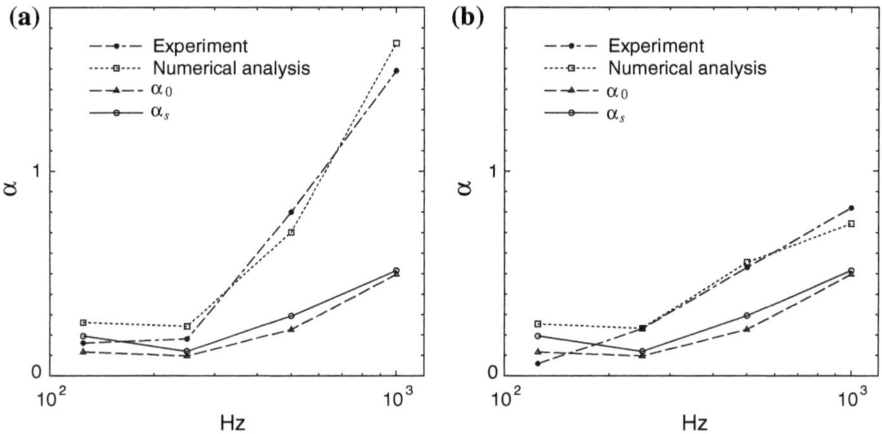

Fig. 8.4 Comparison of measured sound absorption coefficient of finite absorbing surfaces with the theoretical ones: **a** 0.5 × 0.5 m, **b** 3.0 × 4.0 m; α_0 denotes normal incidence sound absorption coefficient, α_s statistical sound absorption coefficient

Table 8.1 Measured values of absorption coefficient α_0, sound pressure reflection coefficient r_p, phase shift δ, and impedance ratio z at normal incidence for glass wool with density of 32 kg/m^3, 25 mm thickness and no air space

Frequency (Hz)	α_0	$\lvert r_p \rvert$	δ (deg)	Re(z)	Im(z)
125	0.116	0.94	7.393	6.557	12.658
250	0.098	0.95	15.086	1.333	7.289
500	0.226	0.88	31.233	0.787	3.410
1000	0.496	0.71	57.646	0.664	1.615

measurement tube) (see Table 8.1). We can see that the area effect becomes evident when the absorption coefficient is large and the dimensions of the specimens are small, and also that the theoretical values are in good agreement with the experimental values. Though reverberation absorption coefficient should have a value close to statistical absorption coefficient with no area effect, it can be seen that even in the results for the size of 3.0 × 4.0 m considerable area effect occurs.

Consider the sound absorption coefficient of an open window as a topic related to the area effect. Such an opening is usually treated as the perfectly absorbing surface [9, 10]. If it is assumed that there is an opening in the plane rigid wall with thin thickness, the relational expression of sound source side semi-infinite space is Eq. (8.14) and that of the other side semi-infinite space is a similar expression with no source term. Solving $\partial\Phi/\partial n$ on the opening by coupling at the opening above two boundary integral equations, we can obtain random-incident absorption coefficient (=random-incidence transmission coefficient) in a manner similar to the above. Figure 8.5 shows the result of an opening with dimensions of 0.5 × 0.5 m. It can be seen that the sound absorption coefficient is greater than 1 at low frequency

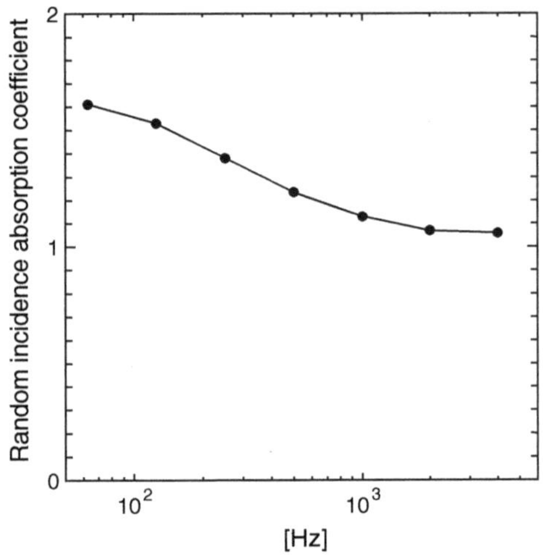

Fig. 8.5 Random-incidence absorption coefficient of an open window with the dimensions of 0.5 ×0.5 m

range. Note, it is not possible to explain such a phenomenon by Kirchhoff's diffraction theory [11] that are often used conventionally.

8.1.3 Simple FDTD Model

8.1.3.1 Rayleigh Model

The Rayleigh model is quite popular as a theoretical and qualitative model for porous materials [10]. The Rayleigh model is defined as simple differential equations which have a flow resistance term in the momentum equation. This model is not enough to simulate real porous material, but an easy-to-use model for the FDTD method. Some examples will be introduced in this section, these enable us to analyze an internal element of acoustic material and extended reaction of acoustic boundary.

In the Rayleigh model, a porous material is simplified as the set of thin and rigid pipes shown in Fig. 8.6. In the pipe, air viscosity causes flow resistance and friction which depends on the flow speed.

In that condition, the momentum equation of the Rayleigh model is defined as:

$$\frac{\partial p}{\partial x} + \rho \frac{\partial u_x}{\partial t} + \sigma u_x = 0, \tag{8.20}$$

and the continuous equation is

$$\frac{\partial p}{\partial t} + \kappa \left(\frac{\partial u_x}{\partial x} + \frac{\partial u_y}{\partial y} \right) = 0, \tag{8.21}$$

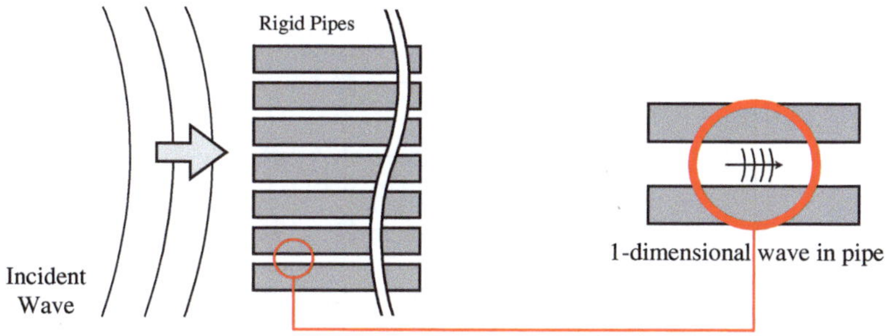

Fig. 8.6 Rayleigh model

where p is the sound pressure, u_x is particle velocity in the x direction, ρ is the density of air, and κ is the bulk modulus of the air. In the momentum equation, the direction x is assumed, because the acoustic wave only propagate in the direction of the pipe.

In a porous material the medium density for acoustic wave is larger than air's value, the effective density is required to perform this property. By the same token, the effective bulk modulus is necessary to describe that the sound speed is low in a material. The flow resistance in the momentum equation describes the resistance of air viscosity in a thin tube. On the other hand, the relaxation coefficient could be introduce to the continuous equation. Finally, the expanded momentum equation is

$$\rho_{\mathrm{e}}\frac{\partial u_x}{\partial t}+\frac{\partial p}{\partial x}+\sigma_x u_x=0, \tag{8.22}$$

$$\rho_{\mathrm{e}}\frac{\partial u_y}{\partial t}+\frac{\partial p}{\partial y}+\sigma_y u_y=0, \tag{8.23}$$

and the expanded continuous equation is

$$\frac{\partial p}{\partial t}+\kappa_{\mathrm{e}}\left(\frac{\partial u_x}{\partial x}+\frac{\partial u_y}{\partial y}+\frac{\partial u_z}{\partial z}\right)+\sigma_{\mathrm{p}}p=0, \tag{8.24}$$

where σ_y is flow resistance in y direction, ρ_{e} is the effective density of the material, κ_{e} is the effective bulk modulus of the material, and σ_p is relaxation coefficient. In the Ralyleigh model, flow resistance is assumed in the direction of pipe, but it could be assumed in any direction, the σ_y is added in this formulation. In a porous material, thermal and viscous diffusion are appeared as relaxation process [12] and it reduces the sound pressure over time. If the plane wave solution $p=e^{(j\omega+\sigma_p)t-kx}$ is assumed in nondissipative wave equation, Eq. (8.24) can be obtained easily. Then it also can be understood that the σ_p performs as a decay of sound pressure in time.

Now we assume a two-dimensional staggered-grid mesh as shown in Fig. 8.7 to investigate the finite difference forms of the proposed model. Each sound pressure

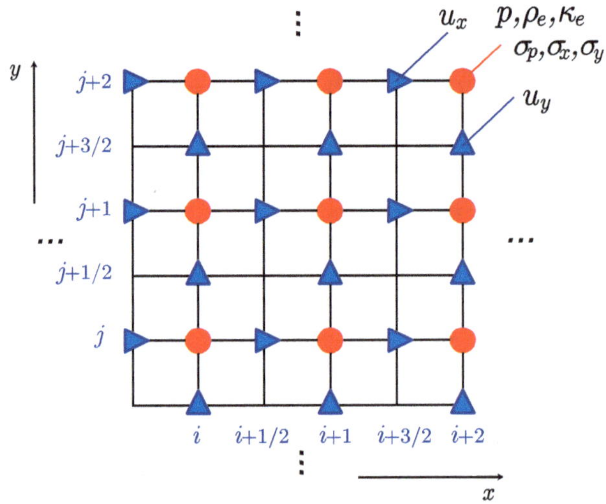

Fig. 8.7 Spacial placement of FDTD cells in two dimensions. p is sound pressure and u_x and u_y are particle velocities in the x, y direction, respectively. ρ_e is the effective density, κ_e is the effective bulk modulus, σ_p is a relaxation coefficient, and σ_x and σ_y are flow resistances in the x, y directions, respectively

and particle velocity is placed alternately, and the medium constants are defined at the same points as the sound pressures.

In this mesh, the finite difference equations of Eqs. (8.22), (8.23) and (8.24) can be written as:

$$u_x^{n+1}(i+\frac{1}{2}, j) = \left(1 - \frac{\Delta t \sigma_x(i+\frac{1}{2}, j)}{\rho_e(i+\frac{1}{2}, j)}\right) u_x^n(i+\frac{1}{2}, j) - \frac{\Delta t}{\Delta x \rho_e(i+\frac{1}{2}, j)} \left(p^{n+\frac{1}{2}}(i+1, j) - p^{n+\frac{1}{2}}(i, j)\right), \tag{8.25}$$

$$u_y^{n+1}(i, j+\frac{1}{2}) = \left(1 - \frac{\Delta t \sigma_y(i, j+\frac{1}{2})}{\rho_e(i, j+\frac{1}{2})}\right) u_x^n(i, j+\frac{1}{2}) - \frac{\Delta t}{\Delta x \rho_e(i, j+\frac{1}{2})} \left(p^{n+\frac{1}{2}}(i, j+1) - p^{n+\frac{1}{2}}(i, j)\right), \tag{8.26}$$

$$p^{n+\frac{3}{2}}(i, j) = \left(1 - \Delta t \sigma_p(i, j)\right) p^{n+\frac{1}{2}}(i, j) - \frac{\Delta t \kappa_e(i, j)}{\Delta x} \left(u_x^{n+1}(i+\frac{1}{2}, j) - u_x^{n+1}(i-\frac{1}{2}, j)\right), \tag{8.27}$$

where $p^n(i+\frac{1}{2})$ is the sound pressure at a discrete time n at the discrete spatial position $i+\frac{1}{2}$, $u_x^n(i)$ is particle velocity in the x direction at time n and position i, Δt is a discrete time width, and Δx is a discrete spatial width in the x direction.

$\sigma_x(i+1/2, j)$, $\sigma_y(i, j+1/2)$, $\rho_e(i+1/2, j)$ and $\rho_e(i, j+1/2)$ are defined as averaged values such as

$$\sigma_x(i+\frac{1}{2}, j) = \frac{\sigma_x(i+1, j)+\sigma_x(i, j)}{2}, \tag{8.28}$$

$$\sigma_y(i, j+\frac{1}{2}) = \frac{\sigma_y(i, j+1)+\sigma_y(i, j)}{2}, \tag{8.29}$$

$$\rho_e(i+\frac{1}{2}, j) = \frac{\rho_e(i+1, j)+\rho_e(i, j)}{2}, \tag{8.30}$$

$$\rho_e(i, j+\frac{1}{2}) = \frac{\rho_e(i, j+1)+\rho_e(i, j)}{2}. \tag{8.31}$$

In these finite-difference forms, it can be seen that the terms $1-\Delta t\sigma_p$, $1-\Delta t\sigma_x/\rho_e$ and $1-\Delta t\sigma_y/\rho_e$ are just added to the finite-difference forms of the air, these forms therefore maintain coding simplicity.

8.1.3.2 Characteristic Impedance

In this section, a one-dimensional sound field is assumed to confirm the normal incident characteristic impedance of the proposed model.

The one-dimensional momentum equation can be written as:

$$\rho_e\frac{\partial u}{\partial t}+\frac{\partial p}{\partial x}+\sigma_x u = 0, \tag{8.32}$$

and the continuous equation is

$$\frac{\partial p}{\partial t}+\kappa_e\frac{\partial u}{\partial x}+\sigma_p p = 0. \tag{8.33}$$

The plane wave solutions are assumed to be

$$p = Pe^{j(\omega t-k'x)}, u_x = U_X e^{j(\omega t-k'x)}, \tag{8.34}$$

where ω is the angular frequency and k' is a complex wave number. Equations (8.32) and (8.33) are simplified by substituting Eq. (8.34) thus

$$\begin{bmatrix} -j\omega+\sigma_p, -\kappa_e jk' \\ -jk', -j\omega\rho_e+\sigma_x \end{bmatrix} \begin{Bmatrix} p \\ u_x \end{Bmatrix} = 0. \tag{8.35}$$

By setting the determinant of coefficient matrix in Eq. (8.35) equal to zero, k' can be solved as:

$$k' = \frac{\omega}{c_\mathrm{e}}\sqrt{1 + \frac{j}{\rho_\mathrm{e}\omega}\left(\rho_\mathrm{e}\sigma_\mathrm{p} + \sigma_x\right) - \frac{\sigma_\mathrm{p}\sigma_x}{\rho_\mathrm{e}\omega^2}}, \tag{8.36}$$

and the characteristic impedance Z is obtained as:

$$Z = \rho_\mathrm{e}c_\mathrm{e}\sqrt{1 + \frac{j}{\rho_\mathrm{e}\omega}\left(\rho_\mathrm{e}\sigma_\mathrm{p} + \sigma_x\right) - \frac{\sigma_\mathrm{p}\sigma_x}{\rho_\mathrm{e}\omega^2}}, \tag{8.37}$$

and no normalization is performed by the characteristic impedance of the air.

8.1.3.3 Stability Analysis

To derive the stability condition of finite-difference forms, we again assume a one-dimensional plane wave solution, because an arbitrary wave can be decomposed into a number of plane waves.

In a one-dimensional scenario, the finite-difference forms at i can be written as:

$$u_x^{n+1}(i) = \left(1 - \frac{\Delta t\sigma_x(i)}{\rho_\mathrm{e}(i)}\right)u_x^n(i) - \frac{\Delta t}{\Delta x\rho_\mathrm{e}(i)}\left(p^{n+\frac{1}{2}}(i+\frac{1}{2}) - p^{n+\frac{1}{2}}(i-\frac{1}{2})\right), \tag{8.38}$$

$$p^{n+\frac{3}{2}}(i) = \left(1 - \Delta t\sigma_\mathrm{p}(i)\right)p^{n+\frac{1}{2}}(i) - \frac{\Delta t\kappa_\mathrm{e}(i)}{\Delta x}\left(u_x^{n+1}(i+\frac{1}{2}) - u_x^{n+1}(i-\frac{1}{2})\right). \tag{8.39}$$

Now if we assume $L_p(i) = 1 - \Delta t\sigma_\mathrm{p}(i)$, $L_v(i) = 1 - \Delta t\sigma_x(i)/\rho_e(i)$, $Z = \rho_e(i)c(i)$, and the Courant number $C(i) = c(i)\Delta t/\Delta x$, then Eqs. (8.38) and (8.39) can be simplified

$$u_x^{n+1} = L_v u_x^n + \frac{C}{Z}2jSp^n, \tag{8.40}$$

$$p^{n+\frac{3}{2}} = L_p p^n + CZ2jSu_x^{n+1}, \tag{8.41}$$

where $S = \sin(\Delta x/2)$.

Finally, the vector matrix form can be obtained by

$$\boldsymbol{x}^{n+1} = \boldsymbol{A}\boldsymbol{x}^n, \tag{8.42}$$

where $\mathbf{x}^n$ and $\mathbf{A}$ are defined

$$\boldsymbol{x}^n = [p^{n+\frac{1}{2}} \ u_x^n]^T, \tag{8.43}$$

$$\boldsymbol{A} = \begin{bmatrix} L_p - 4C^2S^2 & 2jCZL_vS \\ 2jL_v\dfrac{C}{Z}S_x & L_v \end{bmatrix}. \tag{8.44}$$

To obtain stable and nontrivial solutions, all eigenvalues λ of $\mathbf{A}$ have to be $|\lambda| \le 1$. In that condition, the stability condition in one-dimensinal case can be solved as:

$$\Delta t \le \frac{1}{\frac{c_e}{\Delta x} + max(\sigma_p, \sigma_x/\rho)}. \tag{8.45}$$

In two- or three-dimensional cases, the matrices $\mathbf{A}$ can be obtained as

$$\Delta t \le \frac{1}{c_e\sqrt{\frac{1}{\Delta x^2} + \frac{1}{\Delta y^2}} + max(\sigma_p, \frac{\sigma_x}{\rho_e}, \frac{\sigma_y}{\rho_e})}, \tag{8.46}$$

$$\Delta t \le \frac{1}{c_e\sqrt{\frac{1}{\Delta x^2} + \frac{1}{\Delta y^2} + \frac{1}{\Delta z^2}} + max(\sigma_p, \frac{\sigma_x}{\rho_e}, \frac{\sigma_y}{\rho_e}, \frac{\sigma_z}{\rho_e})}. \tag{8.47}$$

8.1.4 Layered Material FE Model

8.1.4.1 Basic Equations for Poroelastic Material

For a multilayer structure consisting of porous materials with elastic frames, a model based on the Biot theory [13, 14] is preferable. The poroelastic model describes the interactive propagation of elastic waves in solid and fluid phases of porous media. There are several representation on the Biot theory with regard to physical variables. For a time-harmonic analysis, a mixed displacement-pressure ($\mathbf{u}^s$-p) formulation [15] is advantageous in terms of computational efficiency. According to the ($\mathbf{u}^s$-p) formulation, poroelastic wave propagation is described by simultaneous equations composed of an equivalent elastodynamic equation and an equivalent Helmholtz equation with volumetric coupling terms as follows:

$$\begin{aligned} &\text{div}\boldsymbol{\sigma}^s(\mathbf{u}^s) + \omega^2\tilde{\rho}_s\mathbf{u}^s + \tilde{\gamma}\text{grad}p = 0, \\ &\nabla^2 p + \omega^2\frac{\tilde{\rho}_{22}}{\tilde{R}}p - \omega^2\frac{\tilde{\gamma}\tilde{\rho}_{22}}{\phi^2}\text{div}\mathbf{u}^s = 0. \end{aligned} \tag{8.48}$$

The following notation is introduced for the physical parameters; in the fluid-phase:

- $\tilde{\rho}_{\mathrm{f}}$ and $\tilde{K}_{\mathrm{f}}$: effective density and bulk modulus of equivalent fluid,
- $\tilde{\rho}_{22}$: Biot's inertial coefficient for fluid-phase defined by: $\tilde{\rho}_{22} = \phi\tilde{\rho}_{\mathrm{f}}$,
- $\tilde{R}$: Biot's elastic coefficient for fluid-phase dilatation defined by: $\tilde{R} = \frac{\phi^2 K_{\mathrm{s}}}{1-\phi-K_{\mathrm{b}}/K_{\mathrm{s}}+\phi K_{\mathrm{s}}/\tilde{K}_{\mathrm{f}}}$,

in the solid-phase:

- ρ_{b} and K_{b}: the bulk density and the bulk modulus of the poroelastic material,
- K_{s}: the bulk modulus of the material of which the skeleton is made,
- $\tilde{\rho}_{\mathrm{s}}$: effective solid-phase density defined by: $\tilde{\rho}_{\mathrm{s}} = \rho_{\mathrm{b}} + \phi\rho_0\left(1 - \frac{\rho_0}{\tilde{\rho}_{\mathrm{f}}}\right)$,
- λ and μ: the Lamé's coefficients,
- $\boldsymbol{\varepsilon}^{\mathrm{s}}(\mathbf{u}^{\mathrm{s}})$: strain tensor of the solid-phase of poroelastic material defined by: $\varepsilon^{\mathrm{s}}_{ij} = \frac{1}{2}\left(\frac{\partial u_i}{\partial x_j} + \frac{\partial u_j}{\partial x_i}\right)$,
- $\boldsymbol{\sigma}^{\mathrm{s}}(\mathbf{u}^{\mathrm{s}})$: stress tensor of the poroelastic material *in vacuo*; It is related to the strain tensor by Hooke's law together with the unit tensor $\mathbf{I}$ by $\boldsymbol{\sigma}^{\mathrm{s}}(\mathbf{u}^{\mathrm{s}}) = \lambda \mathrm{div}\mathbf{u}^{\mathrm{s}}\mathbf{I} + 2\mu\boldsymbol{\varepsilon}^{\mathrm{s}}(\mathbf{u}^{\mathrm{s}})$,
- $\boldsymbol{\sigma}^{\mathrm{t}}$: total stress tensor of the solid-phase of the poroelastic material defined by: $\boldsymbol{\sigma}^{\mathrm{t}} = \boldsymbol{\sigma}^{\mathrm{s}}(\mathbf{u}^{\mathrm{s}}) - (1 - \phi - \frac{\boldsymbol{K_b}}{\boldsymbol{K_s}})p\mathbf{I}$,

and in both phases:

- ϕ: porosity of the poroelastic material,
- $\tilde{\gamma}$: volumetric coupling coefficient defined by: $\tilde{\gamma} = \phi\frac{\rho_0}{\tilde{\rho}_{\mathrm{f}}} + \frac{\boldsymbol{K_b}}{\boldsymbol{K_s}} - 1$

Note that the values with tilde are complex and frequency dependent. $\tilde{\rho}_{\mathrm{f}}$ and $\tilde{K}_{\mathrm{f}}$ may be estimated by different general semi-phenomenological models taking into account the viscous and thermal losses of the acoustic compression wave. In the following, isotropic and homogenous materials are discussed.

8.1.4.2 Finite Element Formulation for Poroelastic Material

The weak forms of the $\mathbf{u}^{\mathrm{s}}$-p equations are written as

$$\int_{\Omega^{\mathrm{P}}} \boldsymbol{\varepsilon}^{\mathrm{s}}(\delta\mathbf{u}^{\mathrm{s}}) : \boldsymbol{\sigma}^{\mathrm{s}}(\mathbf{u}^{\mathrm{s}}) dV - \omega^2 \int_{\Omega^{\mathrm{P}}} \tilde{\rho}_{\mathrm{s}} \delta\mathbf{u}^{\mathrm{s}} \cdot \mathbf{u}^{\mathrm{s}} dV$$

$$- \int_{\Omega^{\mathrm{P}}} \tilde{\gamma}\delta\mathbf{u}^{\mathrm{s}} \cdot \mathrm{grad} p dV - \int_{\Gamma^{\mathrm{P}}} \delta\mathbf{u}^{\mathrm{s}} \cdot \boldsymbol{\sigma}^{\mathrm{s}} \cdot \mathbf{n} dS = 0, \tag{8.49}$$

$$\frac{1}{\omega^2}\int_{\Omega^{\mathrm{P}}} \frac{\phi^2}{\tilde{\rho}_{22}} \mathrm{grad}\delta p \cdot \mathrm{grad} p dV - \int_{\Omega^{\mathrm{P}}} \frac{\phi^2}{\tilde{R}} \delta p \cdot p dV$$

$$- \int_{\Omega^{\mathrm{P}}} \tilde{\gamma} \mathrm{grad}\delta p \cdot \mathbf{u}^{\mathrm{s}} dV + \int_{\Gamma^{\mathrm{P}}} \delta p \left(\tilde{\gamma}\mathbf{u}^{\mathrm{s}} \cdot \mathbf{n} - \frac{\phi^2}{\omega^2\tilde{\rho}_{22}}\frac{\partial p}{\partial n}\right) dS = 0, \tag{8.50}$$

where $\delta \mathbf{u}^{\mathrm{s}}$ and δp are admissible variations of the solid-phase displacement and the fluid-phase pressure, respectively. Based on the Galerkin's method, the above integrals are discretized with the interpolation

$$\mathbf{u}^{\mathrm{s}} = [\mathbf{N}^{\mathrm{s}}]\{\mathbf{u}_e^{\mathrm{s}}\} \quad \text{and} \quad p = [\mathbf{N}^{\mathrm{f}}]\{\mathbf{p}_e\}, \tag{8.51}$$

for volume integrals in the solid-phase,

$$\begin{aligned}\int_{\Omega^{\mathrm{P}}} \boldsymbol{\varepsilon}^{\mathrm{s}}(\delta \mathbf{u}^{\mathrm{s}}) : \boldsymbol{\sigma}^{\mathrm{s}}(\mathbf{u}^{\mathrm{s}}) dV &= \sum_{e \in \Omega^{\mathrm{P}}} \left(\{\delta \mathbf{u}_e^{\mathrm{s}}\} \int_{\Omega_e^{\mathrm{P}}} [\mathbf{B}^{\mathrm{s}}]^T [\mathbf{D}][\mathbf{B}^{\mathrm{s}}] dV \{\mathbf{u}_e^{\mathrm{s}}\} \right) \\ &= \{\delta \mathbf{u}^{\mathrm{s}}\}[\mathbf{K}^{\mathrm{s}}]\{\mathbf{u}^{\mathrm{s}}\}, \end{aligned} \tag{8.52}$$

$$\begin{aligned}\int_{\Omega^{\mathrm{P}}} \tilde{\rho}_{\mathrm{s}} \delta \mathbf{u}^{\mathrm{s}} \cdot \mathbf{u}^{\mathrm{s}} dV &= \sum_{e \in \Omega^{\mathrm{P}}} \left(\{\delta \mathbf{u}_e^{\mathrm{s}}\} \int_{\Omega_e^{\mathrm{P}}} \tilde{\rho}_{\mathrm{s}} [\mathbf{N}^{\mathrm{s}}]^T [\mathbf{N}^{\mathrm{s}}] dV \{\mathbf{u}_e^{\mathrm{s}}\} \right) \\ &= \{\delta \mathbf{u}^{\mathrm{s}}\}[\mathbf{M}^{\mathrm{s}}]\{\mathbf{u}^{\mathrm{s}}\}, \end{aligned} \tag{8.53}$$

$$\begin{aligned}\int_{\Omega^{\mathrm{P}}} \tilde{\gamma} \delta \mathbf{u}^{\mathrm{s}} \cdot \mathrm{grad} p dV &= \sum_{e \in \Omega^{\mathrm{P}}} \left(\{\delta \mathbf{u}_e^{\mathrm{s}}\} \int_{\Omega_e^{\mathrm{P}}} \tilde{\gamma} [\mathbf{N}^{\mathrm{s}}]^T [\mathbf{B}^{\mathrm{f}}] dV \{\mathbf{p}_e\} \right) \\ &= \{\delta \mathbf{u}^{\mathrm{s}}\}[\mathbf{C}]\{\mathbf{p}\}, \end{aligned} \tag{8.54}$$

and for volume integrals in the fluid-phase,

$$\begin{aligned}\int_{\Omega^{\mathrm{P}}} \frac{\phi^2}{\tilde{\rho}_{22}} \mathrm{grad} \delta p \cdot \mathrm{grad} p dV &= \sum_{e \in \Omega^{\mathrm{P}}} \left(\{\delta \mathbf{p}_e\} \int_{\Omega_e^{\mathrm{P}}} \frac{\phi^2}{\tilde{\rho}_{22}} [\mathbf{B}^{\mathrm{f}}]^T [\mathbf{B}^{\mathrm{f}}] dV \{\mathbf{p}_e\} \right) \\ &= \{\delta \mathbf{p}\}[\mathbf{K}^{\mathrm{f}}]\{\mathbf{p}\}, \end{aligned} \tag{8.55}$$

$$\begin{aligned}\int_{\Omega^{\mathrm{P}}} \frac{\phi^2}{\tilde{R}} \delta p \cdot p dV &= \sum_{e \in \Omega^{\mathrm{P}}} \left(\{\delta \mathbf{p}_e\} \int_{\Omega_e^{\mathrm{P}}} \frac{\phi^2}{\tilde{R}} [\mathbf{N}^{\mathrm{f}}]^T [\mathbf{N}^{\mathrm{f}}] dV \{\mathbf{p}_e\} \right) \\ &= \{\delta \mathbf{p}\}[\mathbf{M}^{\mathrm{f}}]\{\mathbf{p}\}, \end{aligned} \tag{8.56}$$

$$\int_{\Omega^{\mathrm{P}}} \tilde{\gamma}\mathrm{grad}\delta p \cdot \mathbf{u}^{\mathrm{s}} dV = \sum_{e\in\Omega^{\mathrm{P}}} \left(\{\delta \mathbf{p}_e\} \int_{\Omega_e^{\mathrm{P}}} \tilde{\gamma}[\mathbf{B}^{\mathrm{f}}]^T [\mathbf{N}^{\mathrm{s}}] dV \{\mathbf{u}_e\} \right)$$
$$= \{\delta \mathbf{p}\}[\mathbf{C}]^T \{\mathbf{u}^{\mathrm{s}}\}, \tag{8.57}$$

with the introduction of strain matrices

$$[\mathbf{B}^{\mathrm{s}}] = \begin{bmatrix} \frac{\partial}{\partial x_1} & 0 & 0 & \frac{\partial}{\partial x_2} & 0 & \frac{\partial}{\partial x_3} \\ 0 & \frac{\partial}{\partial x_2} & 0 & \frac{\partial}{\partial x_1} & \frac{\partial}{\partial x_3} & 0 \\ 0 & 0 & \frac{\partial}{\partial x_3} & 0 & \frac{\partial}{\partial x_2} & \frac{\partial}{\partial x_1} \end{bmatrix}^T [\mathbf{N}^{\mathrm{s}}], \text{ and } [\mathbf{B}^{\mathrm{f}}] = \begin{bmatrix} \frac{\partial}{\partial x_1} \\ \frac{\partial}{\partial x_2} \\ \frac{\partial}{\partial x_3} \end{bmatrix} [\mathbf{N}^{\mathrm{f}}], \tag{8.58}$$

and the elastic moduli matrix,

$$[\mathbf{D}] = \begin{bmatrix} \lambda+2\mu & \lambda & \lambda & 0 & 0 & 0 \\ \lambda & \lambda+2\mu & \lambda & 0 & 0 & 0 \\ \lambda & \lambda & \lambda+2\mu & 0 & 0 & 0 \\ 0 & 0 & 0 & \mu & 0 & 0 \\ 0 & 0 & 0 & 0 & \mu & 0 \\ 0 & 0 & 0 & 0 & 0 & \mu \end{bmatrix}, \tag{8.59}$$

where $[\mathbf{N}^{\mathrm{s}}]$ and $[\mathbf{N}^{\mathrm{f}}]$ are the corresponding shape functions, $\{\mathbf{u}_e^{\mathrm{s}}\}$ and $\{\mathbf{p}_e\}$ are element nodal displacement and pressure vectors, and $\{\mathbf{u}^{\mathrm{s}}\}$ and $\{\mathbf{p}\}$ are global nodal displacement and pressure vectors respectively. It is noted that Eq. (8.52) is arranged according to the Voigt notation of stress and strain tensor and Hooke's law for isotropic materials.

Finally, the following global matrix equation is obtained:

$$\begin{bmatrix} [\mathbf{K}^{\mathrm{s}}] - \omega^2[\mathbf{M}^{\mathrm{s}}] & -[\mathbf{C}] \\ -[\mathbf{C}]^T & \frac{1}{\omega^2}[\mathbf{K}^{\mathrm{f}}] - [\mathbf{M}^{\mathrm{f}}] \end{bmatrix} \begin{Bmatrix} \{\mathbf{u}^{\mathrm{s}}\} \\ \{\mathbf{p}\} \end{Bmatrix} = \begin{Bmatrix} \{\mathbf{F}^{\mathrm{s}}\} \\ \{\mathbf{F}^{\mathrm{f}}\} \end{Bmatrix}, \tag{8.60}$$

where $\begin{Bmatrix} \{\mathbf{F}^{\mathrm{s}}\} \\ \{\mathbf{F}^{\mathrm{f}}\} \end{Bmatrix}$ denotes an external force vector which depends on how the poroelastic material is excited; the details can be found in reference [16]. The surface integrals in Eqs. (8.49) and (8.50) are omitted here for simplicity.

8.1.4.3 Coupling Conditions on Interfaces Between Different Media

This subsection briefly presents coupling conditions on the interfaces between different media. In the following, it is assumed that the material of which the frame is made is incompressible, accordingly, $\tilde{\gamma}$ is reduced, by taking the limit as K_{s} approaches to infinity, into

$$\tilde{\gamma} \approx \phi \frac{\rho_0}{\tilde{\rho}_{\mathrm{f}}} - 1. \tag{8.61}$$

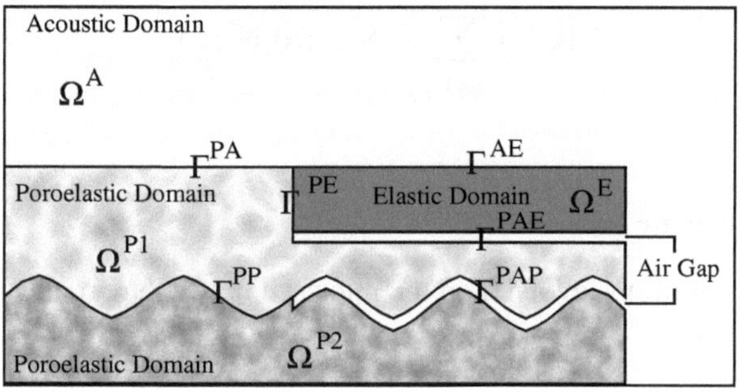

Fig. 8.8 Domain notation for a coupled problem

Moreover, in the finite element implementation, coupling treatments on interfaces depicted in Fig. 8.8 also become simple.

It is noted that coupling conditions involved in poroelastic media are expressed together with fluid-phase displacement in order to make the physical meaning of the condition clear; the fluid-phase displacement is related to solid-phase displacement and fluid-phase pressure by

$$\mathbf{u}^{\mathrm{f}} = \frac{\phi}{\tilde{\rho}_{22}\omega^2}\operatorname{grad} p - \left(\frac{\rho_0}{\tilde{\rho}_{\mathrm{f}}} - 1\right)\mathbf{u}^{\mathrm{s}}. \tag{8.62}$$

Let $\boldsymbol{\sigma}^{\mathrm{E}}$ and $\mathbf{u}^{\mathrm{E}}$ denote stress tensor and displacement of elastic domain, p^{A} denote sound pressure of acoustic domain, and $\mathbf{n}$ denotes the normal direction on the interface, respectively.

In-Contact Condition

In-contact condition can be considered as a situation where two materials are bonded at the interface. The surface integrals in weak form are rearranged in accordance with following coupling conditions.

Elastic–Acoustic Interface: Γ^{AE}

The coupling conditions involved on the interface between elastic and acoustic domains are given as:

$$\begin{cases} \boldsymbol{\sigma}^{\mathrm{E}} \cdot \mathbf{n} = -p^{\mathrm{A}}\mathbf{n}, \\ \dfrac{1}{\rho_0\omega^2}\dfrac{\partial p^{\mathrm{A}}}{\partial n} = \mathbf{u}^{\mathrm{E}} \cdot \mathbf{n}. \end{cases} \tag{8.63}$$

First equation ensures the continuity of normal stress and second equation does the continuity of normal displacement. In the finite element implementation, these conditions are attained by using coupling matrix obtained as following.

$$[\mathbf{Q}] = \sum_{e \in \Gamma} \int_{\Gamma_e} [\mathbf{N}^{\mathrm{s}}]^T [\mathbf{n}][\mathbf{N}^{\mathrm{f}}] dS. \tag{8.64}$$

Poroelastic–Acoustic Interface: Γ^{PA}

The coupling conditions involved on the interface between poroelastic and acoustic domains are given as

$$\begin{cases} \boldsymbol{\sigma}^{\mathrm{t}} \cdot \mathbf{n} = -p^{\mathrm{A}}\mathbf{n}, \\ p = p^{\mathrm{A}}, \\ \dfrac{1}{\rho_0 \omega^2} \dfrac{\partial p^{\mathrm{A}}}{\partial n} = (1-\phi)\mathbf{u}^{\mathrm{s}} \cdot \mathbf{n} + \phi \mathbf{u}^{\mathrm{f}} \cdot \mathbf{n}. \end{cases} \tag{8.65}$$

First and second equations ensure the continuity of normal stresses between the solid and fluid phases, respectively, and third equation refers to the continuity between the acoustic displacement, and the effective displacement of the poroelastic medium. Under the incompressible frame assumption, poroelastic, and acoustic domains are coupled naturally by using common unknowns in accordance with the second equation at the assembly step of the finite element implementation.

Poroelastic–Elastic Interface: Γ^{PE}

The coupling conditions involved on the interface between poroelastic and elastic domains are given as

$$\begin{cases} \boldsymbol{\sigma}^{\mathrm{t}} \cdot \mathbf{n} = \boldsymbol{\sigma}^{\mathrm{E}} \cdot \mathbf{n}, \\ \mathbf{u}^{\mathrm{s}} = \mathbf{u}^{\mathrm{E}}, \\ \mathbf{u}^{\mathrm{f}} \cdot \mathbf{n} - \mathbf{u}^{\mathrm{s}} \cdot \mathbf{n} = 0. \end{cases} \tag{8.66}$$

First and second equations ensure the continuity of normal stress and three-dimensional displacement between the solid-phase of poroelastic domain and elastic domain, respectively. Third equation states that relative mass flux in normal direction does not appear due to the impervious nature of the elastic domain. In the finite element implementation, two procedures are carried out to impose these conditions. First, the common unknowns are eliminated in accordance with the second equation. Next, a similar coupling matrix to Eq. (8.64) is considered for the fluid-phase pressure and solid-phase displacement ($\mathbf{u}^{\mathrm{E}}$ or $\mathbf{u}^{\mathrm{s}}$).

Poroelastic–Poroelastic Interface: Γ^{PP}

The coupling conditions involved on the interface between different poroelastic domains are given as

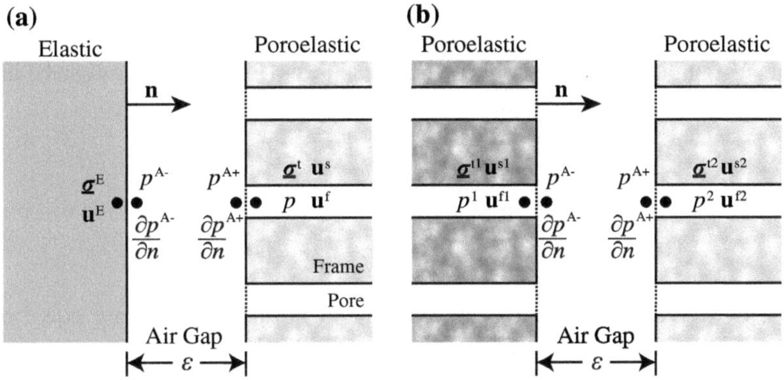

Fig. 8.9 Out-of-contact conditions: **a** Poroelastic–Elastic interface and **b** Poroelastic–Poroelastic interface

$$\begin{cases} \boldsymbol{\sigma}^{t1} \cdot \mathbf{n} = \boldsymbol{\sigma}^{t2} \cdot \mathbf{n}, \\ p^1 = p^2, \\ \mathbf{u}^{s1} = \mathbf{u}^{s2}, \\ \phi^1(\mathbf{u}^{f1} \cdot \mathbf{n} - \mathbf{u}^{s1} \cdot \mathbf{n}) = \phi^2(\mathbf{u}^{f2} \cdot \mathbf{n} - \mathbf{u}^{s2} \cdot \mathbf{n}), \end{cases} \tag{8.67}$$

where superscripts 1 and 2 are used in order to represent attribution of the value to the primary or secondary poroelastic domain. First two equations reflect the normal stress continuity of each phase between poroelastic domains. Third equation ensures the continuity of solid displacements, and fourth equation does the continuity of the relative mass flux. Poroelastic domains are naturally coupled by using common unknowns in accordance with the second and third equations at the assembly step of the finite element implementation.

Out-of-Contact Condition

Out-of-contact condition can be considered as a situation where two materials are coupled through thin air gap as depicted in Fig. 8.9. In case the thin air gap exists between a poroelastic material and another material, it is convenient to handle such a situation as the boundary condition without creating any volumetric meshes for the thin gap.

In the thin air gap, sound pressures and their normal derivatives on the forward and the backward face would coincide, respectively, if the thickness of air gap, ε, tends to zero.

$$\begin{cases} p^{A-} \approx p^{A+}, \\ \dfrac{\partial p^{A-}}{\partial n} \approx \dfrac{\partial p^{A+}}{\partial n}. \end{cases} \tag{8.68}$$

Poroelastic-Elastic Out-of-Contact Interface: Γ^{PAE}

The out-of-contact condition between poroelastic and elastic domains can be induced with the combination of Eqs. (8.63), (8.65) and (8.68) as

$$\begin{cases} \boldsymbol{\sigma}^{\mathrm{t}} \cdot \mathbf{n} = -p\mathbf{n}, \\ \boldsymbol{\sigma}^{\mathrm{E}} \cdot \mathbf{n} = -p\mathbf{n}, \\ \mathbf{u}^{\mathrm{E}} \cdot \mathbf{n} = (1-\phi)\mathbf{u}^{\mathrm{s}} \cdot \mathbf{n} + \phi\mathbf{u}^{\mathrm{f}} \cdot \mathbf{n}. \end{cases} \tag{8.69}$$

First two equations state that all associated normal stresses are equal on this interface. Third equation states that the normal displacement of elastic domain is continuous to the normal effective displacement on the interface of the poroelastic domain. In the finite element implementation, a similar coupling matrix to Eq. (8.64) is considered for the pressure of the poroelastic domain p and the displacement of the elastic domain $\mathbf{u}^{\mathrm{E}}$.

Poroelastic–Poroelastic Out-of-Contact Interface: Γ^{PAP}

Similarly, the out-of-contact condition between two poroelastic media can be induced with the combination of Eqs. (8.65) and (8.68) as

$$\begin{cases} \boldsymbol{\sigma}^{\mathrm{t1}} \cdot \mathbf{n} = -p^1\mathbf{n}, \\ \boldsymbol{\sigma}^{\mathrm{t2}} \cdot \mathbf{n} = -p^2\mathbf{n}, \\ p^1 = p^2, \\ (1-\phi^1)\mathbf{u}^{\mathrm{s1}} \cdot \mathbf{n} + \phi^1\mathbf{u}^{\mathrm{f1}} \cdot \mathbf{n} = (1-\phi^2)\mathbf{u}^{\mathrm{s2}} \cdot \mathbf{n} + \phi^2\mathbf{u}^{\mathrm{f2}} \cdot \mathbf{n}. \end{cases} \tag{8.70}$$

First three equations state that all associated normal stresses are equal on this interface. Fourth equation ensures the continuity of the normal effective displacements on the interface between two poroelastic domains. Differently for the in-contact condition, two poroelastic domains are coupled by using common unknown only for the pressures at the assembly step of the finite element implementation.

8.1.4.4 Calculation Scheme of Sound Absorption Coefficient

This section presents a model for predicting the oblique incidence absorption coefficient of absorbers with arbitrary shape, size, and material composition. A schematic of the model is depicted in Fig. 8.10 and the features are summarized as follows:

- a test sample is mounted in a cavity on the rigid baffle,
- the FEM is employed for the materials and air space in the cavity, and coupled with sound fields out of the baffle by the BEM on the imaginary interface,

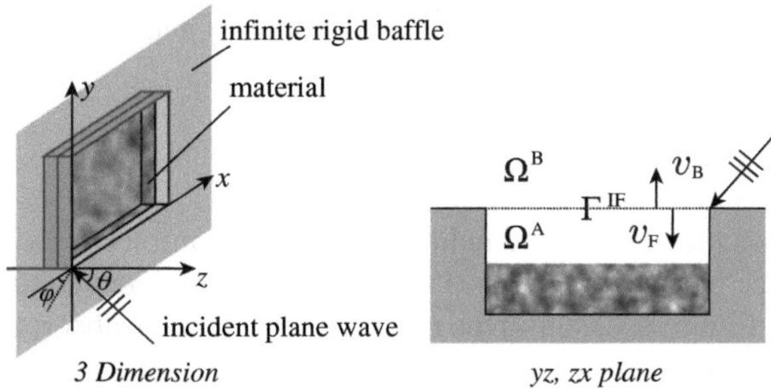

Fig. 8.10 Schematic of a calculation model for oblique incidence absorption coefficient

- absorption coefficient is calculated from the geometrical incidence power and the absorption power on the imaginary interface.

Acoustic FEM-BEM Coupling

For the acoustic field above the rigid baffle, boundary integral equation on the imaginary interface Γ^{IF} is given by

$$p_{\mathrm{B}}(\mathbf{r}_{\mathrm{p}}) + 2j\omega\rho_0 \int_{\Gamma^{\mathrm{IF}}} v_{\mathrm{B}}(\mathbf{r}_{\mathrm{q}}) \frac{\exp(jk|\mathbf{r}_{\mathrm{p}} - \mathbf{r}_{\mathrm{q}}|)}{4\pi|\mathbf{r}_{\mathrm{p}} - \mathbf{r}_{\mathrm{q}}|} dS = 2p_{\mathrm{D}}(\mathbf{r}_{\mathrm{p}}). \tag{8.71}$$

For the acoustic field within the cavity, integral equation in weak form is given by

$$\int_{\Omega^{\mathrm{A}}} \nabla\delta p_{\mathrm{F}} \cdot \nabla p_{\mathrm{F}} dV - k^2 \int_{\Omega^{\mathrm{A}}} \delta p_{\mathrm{F}} p_{\mathrm{F}} dV + j\omega\rho_0 \int_{\Gamma^{\mathrm{IF}}} \delta p_{\mathrm{F}} v_{\mathrm{F}} dS = 0. \tag{8.72}$$

There are two approaches to couple acoustic fields formulated by the BEM and the FEM according to the interpolation scheme of physical variables in the BEM elements. One is interpolation element formulation. Coupling condition in this approach is attained by the elimination of degree of freedom in common on the interface boundary. The other approach is constant element formulation. In this approach, unknowns are generally defined at different points in the BEM and the FEM, thus the elimination of nodal unknowns cannot be applied. Alternatively, the following coupling conditions are imposed to each element.

$$\begin{cases} \displaystyle\int_{\Gamma_e^{\mathrm{IF}}} p_{\mathrm{B}} dS = \int_{\Gamma_e^{\mathrm{IF}}} p_{\mathrm{F}} dS, \\ \displaystyle\int_{\Gamma_e^{\mathrm{IF}}} v_{\mathrm{B}} dS = -\int_{\Gamma_e^{\mathrm{IF}}} v_{\mathrm{F}} dS. \end{cases} \tag{8.73}$$

These equations ensure the continuity of the forces and the volumetric velocity across an interface element, respectively.

Example 1: Single Layer Porous Material

Figure 8.11 shows the random and field incidence absorption coefficients calculated for a single layer porous material set on the bottom of a rectangular cavity with different depths. Figure 8.12 shows the polar angle-dependent absorption coefficients in the same conditions. The Craggs' method [17] was employed for the porous material domain, thus the frame vibration of the porous material was not taken into account here. The effective density and bulk modulus of porous material was evaluated by Kato's model [18] with the material density $\rho_M = 1\,186\,[\text{kg/m}^3]$, the bulk density $\rho_b = 50\,[\text{kg/m}^3]$ and the fiber diameter $D = 21\,[\mu\text{m}]$. In Fig. 8.11, there are also plotted one-third octave band values measured at two small reverberation rooms with the volume 6.5 and 36 $[\text{m}^3]$. In the measurement, different height fences are placed in the perimeter of the material. Furthermore, transfer matrix method (TMM) [14] is employed for the calculation of theoretical values to the infinite area material.

Calculated absorption coefficients remarkably exceed the theoretical values at mid-frequency range due to the area effect. As the cavity becomes deeper, the values approach the theoretical one from high frequency range, which support the validity of a deep-well approach in the measurement by reverberation room method. However for deeper cavities, fluctuations also appear in the absorption coefficients due to the resonance to opposed faces of the cavity. In general, absorption coefficients in random incidence become higher than those in field incidence. Figure 8.12 shows the reason that the discrepancies between the calculated and theoretical values become remarkable at grazing incidence in particular.

Compared to the measurement values, calculated values in field incidence give satisfactory results particularly for the material set in the shallow cavity. The primary factor of the discrepancy seen in the deeper well is considered to be the difference of the geometrical condition around the absorbing material.

Example 2: Layered Poroelastic Materials

Figure 8.13 shows the normal, random and field incidence absorption coefficients calculated for double and triple-layer materials set on the bottom of the cavity. The material properties are listed in Table 8.2. Kato's model was used to evaluate the effective density and bulk modulus of the poroelastic materials. The membrane is limp and impervious. In the finite element formulation, the coupling condition between a poroelastic material and a membrane can be attained by an analogy with the poroelastic–elastic interface. There is thin air gap between the layered material and bottom of the cavity. This boundary condition is attained naturally in the $\mathbf{u}^s - p$ formulation. For P_2AMP_1 and P_2MAP_1, there are also plotted one-third octave band values measured at the same reverberation rooms as mentioned in the Example 1.

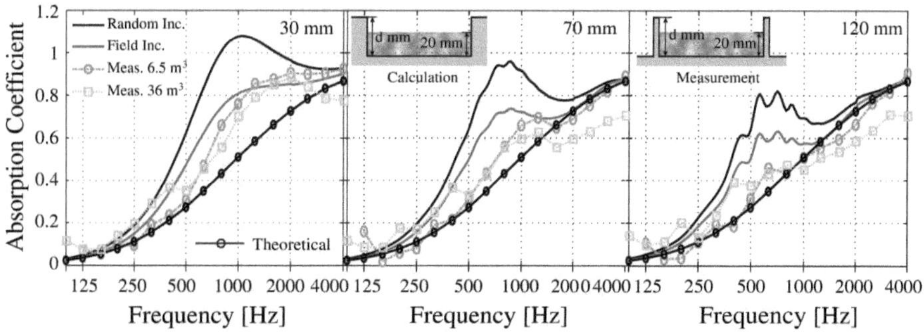

Fig. 8.11 Random and field incidence absorption coefficient calculated for a single layer porous material set on the bottom of the rectangular cavity with different depths. The material thickness is 20 mm, and the cavity and material area is 1 m^2

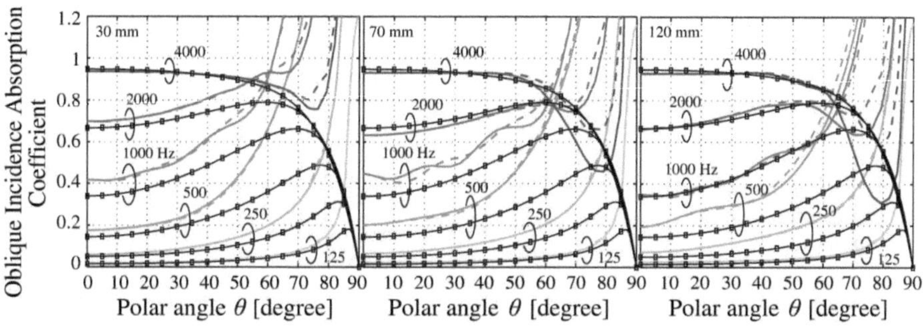

Fig. 8.12 Polar angle-dependent absorption coefficient at 1/1 octave center frequencies calculated for the same condition as in Fig. 8.11. The incidence azimuths φ are 0° and 45° shown in solid and dotted lines, respectively

Fences with 30mm height are placed in the perimeter of the layered material. TMM is also employed for the calculation of theoretical values to the infinite area material.

Remarkable peaks appear at the low- and mid-frequency range when a membrane is stacked between two poroelastic materials. First peaks in triple-layer materials are caused by the mass-spring resonance; the spring is that of the air saturated in PEM1. The second peak in P_2MP_1 is also caused by the mass-spring resonance; however, in this case the spring is that of the frame of PEM1. In case the membrane is in contact with the poroelastic material on the incidence side (PEM2), absorption coefficients decrease at high frequency range because the impervious layer becomes heavier and stiffer.

In general, tendencies of discrepancy between calculated and theoretical values are similar to those of the single layer material. Accordingly, normal incidence absorption coefficients agree well with the theoretical ones, whereas field and random incidence values exceed the theoretical ones due to the remarkable area effect at the grazing incidence. Compared to measured values, calculated values show similar trend for

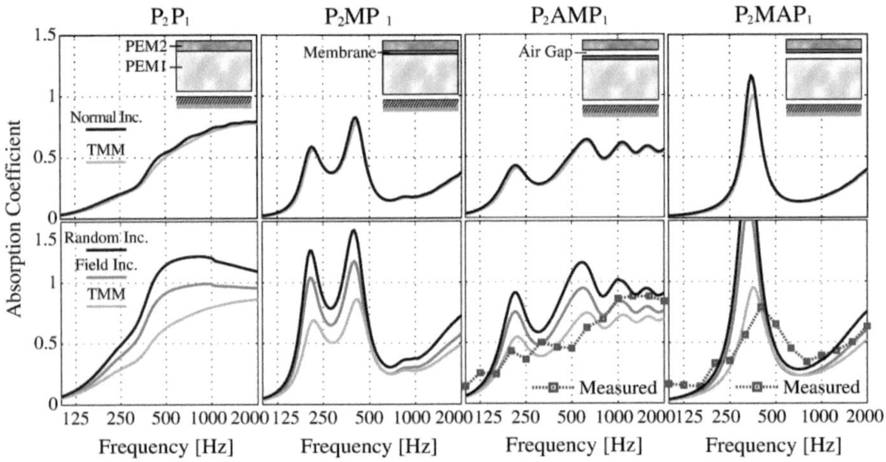

Fig. 8.13 Normal, random and field incidence absorption coefficient calculated for the layered poroelastic materials. The total material thickness is 25 mm, the cavity depth is 30 mm and the cavity and material area is 1 m^2

Table 8.2 Physical properties and dimensions of layered materials for the calculation

PEM 1 and 2	Material Density	$\rho_M = 1186\,(\text{kg/m}^3)$
	Fiber Diameter	$D = 21\,(\mu\text{m})$
	Poisson's Ratio	$\nu = 0$
PEM 1	Bulk Density	$\rho_b = 50\,(\text{kg/m}^3)$
	Young's Modulus	$E = 2.2 \times 10^4\,(\text{N/m}^2)$
	Loss Factor	$\eta = 0.24$
	Thickness	$t = 20\,(\text{mm})$
PEM 2	Bulk Density	$\rho_b = 200\,(\text{kg/m}^3)$
	Young's Modulus	$E = 1.5 \times 10^5\,(\text{N/m}^2)$
	Loss Factor	$\eta = 0.45$
	Thickness	$t = 5\,(\text{mm})$
Membrane	Area Density	$\rho_m = 0.04\,(\text{kg/m}^2)$

both of P_2AMP_1 and P_2MAP_1. Quantitative discrepancies may be caused by the lack of diffusivity in the measurement field and/or the difference of contact condition.

8.2 Diffusers

In room acoustics design, scattering properties of architectural walls and components are important factors. However, measurement of the scattering properties is not easy because it is required to observe a complicated scattering phenomenon in spatial and temporal aspects. Numerical simulation enables detailed examinations on transient scattering, frequency characteristics of reflection directivity, and so on. Furthermore,

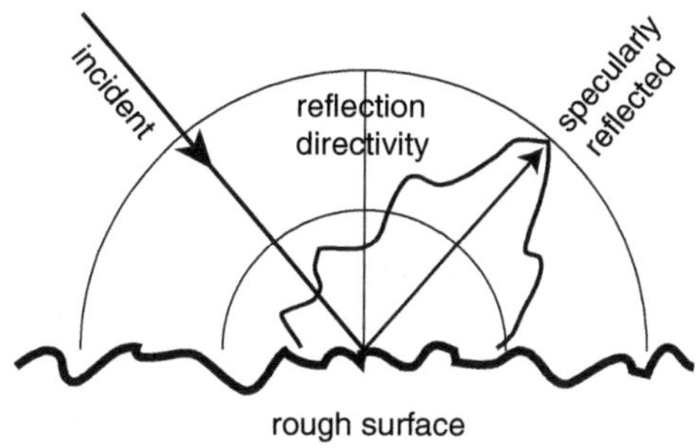

Fig. 8.14 Reflection directivity and specular reflection

it is applicable to estimate the two typical indicators, diffusion coefficient and scattering coefficient. This section introduces application of the boundary element method to the estimation of scattering properties [19].

8.2.1 BE Analysis of Surface Scattering

8.2.1.1 Indicators of Surface Scattering

When a sound wave impinges on a surface, reflected waves spread over the space with directivity, as shown in Fig. 8.14. The reflected waves are divided into a specularly reflected component that obeys Snell's law, and a nonspecularly reflected component that is scattered in other directions. To characterize surface scattering, two kinds of indicators were proposed: diffusion coefficient as a measure of uniformity of reflection directivity and scattering coefficient as the ratio of nonspecularly reflected energy to total reflected energy. The diffusion coefficient is defined as the average of directional autocorrelation coefficients in reflection directivity under arbitrary plane wave incidence [20, 21], which is represented by

$$d_\theta = \frac{\left(\sum_{i=1}^{n} E_i\right)^2 - \sum_{i=1}^{n} E_i^2}{(n-1)\sum_{i=1}^{n} E_i^2}, \tag{8.74}$$

where E_i is the reflected energy to the ith direction, and n is the total number of directions. In the case of perfectly specular reflection, $d_\theta = 0$, while, in the case of perfectly diffuse reflection, $d_\theta = 1$. The scattering coefficient is defined as the ratio of nonspecularly reflected energy to total reflected energy [22, 23], which is represented by

$$s_\theta = 1 - \frac{E_{\text{spec}}}{E_{\text{total}}} = \frac{\alpha_{\text{spec}} - \alpha}{1-\alpha}, \tag{8.75}$$

where E_{total} is the total reflected energy, E_{spec} is the specularly reflected energy, α is the absorption coefficient of a surface, and α_{spec} is the specular absorption coefficient that considers nonspecularly reflected energy to be apparently dissipated. In the case of perfectly specular reflection, $s_\theta = 0$, whereas $s_\theta = 1$ merely represents no energy in specular reflection direction. For a surface of finite size, the above two indicators can be numerically estimated in a free field. Regarding scattering coefficient, specularly reflected energy cannot be directly determined due to diffracted waves from the edge of the surface. As an alternative way, the directivity correlation method [24] was proposed to determine a scattering coefficient from two reflection directivities for a test sample and for a flat reference surface. Under arbitrary plane wave incidence, the directional scattering coefficient is given by

$$s_\theta = 1 - \frac{\left| \sum_{i=1}^{n} p_i \cdot \hat{p}_i^* \right|^2}{\sum_{i=1}^{n} |p_i|^2 \sum_{i=1}^{n} |\hat{p}_i|^2}, \tag{8.76}$$

where p_i and $\hat{p}_i$ denote the complex sound pressure of reflection wave in the ith direction for the sample and for the flat surface, respectively.

8.2.1.2 Calculation of Reflection Directivity

The boundary element method (BEM) is effective for the free field analysis of reflection directivity of a surface. If a test sample is considered as a rigid thin panel with zero thickness, the indirect BEM using degenerate boundary can be applied more efficiently (see Sect. 4.2.1). Applying the indirect method to a sample with plane wave incidence as shown in Fig. 8.15, the following equation is derived from Eq. (4.60):

$$\boldsymbol{H}' \tilde{\boldsymbol{p}} = -\boldsymbol{p}'_{\text{d}}, \tag{8.77}$$

where $\tilde{\boldsymbol{p}}$ is the sound pressure difference vector for the two sides of the sample, and $\boldsymbol{p}'_{\text{d}}$ is the pressure gradient vector of the incident wave. Considering a plane wave incidence with unit amplitude and wave number vector $\boldsymbol{k}_{\text{d}}$, the entries of $\boldsymbol{p}'_{\text{d}}$ are given by

$$p'_{\text{d},i} = \frac{\partial}{\partial n_i} \exp\left(j \boldsymbol{k}_{\text{d}} \cdot \boldsymbol{r}_i\right). \tag{8.78}$$

After the nodal sound pressure differences are obtained by solving the above linear system, the sound pressure of reflected wave at an observation point p is calculated in the postprocessing by

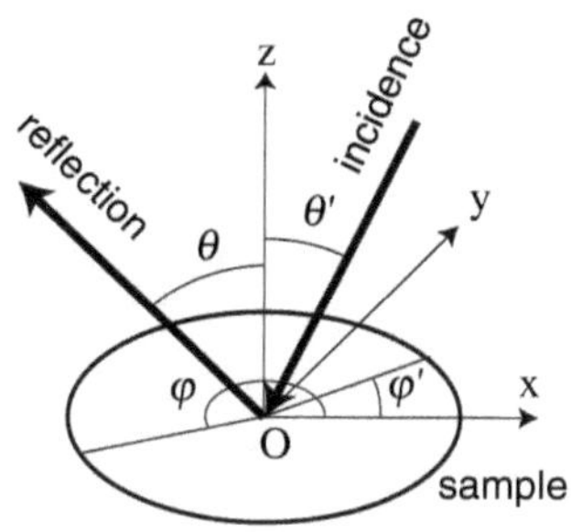

Fig. 8.15 Schematic for reflection directivity analysis in a free field

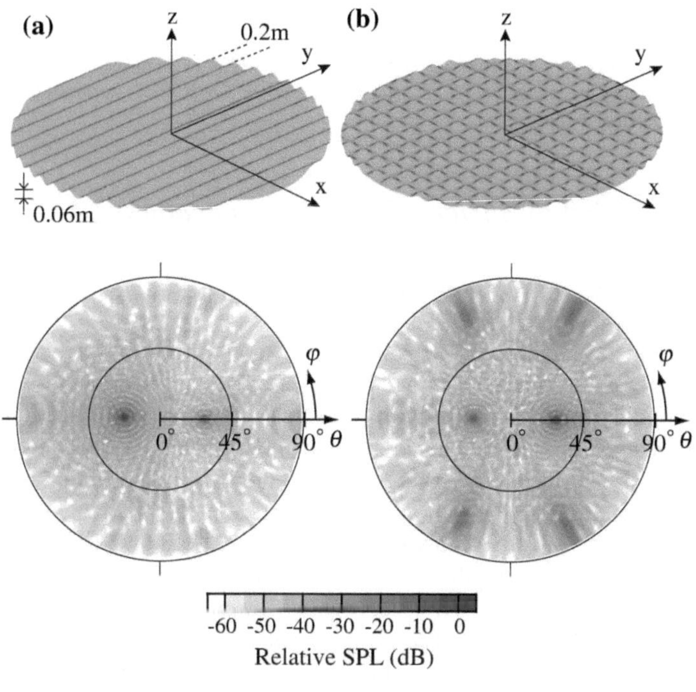

Fig. 8.16 Reflection directivities of the 1D and 2D sinusoidal surfaces (diameter: 3 m, period: 20, height: 6 cm), with plane wave incidence of $\theta = 31.5°$ and $\varphi = 181.5°$, at 2 kHz. **a** 1D sinusoidal, **b** 2D sinusoidal

$$p(r_\mathrm{p}) = \boldsymbol{h}^T(r_\mathrm{p})\tilde{\boldsymbol{p}}. \tag{8.79}$$

In the calculation of reflection directivity, observation points should be arranged on a hemisphere sufficiently far from the center of the sample. For typical samples with a width of 3 m, the angular resolution should be less than 5° for 2 kHz, and 3° for 4 kHz [25]. For example, Fig. 8.16 shows reflection directivities for two kinds of sinusoidal surfaces under a plane wave incidence.

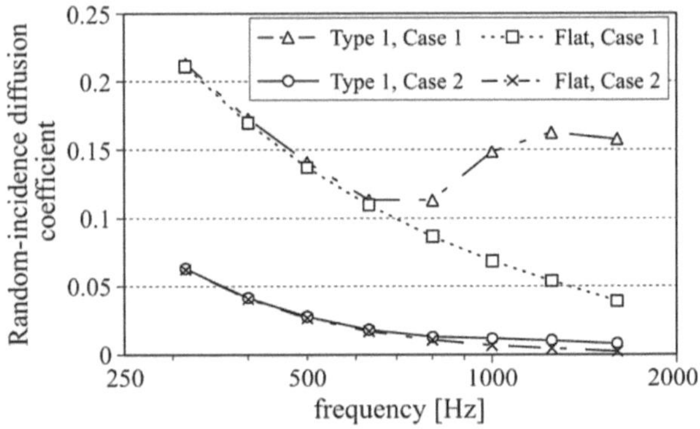

Fig. 8.17 Random-incidence diffusion coefficients of the 1D sinusoidal surface (Type 1: period: 20, height: 6 cm) and the flat panel (Flat) with a diameter of 3 m. Case 1: semicircular evaluation; Case 2: hemispherical evaluation

8.2.2 Calculation of Surface Scattering Indicators

8.2.2.1 Estimation of Diffusion Coefficient

From the reflection directivity calculated as mentioned above, the directional diffusion coefficient can be determined by Eq. (8.75), and then the random-incidence diffusion coefficient is obtained with statistical averaging of directional values. Figure 8.17 shows random-incidence diffusion coefficients for a 1D sinusoidal surface and for a flat surface [26], where the coefficients are determined in two ways: semicircular evaluation in the orthogonal section and hemispherical evaluation. The former evaluation gives values close to the results by 2D BEM [27], whereas the latter evaluation gives very small values, suggesting that it is unsuitable for single-plane surfaces. The diffusion coefficient increases at lower frequencies due to edge diffraction regardless of surface profile, and the scattering effect of surface unevenness appears above about 800 Hz. In the estimation of diffusion coefficient for a sample, the additional effect of edge diffraction can be excluded by normalizing with the value of the flat surface $d_{\theta,\mathrm{r}}$, such as [21]

$$d_{\theta,\mathrm{n}} = \frac{d_\theta - d_{\theta,\mathrm{r}}}{1 - d_{\theta,\mathrm{r}}}. \tag{8.80}$$

8.2.2.2 Estimation of Scattering Coefficient

The directional scattering coefficient can be also determined from the reflection directivity by Eq. (8.74), and the random-incidence scattering coefficient is given with statistical averaging. For the three types of 1D periodical surface with 3 m^2 (Fig. 8.18),

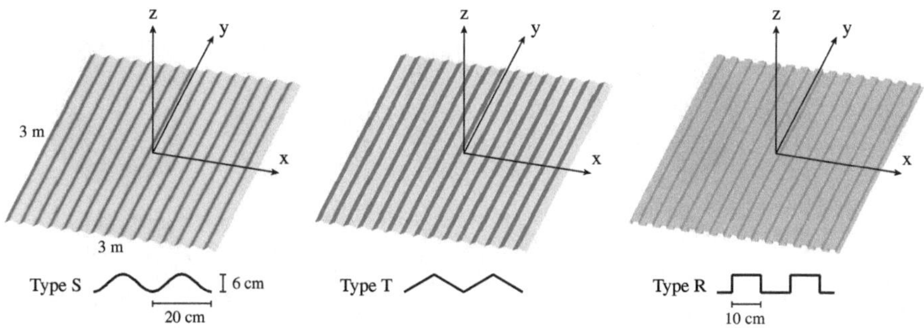

Fig. 8.18 Three types of surface: Type S: sinusoid, Type T: triangular wave, Type R: rectangular wave

distributions of directional scattering coefficients are shown in Fig. 8.19. The effect of surface profile is similar below 1 kHz, whereas above 2 kHz, the rectangular wave type has low values in wide direction compared with the sinusoid and the triangular wave types. Figure 8.20 shows random-incidence scattering coefficients for the two types of sinusoid and rectangular wave, with changing the height to period. In the low frequency range with $d/\lambda < 1/2$, the two types take the greatest values at $h/d = 50$ [%]. In the high frequency, the sinusoid type generally has the greatest values at $h/d = 30$ [%], while the rectangular wave type has peaks and dips due to interference between the top and bottom surfaces, and the greatest peak value at $h/d = 20$ [%]. As demonstrated above, numerical analysis of diffusers is useful to investigate effective surface profiles.

8.3 Insulation of Windows

To obtain the sound insulation performance of wall structures in the buildings, experiments in the laboratory or in-situ measurements are performed following the standardized methods such as ISO. In such a situation, much more effective designing of buildings is possible by applying thc wave-based numerical simulation technique in the planning phase. Many researches are performed as follows; the analytical solution by London [28], four-terminal network theory by Beranek [29], SEA method by Lyon [30]. On the other hand, the vibroacoustical simulation method based on the wave-based numerical schemes are also applied to the sound transmission problems [31–34]. In such a method applying the wave-based scheme, the finiteness of the plates and the vibration reflection and absorption on the boundary of the plates can be easily treated, because the method discretizes the vibration field into multiple meshes and that can model the shape of the field accurately. In this section, as an example of the numerical simulation for the acoustical performance of the building elements

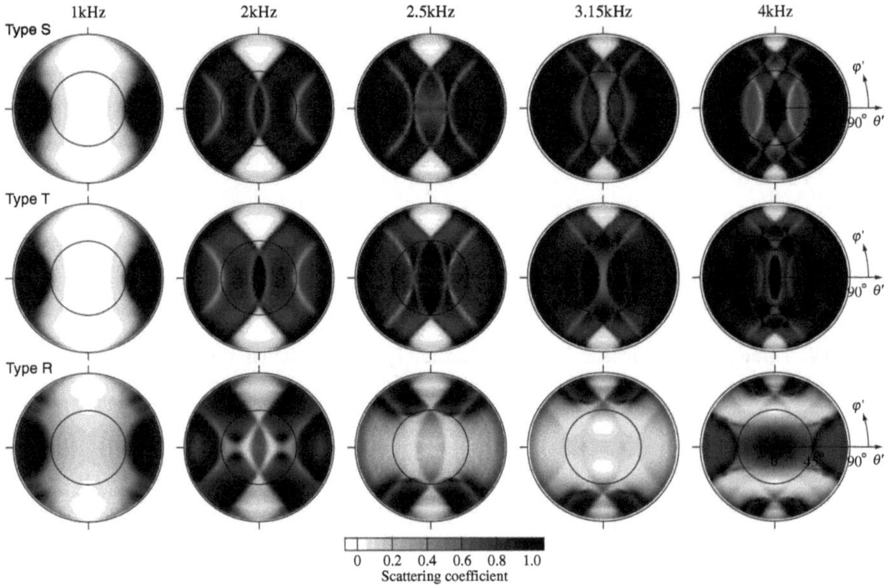

Fig. 8.19 Distribution of directional scattering coefficients for the three types of surface

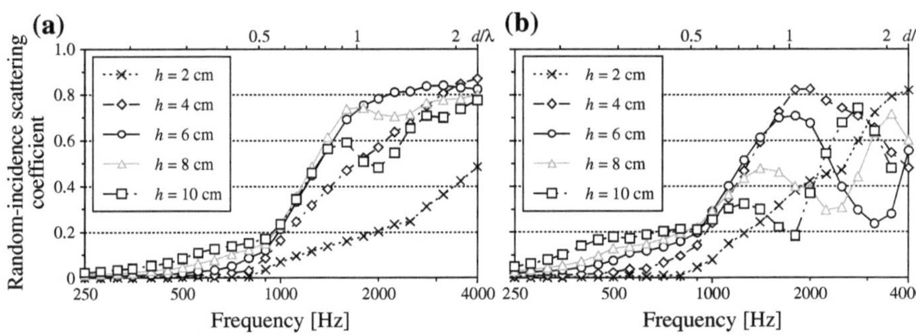

Fig. 8.20 Random-incidence scattering coefficients with changing the height of structure (period: 20 cm): **a** sinusoid, **b** rectangular wave

based on the wave-based scheme, a simulation study using the finite-difference time-domain method and the boundary element method are described.

8.3.1 Analysis Using BEM

8.3.1.1 Detail of Method

A detailed numerical method used in this section is described below. In the case study of this section, the vibration field on the plate element and two semi-free sound fields on the both sides of the plate are simulated by applying a coupling method. As for the sound field, the boundary element method is applied to the integral equation on the sound pressure difference between the sound field on the both sides of the plate. Finally, we obtain

$$\Delta \boldsymbol{p} = 2\boldsymbol{p}_{\mathrm{d}} + 4\omega^2 \rho_0 \boldsymbol{G}\boldsymbol{w}, \tag{8.81}$$

where $\Delta \boldsymbol{p}$ is the vector for the sound pressure difference, $\boldsymbol{p}_{\mathrm{d}}$ is that for the incident sound pressure, $\boldsymbol{w}$ is that for the displacement of the plate vibration, $\boldsymbol{G}$ is the matrix for the influence function, and ρ_0 is the density of the air. As for the vibration field, the finite element method is applied to the vibration equation of the thin plate theory by Kirchhoff, and finally we obtain

$$(\boldsymbol{K} - \omega^2 \boldsymbol{M}) \begin{Bmatrix} \boldsymbol{w} \\ \boldsymbol{\theta}_x \\ \boldsymbol{\theta}_y \end{Bmatrix} = \boldsymbol{Q}\Delta \boldsymbol{p}, \tag{8.82}$$

where $\boldsymbol{\theta}_x, \boldsymbol{\theta}_y$ are the vectors for the inclination angle of the plate in the x and y direction, $\boldsymbol{K}$, $\boldsymbol{M}$, $\boldsymbol{Q}$ are the stiffness, mass, and conforming matrix, respectively. As damping of the plate, the internal damping loss of the plate is considered by setting appropriate values for the stiffness matrix, and the damping loss at the boundary of the plate is considered by the edge damping model described in the following section. Two equations described above are simultaneously solved, and the displacement of the vibrating plate can be obtained. The vibration velocity and the sound pressure level in the near filed of the plate surface are also calculated based on the obtained results.

8.3.1.2 Vibration Energy Loss at Edge of Plate

To predict accurately the sound transmission loss of a glass plate, the vibration and damping characteristics of the edge supporting parts should be appropriately considered in the simulation. Especially, the damping loss of the vibration at the edge of the plate has a great influence on the sound transmission loss. For such a reason, the vibration mechanism of the edge supporting part is replaced by a simple mechanical impedance model. A detailed procedure is described below. Based on the assumption that the sealing materials surrounding and fixing the plate counteract to the displacement and inclination of the edge part, the translational and rotary

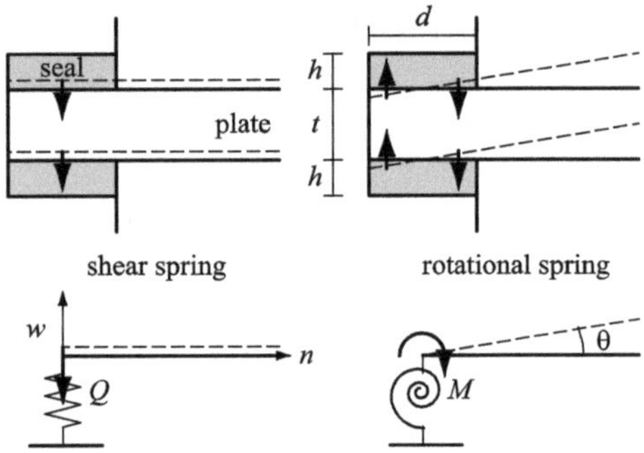

Fig. 8.21 Modeling of the plate edge part by the translational and rotary springs

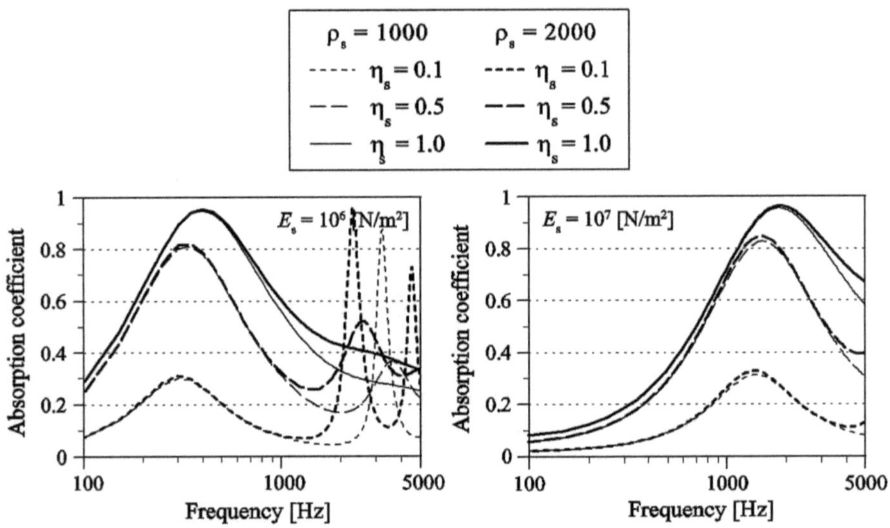

Fig. 8.22 Random incidence absorption coefficient by the elastic continuum mode (Thickness t = 10 [mm])

spring model equivalent to elastic body is considered. A modeling scheme is shown in Fig. 8.21. Here, the damping of the vibration energy at the edge is caused due to the internal damping of the spring. Assuming a one-dimensional longitudinal wave-motion to the translational deformation of the sealing material, the mechanical impedance of the edge can be described as:

$$Z_{\mathrm{q}} = \frac{2\rho_{\mathrm{s}} c_{\mathrm{s}} d}{j \tan \omega h / c_{\mathrm{s}}}, \tag{8.83}$$

where $c_s = \sqrt{E_{\mathrm{s}}(1 + j\eta_{\mathrm{s}})/\rho_{\mathrm{s}}}$ is the velocity of the longitudinal wave of the sealing material, $E_{\mathrm{s}}, \eta_{\mathrm{s}}, d, h, \rho_{\mathrm{s}}$ are the Young's modulus, loss factor, width, thickness, and density of the sealing material. In addition, the moment impedance Z_m is described as

$$Z_m = \frac{M}{j\omega\theta} = \frac{K_\mathrm{m}}{j\omega} = \frac{2}{j\omega}\frac{E_\mathrm{s}(1+j\eta_\mathrm{s})d^3}{12h}, \tag{8.84}$$

where K_m is the stiffness of the rotary spring, and M is the bending moment. The vibration energy loss can be theoretically calculated by the equation for the impedance at the edge part.

To investigate the effect of the glass thickness and the physical parameters of the sealing material, a calculation example is shown. Then, including the sound transmission loss calculation in the next section, the physical parameters of the glass plate are set as follows; the density $\rho = 2,500$ [kg/m^3], the Young's modulus $E = 7.5 \times 10^{10}$ [N/m^2], Poisson's ratio $\nu = 0.22$, the loss factor $\eta = 0.002$. The random incidence absorption coefficient of the vibration at the edge for the situation that the density of the sealing material is $\rho_\mathrm{s} = 1,000, 2,000$ [kg/m^3] is shown in Fig. 8.22. In situation that the sealing material has relatively low elasticity ($E_\mathrm{s} = 10^6$ [N/m^2]), some peaks are seen in relatively higher frequency range. These peaks are caused by the resonance of the sealing material itself. In the following section, the numerical simulation considering the impedance model for the edge part described in this section is performed.

8.3.1.3 Calculation Method for Sound Transmission Loss

Based on the coupling simulation method described above, the distribution of the vibration velocity and the sound pressure near the plate surface is calculated in every incident angle (Fig. 8.23), and the intensity for the transmitted sound wave is obtained. By integration of the transmitted intensity through the plate, transmitted sound power is calculated, and the transmission rate through the plate in every incident angle is obtained.

Consequently, the random incidence sound transmission loss is obtained by weighted summation of the transmission rate in each incident angle.

8.3.1.4 Comparison Between Calculation and Measurement [35]

The calculation results of the random-incidence sound transmission loss considering the energy loss at the edge parts are compared to the measurement results with putty mounting are shown in Fig. 8.24. The loss factor of the sealing material is set as $\eta_\mathrm{s} = 0.5$ which corresponds to that of the putty. In case of 6 mm thickness, the calculation results agree well with the measurement ones regardless of the Young's modulus of the sealing material. In all the cases of 6 and 10 mm thickness, the calculation and measurement results in the condition with $E_\mathrm{s} = 10^7$ [N/m^2] have the highest correlation.

Next, the measurement and calculation results with silicone mounting are shown in Fig. 8.25. The loss factor of the sealing material is set as $\eta_\mathrm{s} = 0.1$ considering

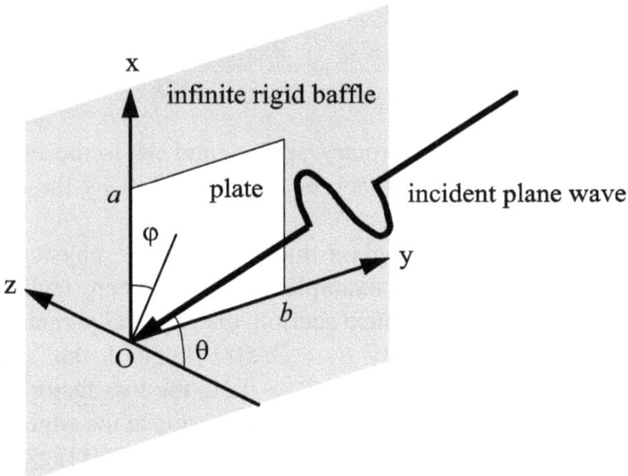

Fig. 8.23 Three-dimensional sound transmission model

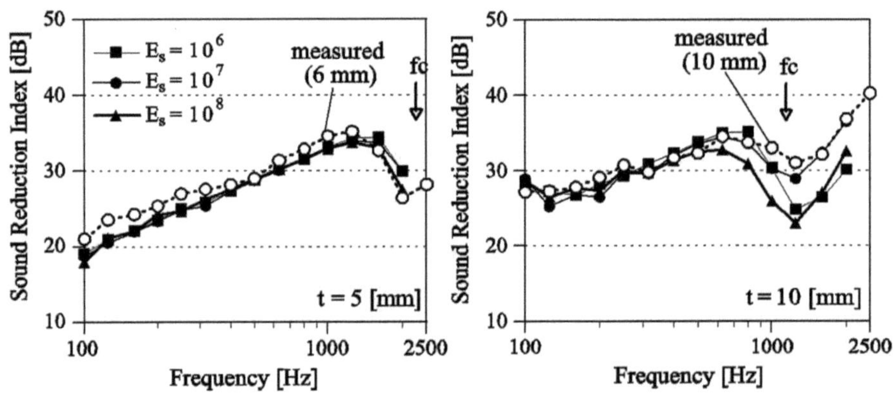

Fig. 8.24 Calculation and measurement results of the random-incidence sound transmission loss in situation of putty mounting in the surrounding parts of the glass plate. (the plate size 1.25 × 1.5 [m^2], the density of the sealing material ρ_s =1 000 [kg/m^3], the loss factor of the sealing material, $\eta_s = 0.5$)

the silicone. The condition with a Young's modulus of $E_s = 10^6$ [N/m^2] has the highest correlation, however they have some discrepancy at around the coincidence frequency. The contacting area between the silicone material and the glass pane is smaller than that in the case with putty mounting [35], and it is considered as the primary factor of the discrepancy that the edge supporting model considered in the simulation is not accurately agreeing with the actual phenomena in that view point.

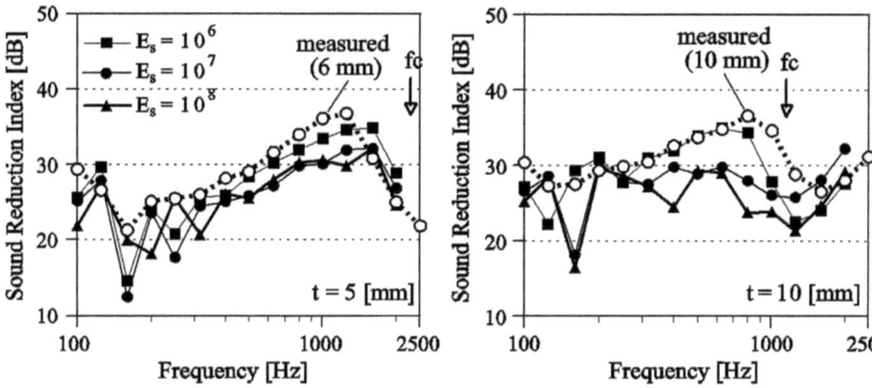

Fig. 8.25 Calculation and measurement results of the random-incidence sound transmission loss in situation of silicone mounting in the surrounding parts of the glass plate. (the plate size 0.9×0.9 [m^2], the density of the sealing material $\rho_s = 1\,000$ [kg/m^3], the loss factor of the sealing material, $\eta_s = 0.1$)

8.3.2 Analysis Using FDTD Method

8.3.2.1 Detail of Method

The momentum equations and the continuity equation for three-dimensional sound field are described as:

$$\frac{\partial p(x, y, z, t)}{\partial x} + \rho_0 \frac{\partial u_x(x, y, z, t)}{\partial t} = 0, \tag{8.85}$$

$$\frac{\partial p(x, y, z, t)}{\partial y} + \rho_0 \frac{\partial u_y(x, y, z, t)}{\partial t} = 0, \tag{8.86}$$

$$\frac{\partial p(x, y, z, t)}{\partial z} + \rho_0 \frac{\partial u_z(x, y, z, t)}{\partial t} = 0, \tag{8.87}$$

$$\frac{\partial p(x, y, z, t)}{\partial t} + \rho_0 {c_0}^2 \left[\frac{\partial u_x(x, y, z, t)}{\partial t} + \frac{\partial u_y(x, y, z, t)}{\partial t} + \frac{\partial u_z(x, y, z, t)}{\partial t} \right] = 0, \tag{8.88}$$

where p is sound pressure, u_x, u_y, u_z are particle velocities for x, y, z directions, ρ_0 and c_0 are the density of the air and speed of sound, respectively. A transient response in a sound field is calculated by solving the discretized equations obtained from the basic equations through finite-difference approximation. A detailed procedure of the calculation is described in the reference [36]. As for the bending wave simulation, the vibration equation based on the thin plate theory by Kirchhoff,

$$D\nabla^4 w + \xi D \frac{\partial}{\partial t} \nabla^4 w + m\mu \frac{\partial w}{\partial t} + m \frac{\partial^2 w}{\partial t^2} = F, \tag{8.89}$$

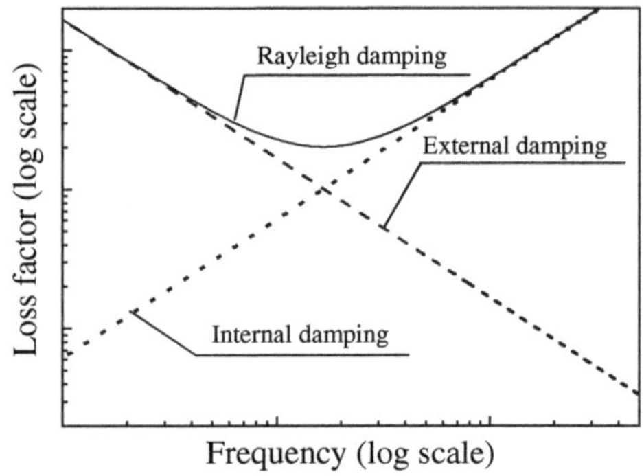

Fig. 8.26 Frequency characteristics of the loss factor simulated in the FDTD analysis

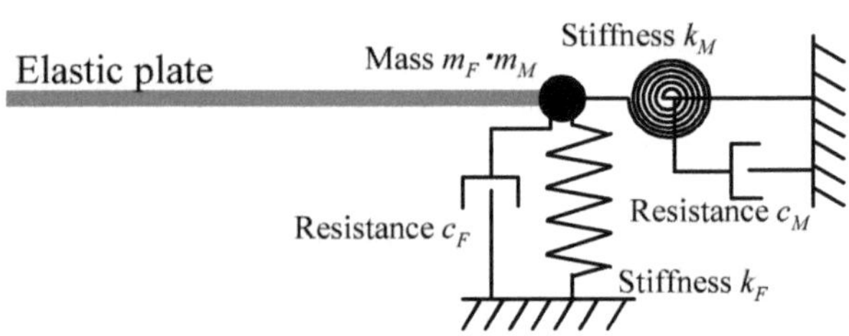

Fig. 8.27 Mechanical equivalent impedance model applied in the vibration analysis

$$D = \frac{Eh^3}{12(1 - \nu^2)}, \tag{8.90}$$

$$F = p_1 - p_2, \tag{8.91}$$

are used. Here, w is the displacement, E is the Young's modulus, h is the thickness, ν is the Poisson's ratio, and p is the external force to the plate. The external force is obtained by the difference between p_1 and p_2, which are the sound pressures on both sides of the plate. In Eq. (8.89), ξ and μ in the second and third term of the left side are the coefficients related to the internal and external damping. The relationship between the coefficients of ξ and μ, and the loss factor is described as

$$\eta = 2\pi f \xi + \frac{\mu}{2\pi f}, \tag{8.92}$$

where η is described as a function of frequency. The frequency characteristics of the modeled loss factor are shown in Fig. 8.26. As shown in the figure, the external/internal damping is in proportion/inverse proportion to the frequency, and the Rayleigh damping obtained as the superposition of the two damping characteristics has curved shape convex downward. To model the actual frequency characteristics of the loss factor, suitable coefficients of ξ and μ are set, and used in the simulation.

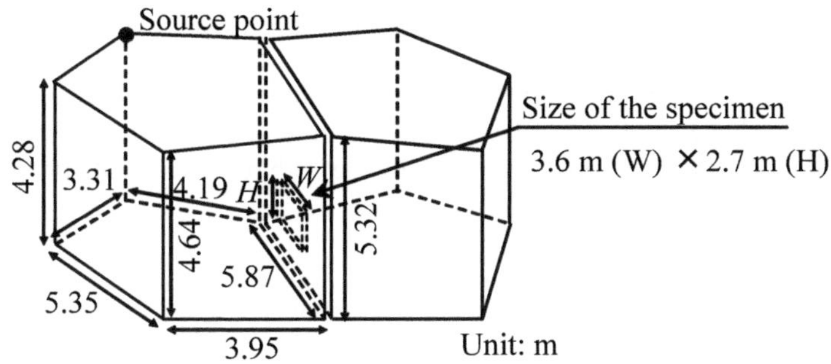

Fig. 8.28 Pair of irregular-shaped reverberation rooms under investigation

8.3.2.2 Vibration Energy Loss at Edge of Plate

As mentioned in Sect. 8.3.1, the energy loss of the bending wave at the elastically supported edge part of the glass pane should be considered to simulate the sound insulation accurately. Therefore, a mechanical equivalent impedance model in which the boundary condition is modeled by an equivalent system comprising masses, springs, and dampers (see Fig. 8.27) is adopted. The assumed mechanical equivalent system vibrates when this system is forced by a shear force F or a flexural moment M, and this vibration is described by the following momentum equations:

$$F = m_F \frac{\partial^2 w_0}{\partial t^2} + c_F \frac{\partial w_0}{\partial t} + k_F w_0, \tag{8.93}$$

$$M = m_M \frac{\partial^2 w_0}{\partial t^2} + c_M \frac{\partial w_0}{\partial t} + k_M w_0, \tag{8.94}$$

where m_F, c_F, k_F, m_M, c_M and k_M are the mass, resistance coefficient, and stiffness of the assumed mechanical system actuated by M or F, respectively. w_0 is the displacement at the edge. The simulation was performed by solving the finite-difference forms of Eqs. (8.89), (8.93), (8.94) simultaneously. Here, F and M are calculated by:

$$F = -D\left[\frac{\partial}{\partial x}\left(\frac{\partial^2 w}{\partial x^2} + \frac{\partial^2 w}{\partial y^2}\right) + (1-\nu)\frac{\partial^3 w}{\partial x \partial y^2}\right], \tag{8.95}$$

$$M = -D\left(\frac{\partial^2 w}{\partial x^2} + \nu \frac{\partial^2 w}{\partial y^2}\right). \tag{8.96}$$

8.3.2.3 Calculation Scheme of Sound Transmission Loss

A pair of irregular-shaped reverberation rooms shown in Fig. 8.28 is simulated. The both rooms have a volume of $164\,\mathrm{m}^3$. In the simulation, 490 receiving points are set, and the averaged sound pressure levels inside both rooms are calculated. Finally, the sound transmission loss is obtained by

$$R = \overline{L_1} - \overline{L_2} + 10\log_{10}\frac{S}{A_2}, \tag{8.97}$$

$$A_2 = 55.3V_2/cT_2, \tag{8.98}$$

where $\overline{L_1}$, $\overline{L_2}$ are the averaged sound energy levels in the rooms of source side and receiving side, S is the total area of the specimen, A_2 is the equivalent sound absorption area, V_2 and T_2 are the volume of the receiving room and the reverberation time of the receiving room, respectively. The reverberation time in each 1/3 Oct. band was calculated by the integrated impulse response method. The absorption coefficients of all surfaces inside the rooms were set to be 0.026.

8.3.2.4 Comparison with Measurement Results [37, 38]

As a case study, the sound transmission loss of the glass pane and the plasterboard wall is calculated. Setting conditions in the simulation is described as follows. As for the glass pane, the vibration energy loss at the elastically supported edge is considered by the method described in the preceding section, and the density $\rho = 2\,500$ [kg/m^3], the Young's modulus $E = 7.16 \times 10^{10}$ [N/m^2], the Poisson's ratio $\nu = 0.22$ are given. The loss factor of the glass pane was neglected. Referring to the measurement data by Yoshimura [35] shown as broken line in Fig. 8.29, the frequency characteristics of the normal incidence absorption coefficient at the edge part was simulated as shown by solid line in the same figure. In the figure, determined parameters are also shown. As investigated conditions, single glazed glass of 6 and 10 mm thickness, double-glazed glass (air layer of 6 mm thickness was sandwiched by two sheets of glass of 6 mm thickness) are treated. As for the plasterboard wall, the boundary condition is set as clamped edge, and the density $\rho = 670$ [kg/m^3], Young's modulus $E = 2.1 \times 10^9$ [N/m^2] and Poisson's ratio $\nu = 0.30$ are given. For the loss factor, the frequency characteristics shown by solid line in Fig. 8.30 were simulated by giving damping coefficients shown in the same figure. A single plasterboard of 12 mm thickness, double wall A (air layer of 80 mm thickness was sandwiched by two sheets of plasterboard of 12 mm thickness), double wall B (air layer and glass wools of 32 kg/m^3 of 80 mm thickness in total was sandwiched by two sheets of plasterboard of 12 mm thickness) are treated. Calculation results are shown in Fig. 8.31. As for the single- and double-glazed glasses, the calculation results agree well with the measurement results. As for the plasterboard, the results of the single wall and the double walls also agree with the measurement.

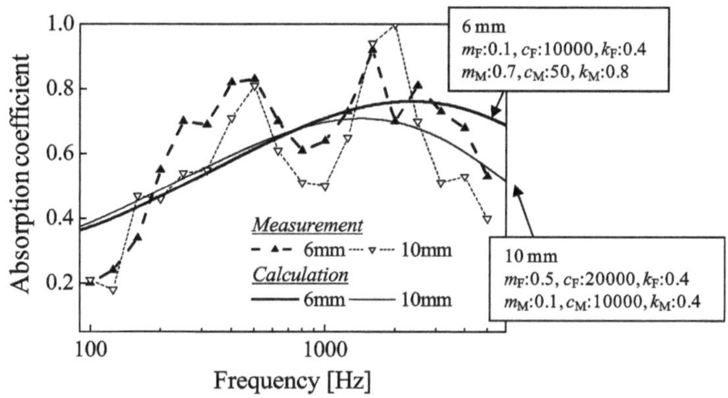

Fig. 8.29 Normal incidence absorption coefficient of the elastically supported edge simulated in the FDTD analysis

Fig. 8.30 Simulated loss factor of the plasterboard and the damping coefficients used in the simulation

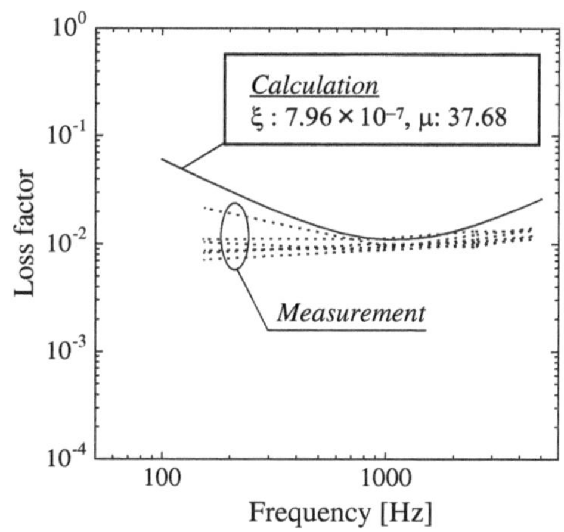

8.4 Acoustic Radiation from Loudspeakers

A great amount of knowledge about loudspeakers has been acquired and has for a long time been utilized in design of loudspeakers [39–41]. However, designing based on this conventional method has recently become more difficult than ever because of increasing of there various applications and the pursuit of design characteristics, functionality, and cost performance. Accordingly, optimization technology for high-sound performance, including installation conditions, is becoming increasingly important. To fulfill these demands and improve sound quality, acoustic simulations based on wave theory can be an effective way. Therefore, development and appli-

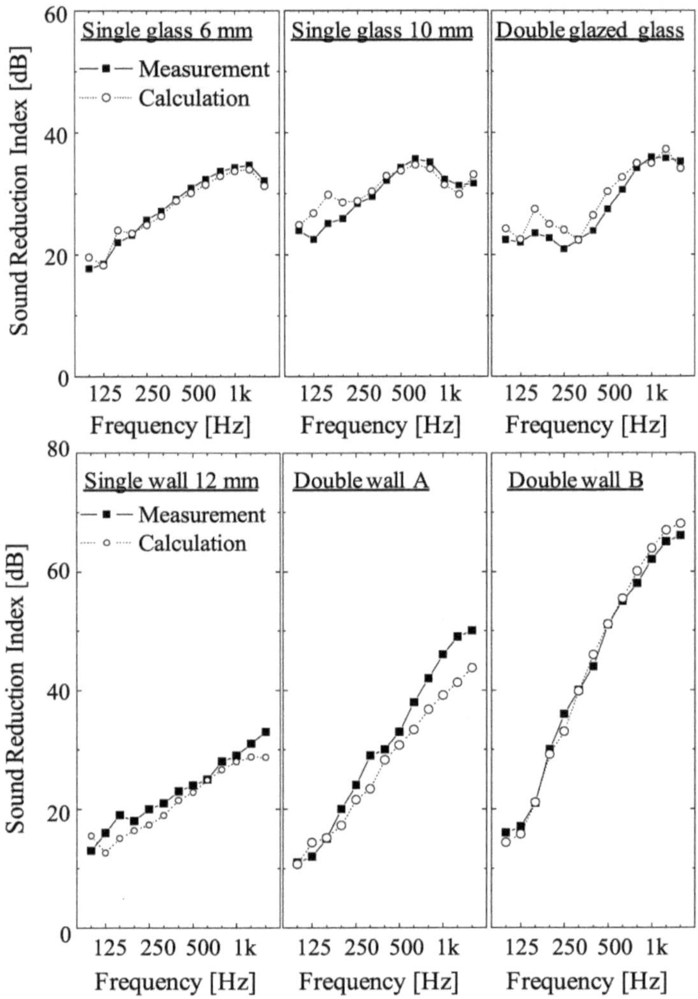

Fig. 8.31 Comparison of the sound transmission loss between calculation and measurement

cation of this technique is now expanding [42, 43]. The calculation procedure of the numerical simulations and its application for electrodynamics type loudspeakers at low frequencies are described herein. Evaluation of the sound characteristics of a horn speaker at the middle to high frequencies is also described.

8.4.1 Calculation Procedure

8.4.1.1 Basic Equations

A dynamic cone speaker consists of three systems: the electrical system that changes the electrical signal to mechanical vibration, the mechanical system that generates the acoustic wave, and the acoustic system that radiates the acoustic wave. These three systems are coupled with each other and feed the sound into space, so a numerical simulation model must be an electrical, mechanical, and acoustic coupled system. The coupled system is as follows. First, the electrical system is assumed to be one-dimensional due to the target frequency being limited to low. As a result, the input force to the mechanical system is also one-dimensional. On the other hand, because the modal coordinates are employed to model the mechanical motion (the three-dimensional finite element method is used to calculate the eigenvalues) and the boundary element method is employed for the acoustic system, the phenomena are modeled as three-dimensional.

–Electrical model

$$T_E(E - E_S) = I. \tag{8.99}$$

Here, T_E is the transfer function between the electric voltage and the current of the electric circuit. It is also equivalent to the inverse of the electric impedance Z_e of the voice coil. E is the electric voltage, I is the electric current, and E_S is the back electromotive force produced by the vibration velocity V of the cone. By applying two equations $E_S = AV$ where A (=BD) is force factor and $F_E = AI$ where F_E is the force produced by the coil Eq. (8.99) is changed to the following form:

$$T_E A(E - AV) = F_E. \tag{8.100}$$

–Mechanical system

$$T_S(F_E - F_A) = V. \tag{8.101}$$

Here, T_S is the structural transfer function between the force acting on the cone and its vibration velocity. F_A is the force applied by the acoustic wave.

–Acoustic model

$$T_A V = F_A. \tag{8.102}$$

Here, T_A is the acoustic transfer function between the vibration velocity of the cone and the force applied by the acoustic wave.

8.4.1.2 Calculation Method

–Structural-acoustic coupled model

Substituting Eq. (8.102) with Eq. (8.101), the following structural-acoustic coupled system is obtained:

$$T_{S-A} F_E = V. \tag{8.103}$$

Here, $T_{S-A} = T_S/(1 - T_S T_A)$ is the transfer function of structural-acoustic coupled system. For this system, the methodology described in the next section is applied. And T_{S-A} is the vibration velocity V obtained from the system whose input force F_E is equal to 1.

–Electrical–mechanical-acoustic-coupled model

By substituting Eq. (8.103) into Eq. (8.100), the electrical-mechanical-acoustic coupled system is obtained:

$$T_{E-S-A} E = F_E. \tag{8.104}$$

Here, $T_{E-S-A} = T_E A/(1 - A^2 T_E T_{S-A})$ is the transfer function of the electrical–mechanical-acoustic coupled system. The transfer function T_{E-S-A} is easily obtained because T_E is one-dimensional and T_{S-A} is obtained by the methodology described in the next section. For the calculation of T_E, T-S parameters opened by the manufacturer are effectively used.

8.4.1.3 Calculation Method for Transfer Function of Structural-Acoustic Coupled System

The vibration of the cone and the acoustic field inside and outside the speaker box have three-dimensional characteristics. Therefore the transfer function T_{S-A} that considers three-dimensional phenomena should be used for the prediction.

–Acoustic model

For the acoustic model, the boundary element method is applied. Applying the new boundary Γ_s that couples the structure to the boundary indicated in Fig. 4.2, Eq. (4.14) is changed to the following form.

$$\left(-\frac{1}{2}\boldsymbol{I} + \boldsymbol{H} + \boldsymbol{D}\right)\boldsymbol{p} - \boldsymbol{G}_s \boldsymbol{v}_s = \boldsymbol{G}\boldsymbol{v} - \boldsymbol{p}_d, \tag{8.105}$$

where $\boldsymbol{v}_s$ is the velocity vector obtained from $\boldsymbol{\varphi q}$ at the boundary Γ_s, $\boldsymbol{q}$ is obtained from Eq. (8.106), and $\boldsymbol{G}_{sij} = j\omega\rho \int_{\Gamma_s} G(\boldsymbol{r}_i, \boldsymbol{r}_q)\varphi_j(\boldsymbol{r}_q) d\Gamma$.

–Structural model

The following formulation is obtained by applying the modal coordinates to the structural vibration equation. Note that eigenvectors must be normalized to the mass matrix.

$$(-\omega^2 + 2j\boldsymbol{h}\omega\omega_0 + \omega_0^2)\boldsymbol{q} = \varphi^T(\boldsymbol{f} + \boldsymbol{s}^T \boldsymbol{p}), \tag{8.106}$$

where $\boldsymbol{h}$ is the diagonal matrix consisting of the viscous damping ratio, ω_0 is the diagonal matrix consisting of eigenvalues, $\boldsymbol{q}$ is the modal participation factor vector, $\boldsymbol{f}$ is the load vector, $\boldsymbol{s}$ is the area vector that the sound pressure acts on, and $\boldsymbol{p}$ on the right-hand side is the unknown in Eq. (8.105).

–Coupled model

The structural-acoustic coupled formulation is obtained by using Eqs. (8.105) and (8.106) simultaneously:

$$\begin{bmatrix} \boldsymbol{E}+\boldsymbol{B}+\boldsymbol{C} & \omega^2 \rho \boldsymbol{A}' \varphi \\ -\varphi^T \boldsymbol{s} & (-\omega^2 + 2j\boldsymbol{h}\omega\omega_0 + \omega_0^2) \end{bmatrix} \begin{Bmatrix} \boldsymbol{p} \\ \boldsymbol{q} \end{Bmatrix} = \begin{Bmatrix} j\omega\rho \boldsymbol{A}\boldsymbol{v} \\ -\varphi^T \boldsymbol{f} \end{Bmatrix}. \tag{8.107}$$

8.4.2 Numerical Examples

8.4.2.1 Characteristics of Closed-Type Loudspeaker at Low Frequency Range

Using a simple model (Fig. 8.32), in which a speaker unit (radius: 120 mm) is attached to the JIS box (940 × 1 240 × 640 mm^3), we conducted acoustic simulations with the boundary element method (Fig. 8.33). The transfer function obtained from these simulations and input condition of electric voltage (2 V) are used for the coupled analysis. The frequency response curve at the point 1.0 m distance in front of the speaker unit and electric impedance are shown in Fig. 8.34. These results correspond well to the measured data.

8.4.2.2 Sound Characteristics of Bass Reflex Type Loudspeaker at Low Frequency Range

The bass reflex type loudspeaker is characterized by high-sound performance at the low frequency range compared with closed-type loudspeaker (Figs. 8.35, 8.36). The radiated sound at the low frequency range is increased by the sound wave that is radiated to the inside of the cabinet by the cone and comes through the port (venting hole). This mechanism makes use of the Helmholtz resonance phenomenon, that can be approximately modeled resonance system by the air in the cabinet as spring (K_c) and the air in the port as mass (M_p) (Fig. 8.37).

The analysis case with a model (cabinet: 318×186×186 mm^3; port radius: 28 mm; port length: 80 mm; speaker unit radius: 130 mm) indicated in Fig. 8.38 is addressed here. The sound pressure response and electric impedance are shown in Fig. 8.39. There are two peaks in the electric impedance curve at the low frequency range, the lower one is due to the effect of the electrical–mechanical resonance system on which the mass of the air in the port has a dominant effect, while the higher is due to the effect of the electrical–mechanical resonance system on which the spring of the air in the cabinet has a dominant effect. The transfer function between the force acting on the cone and the radiated sound pressure on the cone is indicated in

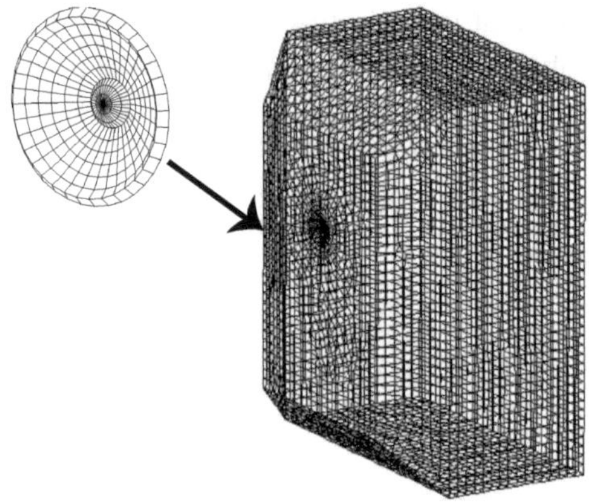

Fig. 8.32 Analysis model

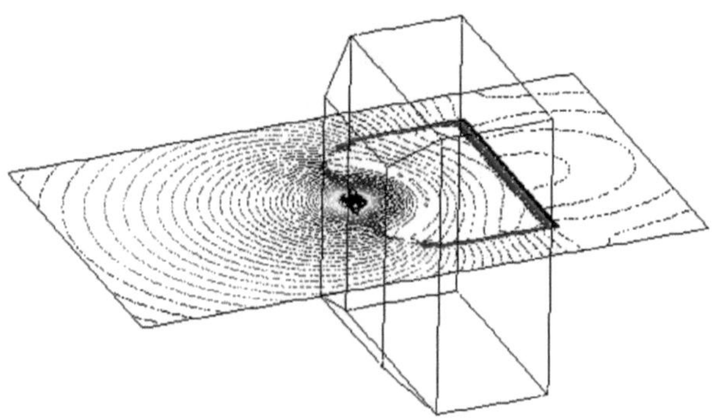

Fig. 8.33 Spatial distribution of sound pressure

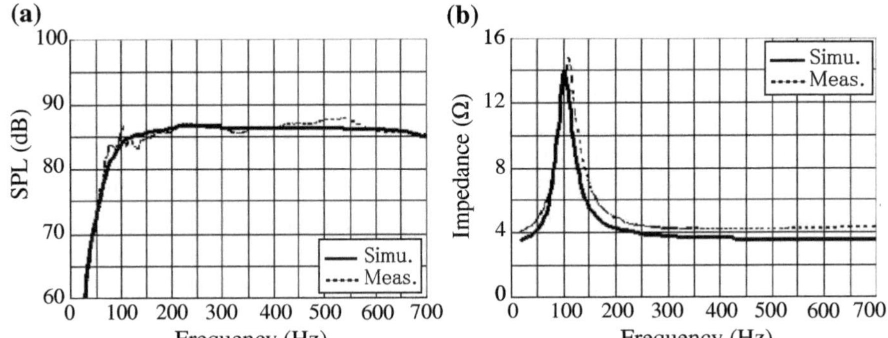

Fig. 8.34 Results of coupled analysis. **a** Radiated sound pressure, **b** Electrical impedance

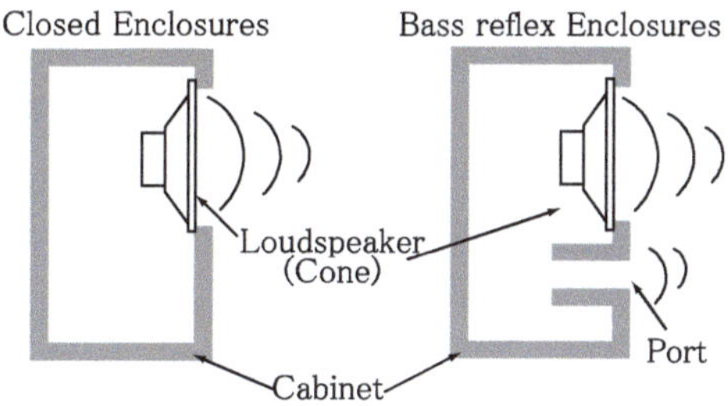

Fig. 8.35 Closed-type loudspeaker and bass reflex type loudspeaker

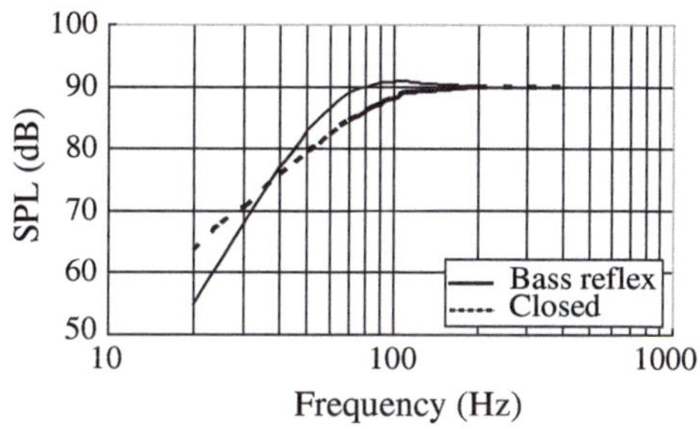

Fig. 8.36 Difference in frequency characteristics

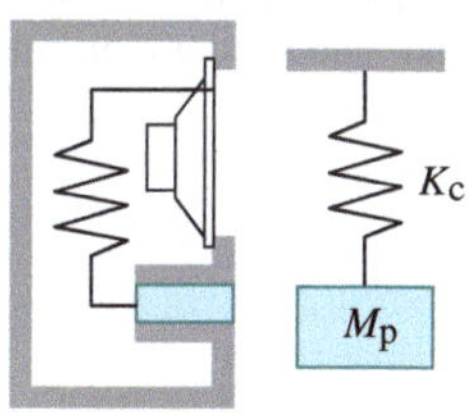

Fig. 8.37 Helmholtz resonance

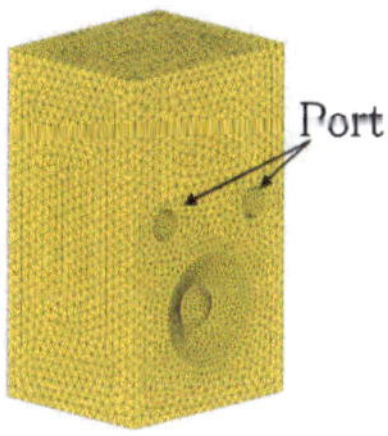

Fig. 8.38 Analysis model

Fig. 8.40. There are also two peaks at the same frequencies and at three frequencies including Helmholtz resonance one phase reversal occur. The force acting on the cone from the electric system and the vibration velocity of the cone in the case of the constant voltage driving is indicated in Fig. 8.41. At the peak frequencies of the electric impedance, the force has minimum value and the velocity has the maximum. This is due to the reverse electromotive force produced by the cone vibration of large amplitude at the two resonance frequencies.

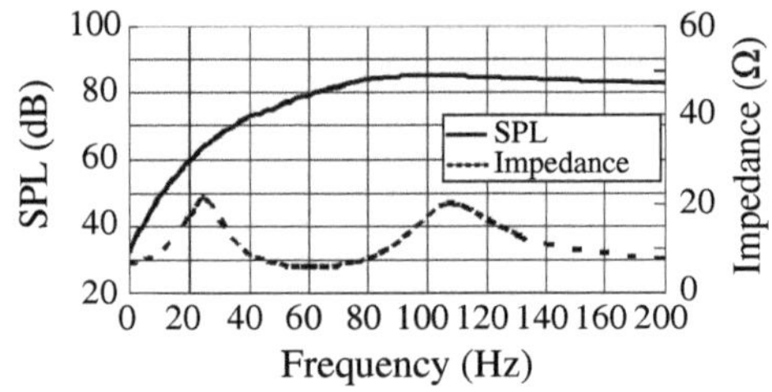

Fig. 8.39 Force acting on the cone and its vibration velocity

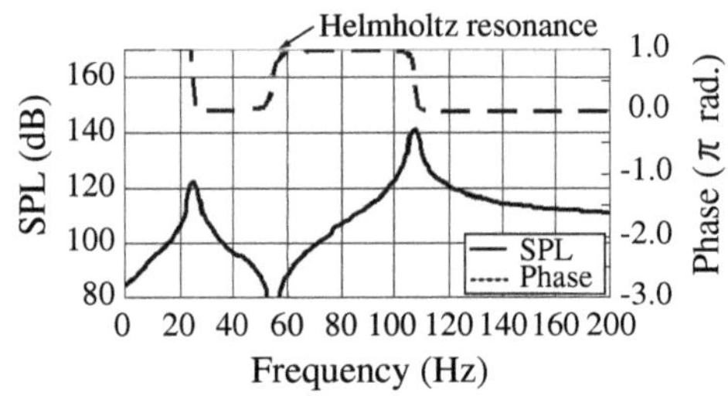

Fig. 8.40 Frequency characteristics of radiated sound and electrical impedance

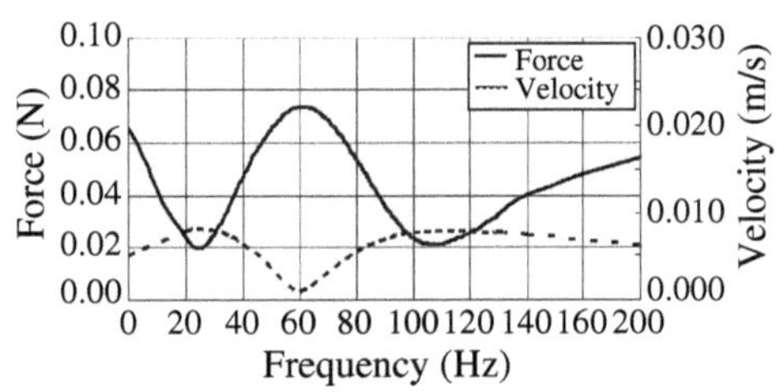

Fig. 8.41 Force acting on the cone and its vibration velocity

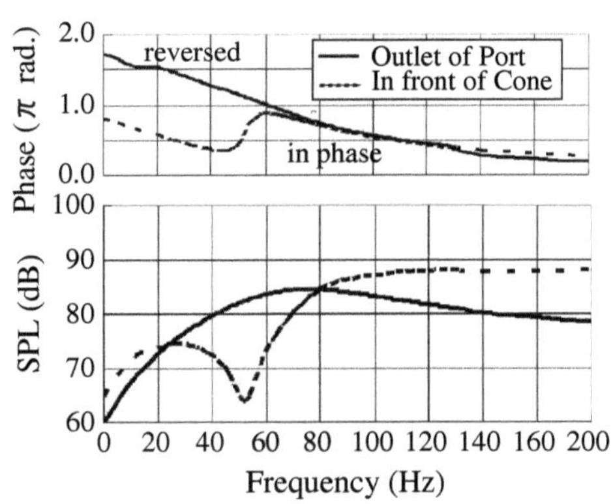

Fig. 8.42 Characteristics of radiated noise from the corn and the port

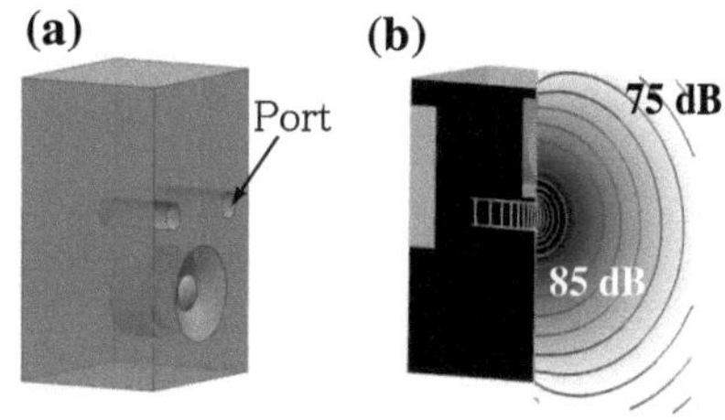

Fig. 8.43 Spatial distribution of radiated sound at Helmholtz resonance frequency. **a** Loudspeaker model, **b** Spatial distribution of pressure

Figure 8.42 shows the frequency response curve of the sound pressure level and phase at the points in front of the cone and the exit of the port. The sound pressures radiated from the cone and the port have reversal phase at the Helmholtz resonance frequency. At frequencies above the resonance frequency, the phase of both are the same, so these two sound components reinforce each other. While at frequencies below it, the phase is opposite, and then radiated pressure is eventually canceled and the total sound power becomes poor.

The distribution of the sound pressure at the resonance is shown in Fig. 8.43. This indicates that the contribution of the cone to the radiated sound is considerably small because a large amount of energy is radiated from the port, and at the same time the vibration velocity of the cone is small.

Like the above, by examining the acoustic phenomenon of the bass reflex type loudspeaker using these numerical simulations, we can predict and evaluate the contribution of the radiated sound from the port and the difference in the acoustic characteristics due to the location or direction of the port.

8.4.2.3 Acoustic Characteristics of Horn Speakers at Middle to High Frequency Range

This section addresses the case of evaluating the directivity at middle to high frequency range by wave-based acoustic analysis. A horn speaker(radius: $200 \times 350\,\text{mm}^2$; length: 155 mm) indicated in Fig. 8.44 is used. For boundary conditions, the vibration velocity is applied to the vibrating surface located inside the horn.

The sound pressure distribution in the horizontal plane is shown in Fig. 8.45, and the directivity of the radiated pressure evaluated on the arc, which has a radius of 1.0 m, is shown in Fig. 8.46. There is good correlation between the simulations and experimental results.

The frequency-dependent directivity angle (angle of directions in which the pressure drops less than 6.0 dB compared with pressure of front direction) curves of different horn shapes are plotted in Fig. 8.47. They are evaluated numerically. The evaluation of the frequency-dependent directivity angle of loudspeakers with different specification is quite effective in having design optimization.

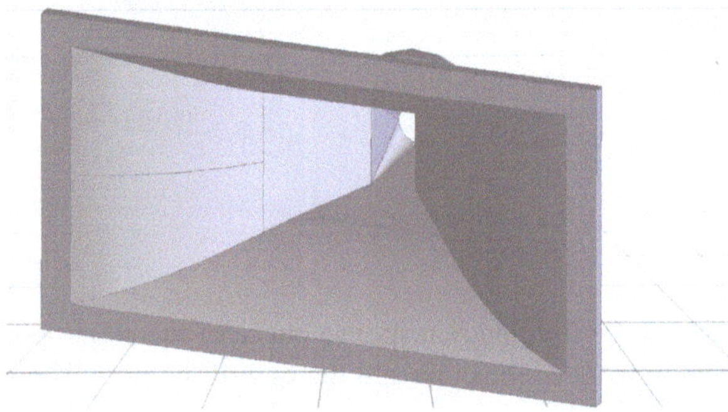

Fig. 8.44 Horn speaker

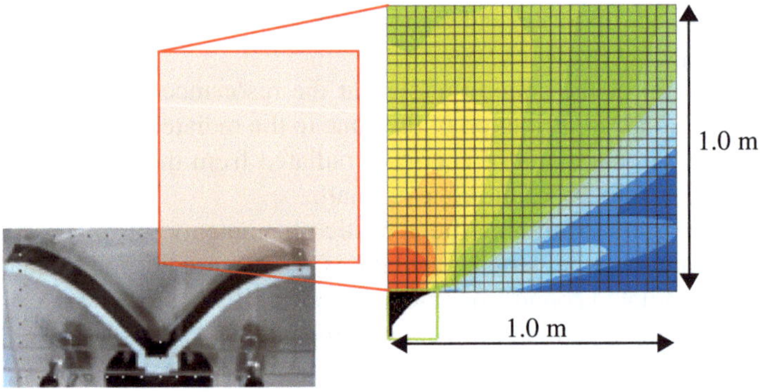

Fig. 8.45 Spatial distribution of sound pressure

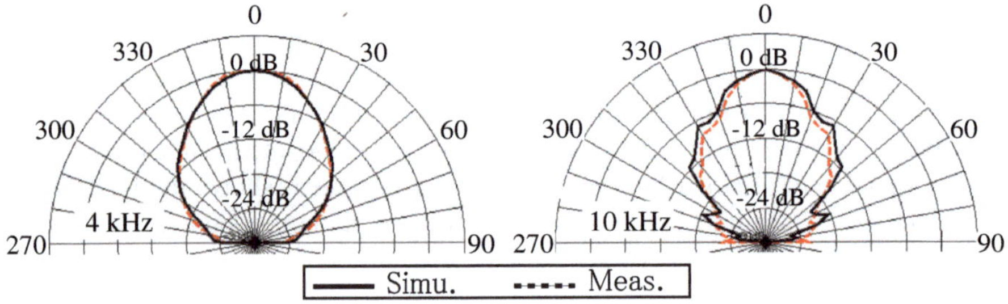

Fig. 8.46 Poler pattern

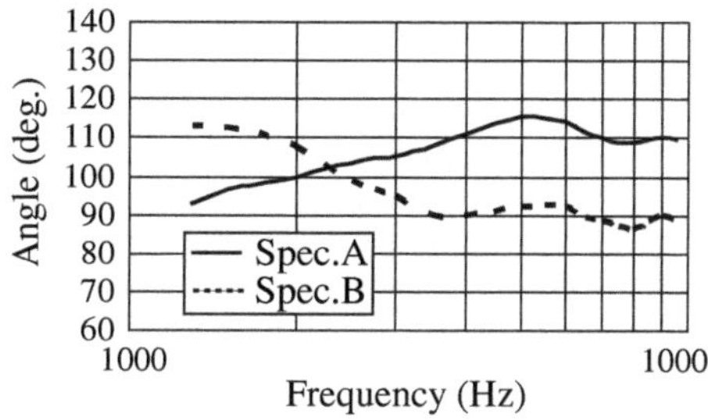

Fig. 8.47 Frequency dependence of directional angle

References

1. T. Terai, On calculation of sound fields around three dimensional objects by integral equation methods. J. Sound Vib. **69**, 71–100 (1980)
2. T. Terai, Method for measurement of sound absorption coefficients in a reverberation room. Architect. Acoust. Noise Control **111**, 15–18 (2000). (in Japanese)
3. Y. Horinouchi, Y. Kawai, H. Meotoiwa, M. Toyoda, An analysis of the area effect of a sound absorbent surface. Transactions of Technical Committee on Architectural Acoustics. The Acoustical Society of Japan, (AA2002-22) (2002) (in Japanese).
4. N. Ogami, M. Tanaka, Y. Kawai, H. Meotoiwa, T. Murakami, Prediction of the area effect of a sound absorbent patch by using a baoundary integral equation. Transactions of Technical Committee on Architectural Acoustics, The Acoustical Society of Japan, (AA2003-08) (2003) (in Japanese).
5. Y. Kawai, H. Meotoiwa, Estimation of the area effect of sound absorbent surfaces by using a boundary integral equation. Acoust. Sci. Tech. **26**, 123–127 (2005)
6. B.B. Baker, E.T. Copson, *The Mathematical Theory of Huygens' Principle* (Chelsea Publishing, Chelsea, 1987)
7. P.M. Morse, K.U. Ingard, *Theoretical Acoustics* (McGraw-Hill, New York, 1968)
8. S.I. Thomasson, On the absorption coefficient. Acustica **44**, 265–273 (1980)
9. Z. Maekawa, J.H. Rindel, P. Lord, *Environmental and Architectural Acoustics* (Spon Press, New York, 2000)
10. H. Kuttruff, *Room Acoustics*, 4th edn. (Spon Press, New York, 2000)
11. M. Born, E. Wolf, *Principles of Optics* (Pergamon Press, London, 1975)
12. D.K. Wilson, Relaxation-matched modeling of propagation through porous media, including fractal pore structure. J. Acoust. Soc. Am. **94**(2), 1136–1145 (1993)
13. M.A. Biot, The theory of propagation of elastic waves in a fluid- saturated porous solid. J. Acoust. Soc. Am. **28**, 168–191 (1956)
14. J.F. Allard, N. Atalla, *Propagation of Sound in Porous Media: Modelling Sound Absorbing Materials*, 2nd edn. (Wiley, Chichester, 2009)
15. N. Atalla, R. Panneton, P. Debergue, A mixed displacement pressure formulation for poroelastic materials. J. Acoust. Soc. Am. **104**, 1444–1452 (1998)
16. P. Debergue, R. Panneton, N. Atalla, Boundary condition for the weak formulation of the mixed (u-p) poroelasticity problems. J. Acoust. Soc. Am. **106**, 2383–2390 (1999)
17. A. Craggs, Coupling of finite element acoustic absorption models. J. Sound Vib. **66**, 605–613 (1979)
18. D. Kato, Predictive model of sound propagation in porous materials: extension of applicability in kato model. J. Acoust. Soc. Jpn. **64**, 339–347 (2008). (in Japanese)
19. T. Sakuma, Analysis and evaluation methodology on acoustic scattering performance of architectural surfaces. J. Acoust. Soc. Jpn. **61**, 39–44 (2005). (in Japanese)
20. T.J. Hargreaves, T.J. Cox, Y.W. Lam, P. D'Antonio, Surface diffusion coefficients for room acoustics: free field measures. J. Acoust. Soc. Am. **108**, 1710–1720 (2000)

21. Acoustics–Sound-scattering properties of surfaces–Part 2: measurement of the directional diffusion coefficient in a free field. ISO 17497–2:2012.
22. M. Vorländer, E. Mommertz, Definition and measurement of random-incidence scattering coefficients. Appl. Acoust. **60**, 187–199 (2000)
23. Acoustics–Sound-scattering properties of surfaces–Part 1: measurement of the random-incidence scattering coefficient in a reverberation room. ISO 17497–1:2004.
24. E. Mommertz, Determination of scattering coefficients from the reflection directivity of architectural surfaces. Appl. Acoust. **60**, 201–203 (2000)
25. Y. Kosaka, T. Sakuma, Numerical examination on the scattering coefficients of architectural surfaces using the boundary element method. Acoust. Sci. Tech. **26**, 136–144 (2005)
26. Y. Kosaka, T. Sakuma, Comparison of scattering coefficients and diffusion coefficients of architectural periodic surfaces. in Proceedings of ASJ Spring Meeting, pp. 855–856 (2005) (in Japanese).
27. T.J. Cox, P. D'Antonio, *Acoustic Absorbers and Diffusers: Theory, Design and Application* (Spon Press, New York, 2004)
28. A. London, Transmission of reverberant sound through single walls. J. Res. Nat. Bur. Stand. **42**, 605–615 (1949)
29. L.L. Beranek, *The Transmission and Radiation of Acoustic Waves by Solid Structures* (McGraw-Hill, New York, 1962)
30. R.H. Lyon, G. Maidanik, Power flow between linearly coupled oscillators. J. Acoust. Soc. Am **34**, 623–629 (1962)
31. A. Craggs, An acoustic finite element approach for studying boundary flexibility and sound transmission between irregular enclosures. J. Sound Vib. **30**, 343–357 (1973)
32. T. Otsuru, H. Yamamoto, Sound transmission using computational mechanics. J. Acoust. Soc. Jpn. **44**, 293–299 (1988). (in Japanese)
33. T. Otsuru, Vibro/acoustic analysis of elastic plates by finite element method. J. Arch. Plan. **461**, 1–8 (1994). (in Japanese)
34. T. Sakuma, K. Egawa, Y. Yasuda, Numrical analysis of sound loss of glass pane-energy loss by edge supports, in Proceedings of ASJ Spring Meeting, pp. 1131–1132 (2009) (in Japanese).
35. J. Yoshimura, S. Sugie, E. Toyoda, Internal and edge damping effects on sound reduction index measurements of a glass pane, in Proceedings of the Inter-Noise 2007 (Istanbul), No. 07–427 (2007).
36. S. Sakamoto, Phase-error analysis of high-order finite difference time domain scheme and its influence on calculation results of impulse response in closed sound field. Acoust. Sci. Tech. **28**, 295–309 (2007)
37. H. Watanabe, K. Miyao, Sound insulation of gypsum-board double wall, in Transactions of Technical Committee on Architectural Acoustics. The Acoustical Society of Japan, (AA88-24) (1988) (in Japanese).
38. H. Yano, H. Tachibana, S. Sakamoto, T. Matsumoto, Alleviation of the coincidence effect in double-layered plasterboards composing multiple dry wall system, in Proceedings of the Inter-Noise 2005 (Rio de Janeiro), No. 1808 (2005).
39. H.F. Olson, *Acoustical Engineering* (D. VAN NOSTRAND COMPANY Inc, New York, 1975)
40. J. Borwick, *Loudspeaker and Headphone Handbook* (Focal Press, Oxford, 1998)
41. L.L. Beranek, *Acoustics* (Mac Graw-Hill Co., New York, 1954)
42. N. Kyouno, T. Yamabuchi, Y. Kagawa, Acoustic response analysis of a cone-type loudspeaker by the finite element method. J. Acoust. Soc. Jpn. **61**(6), 312–319 (2005)
43. D.J. Murphy, R. Morgans, Modelling compression drivers using T-matrices and Finite element analysis in AES 119th Convention Paper, number 6580, pp. 1–14 (2005).

Chapter 9
Auralization

Takatoshi Yokota, Takumi Asakura and Takayuki Masumoto

Abstract This chapter introduces some techniques for the auralization of numerical simulations and provides an example of calculating the head-related transfer function (HRTF) which is a function that is fundamental to reproduce stereophonic sound field. In the first section, we introduce a sound field simulation technique combines a numerical simulation with a multichannel reproduction technique, and we present applications of this method to the problem on room acoustics. In the second section, we discuss an auralization technique that simulates the sound insulation characteristics of the façade of a building. In the final section, we present the results of calculating the HRTF and discuss the applicability and efficiency of the analysis method.

Keywords Room acoustics · Building façades · Insulation characteristics · Vehicle noise · Head-related transfer function (HRTF) · Multichannel reproduction · Finite-difference time-domain (FDTD) method · Boundary element method (BEM) · Fast multipole boundary element method (FMBEM)

T. Yokota (✉)
Kobayasi Institute of Physical Research, 3-20-41, Higashi-motomachi,
Kokubunji, Tokyo 185-0022, Japan
e-mail: t-yokota@kobayasi-riken.or.jp

T. Asakura
Institute of Technology, Shimizu Corporation, 3-4-17, Etchujima,
Koto-ku, Tokyo 135-8530, Japan
e-mail: t_asakura@shimz.co.jp

T. Masumoto
Cybernet Systems Co., Ltd., Fujisoft Bldg. 3 Kanda-neribeicho,
Chiyoda-ku, Tokyo 101-0022, Japan
e-mail: masumoto@cybernet.co.jp

T. Sakuma et al. (eds.), *Computational Simulation in Architectural, and Environmental Acoustics*, DOI: 10.1007/978-4-431-54454-8_9,

9.1 Room Acoustics

The auralization of simulation results is one of the most important topics in room acoustics. The techniques for the auralization of the results obtained from scale model experiments or simulations based on the geometrical acoustics have been developed and have already been put into practical use in the field of acoustical design. In these techniques, the impulse responses at target positions in the sound field under consideration are obtained by scale model experiments or by geometrical acoustic simulations, and they are then convolved with a dry source. The resultant signals can be listened to through headphones or loudspeakers. In recent years, outstanding progress in computer hardware has made it possible to auralize the results obtained from wave-based numerical simulations. In the finite-difference time-domain (FDTD) calculations, the impulse response at an arbitrary receiver position can be obtained directly, and it can then be listened by simply converting the digital signals to analog signals.

In the following section, we introduce a multichannel sound field simulation system for auralizing the results of the FDTD method. As applications of the auralization technique, we conducted a subjective comparison on the effects of sound diffusers and we also reproduced a particular fluttering echo in a shrine (the world-famous phenomenon known as the "Roaring Dragon").

9.1.1 Multichannel Reproduction System with FDTD Method

Figure 9.1 shows the outline of a six-channel sound field simulation system that combines three-dimensional numerical analysis using the FDTD calculation, with a six-channel sound reproduction system [1]. This system was originally developed to simulate the sound field of actual concert halls in an anechoic room [2] and the technique has been applied to the auralization of numerical simulation results. In this system, the directional impulse responses in the directions of every 90° at a receiving point in three-dimensional sound field are first calculated by the FDTD method. Next, the calculated impulse response signals are reproduced directly through the six loudspeakers arranged at every 90° in an anechoic room. In this way, the acoustical properties at an arbitrary receiving point set in the virtual space assumed in the FDTD calculation can be reproduced at the center of the sound reproduction system, and one can experience them.

In order to calculate the directional impulse responses for the six orthogonal directions, an arbitrary directivity factor is assumed and set to every 90° by rotating its direction at the receiving point. By multiplying the directivity factor and the instantaneous sound pressure at the receiving point, each of the directional impulse responses at the receiving point is calculated. Two kinds of directivity factor have been proposed, as follows:

Type-I (Cardioid):

$$D_{1,i} = \frac{1 + \cos\theta_i}{2}, \tag{9.1}$$

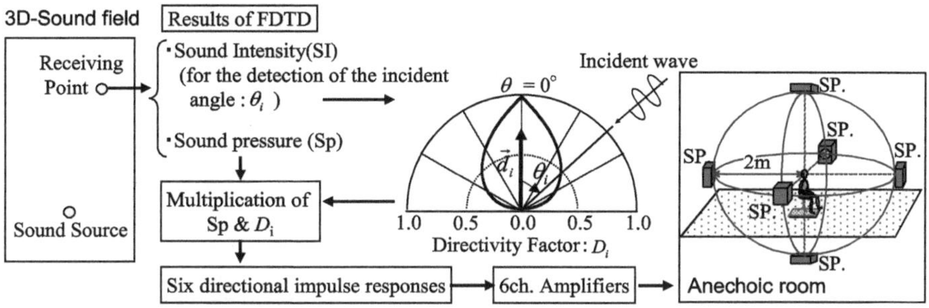

Fig. 9.1 Outline of a six-channel sound simulation system

Type-II:

$$D_{2,i} = \begin{cases} \frac{\cos\theta_i}{A} & \cos\theta_i \geq 0 \\ 0 & \cos\theta_i < 0 \end{cases}, \tag{9.2}$$

$$A = \frac{1}{2}\sum_{i=1}^{6}\left|\cos\theta_i\right|, \tag{9.3}$$

where $D_{1,i}$ or $D_{2,i}$ is the directivity factor (i = 1–6), θ_i is the angle between the front direction of each directivity factor and the incident angle of the sound at the receiving point. The incident angle is obtained from the sound intensity vector, which is calculated by the FDTD method as follows:

$$\cos\theta_i = \frac{-d_{ix}\cdot I_x - d_{iy}\cdot I_y - d_{iz}\cdot I_z}{\sqrt{(I_x)^2 + (I_y)^2 + (I_z)^2}}, \tag{9.4}$$

where $\boldsymbol{I} = (I_x, I_y, I_z)$, $\boldsymbol{d}_i = (d_{ix}, d_{iy}, d_{iz})$, $|\boldsymbol{d}_i| = 1$, $\boldsymbol{d}_i$ is the front direction of each directivity factor, and $\boldsymbol{I}$ is the instantaneous sound intensity vector, which is calculated by the FDTD method. By reproducing the resultant six-directional impulse responses, the amplitude of the sound pressure and the direction of the sound intensity vector at the receiving point can be simulated accurately at the center of the reproduction system.

9.1.2 Simulation of Room Impulse Responses

In this section, we consider the auralization of the impulse responses for rooms with different shapes. Figure 9.2 shows the two-dimensional sound fields for the three different room shapes. The directional impulse responses were calculated by

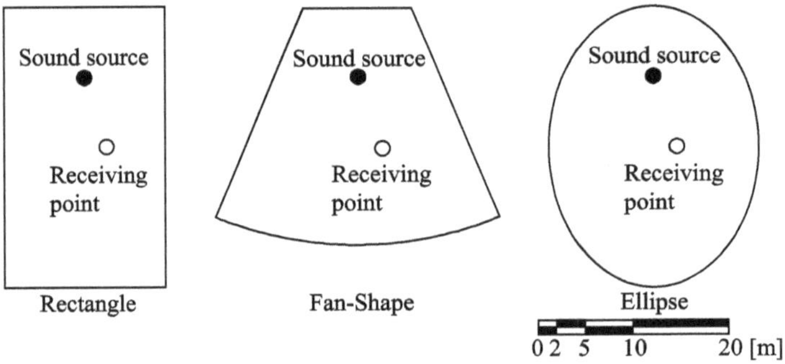

Fig. 9.2 Outline of the sound field

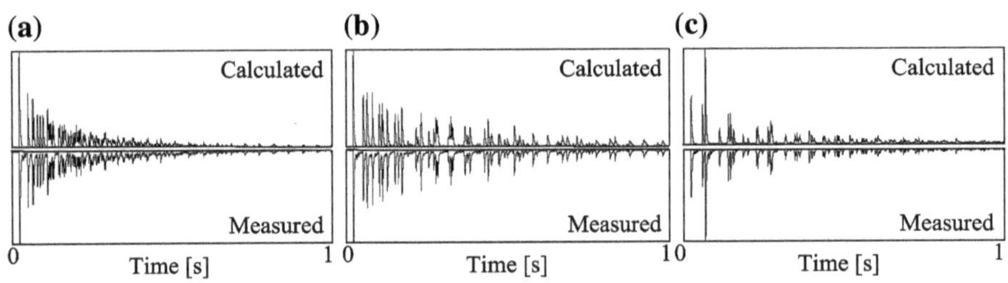

Fig. 9.3 Comparisons of the echo diagrams of the calculations and those of the measurements. **a** Rectangle. **b** Fan-shape. **c** Ellipse

the two-dimensional FDTD method, and the signals were reproduced through four loudspeakers that were set at right angles on an arc of 2 m radius in an anechoic room. In the calculations, the spatial grid size and the time interval were set to be 0.01 m and 0.02 ms, respectively. It was assumed that the normal acoustic impedance of the room boundaries consisted only of the real part, and a constant sound absorption coefficient ($\alpha = 0.2$) was assumed for all boundaries. We used the Type-I directivity factor, defined in Eq. (9.1), to calculate the directional impulse responses.

The calculated directional impulse response signals were reproduced through the four-channel loudspeaker system, and the omnidirectional impulse response was measured at the center of the reproduced sound field. Figure 9.3 compares the echo diagram of the calculation with that of the measurement. These echo diagrams were obtained by passing the omnidirectional impulse response signal through a numerical RMS detector with a 1 ms time constant. As seen in the figure, the calculated results and the measured ones are in very good agreement. The instantaneous sound intensities at the center of the reproduction system were also measured. Figure 9.4 uses radar charts to compare the sound intensity vector of the calculation with that of the measurement. In these figures, it can be seen that in all three cases, the calculated intensity vectors are in fairly good agreement with the measured ones.

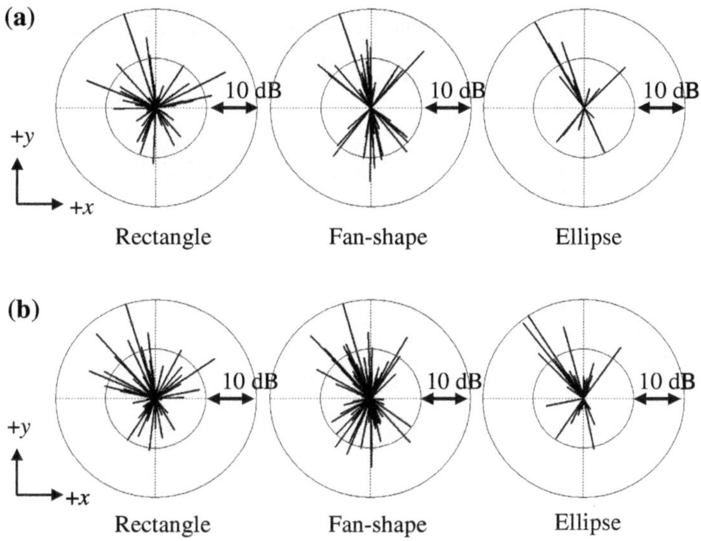

Fig. 9.4 Comparisons between the calculated and measured instantaneous sound intensities. **a** Calculated, **b** measured

9.1.3 Effects of Sound Diffusers in Rooms

As an application to room acoustics of the multichannel sound field simulation method, we conducted an experiment in which we subjectively evaluated the effect of sound diffusers to prevent fluttering echoes. Figure 9.5 shows the rectangular room and two types of sound diffusers attached at the room boundaries (a: Triangular and b: Column). For each type of diffuser, we used four different sets of dimensions, as shown in the figure. For the boundary condition, it was assumed that the normal acoustic impedance at the room boundaries consisted of only the real part, and a constant sound absorption coefficient was assumed for all the boundaries. In the case of a bare room (without diffusers), the absorption coefficient was set to 0.2. For the cases with diffusers, the sound absorption coefficient was assumed to be such that the equivalent sound absorption length (which, in a three-dimensional room, corresponds to sound absorption area) was equal to that of the bare room. In the FDTD calculation, we again used a spatial grid size of 0.01 m and a time step of 0.02 ms. In the experiment, the impulse responses were presented to the subject in a random order through the four-channel loudspeaker system. After hearing each impulse response, the subject judged the strength of the fluttering echo and assigned it to one of five categories as shown in Fig. 9.6. Figure 9.6 shows the results of the subjective experiment. In these figures, each plot shows the arithmetic average of the category number assigned by all of the subjects for each diffuser condition at each receiving point. In the case of the triangular diffusers, it can be seen that the larger the diffusers, the better they prevent the fluttering echoes. On the other hand, in the case of the

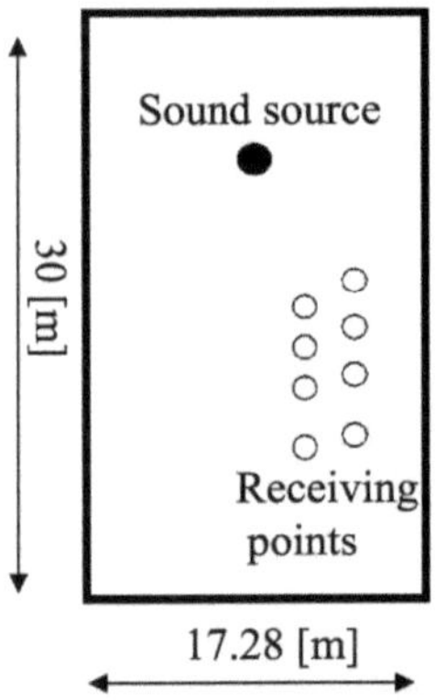

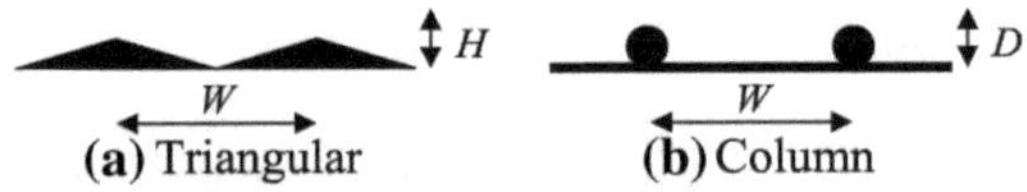

Type	Sizes (unit: m)		Absorption Coefficient
Triangular	1	$W=0.4$ $H=0.06$	0.19
	2	$W=0.75$ $H=0.11$	0.19
	3	$W=1.5$ $H=0.23$	0.19
	4	$W=3.0$ $H=0.45$	0.19
Column	1	$W=0.4$ $D=0.4$	0.13
	2	$W=0.75$ $D=0.4$	0.08
	3	$W=1.5$ $D=0.4$	0.11
	4	$W=3.0$ $D=0.4$	0.15

Fig. 9.5 Two-dimensional rectangular room and the two types of sound diffusers that were investigated

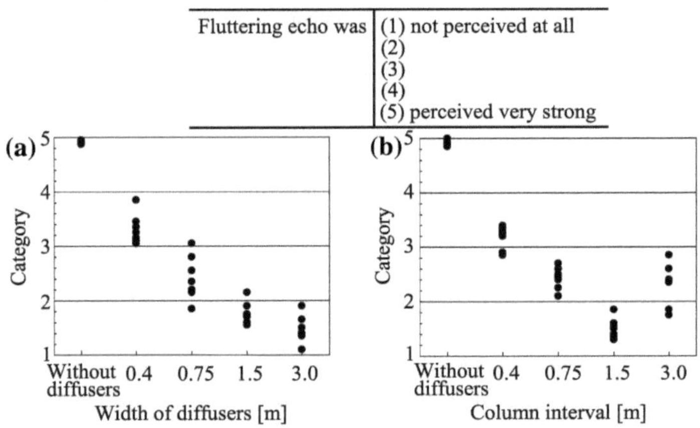

Fig. 9.6 Results of the judgment test on the strength of fluttering echo: **a** triangular diffusers; **b** column diffusers

column diffusers, the effect was greatest when the column interval was 1.5 m. This indicates that an optimum scale exists for the prevention of a fluttering echo by a column diffuser.

9.1.4 Auralization of "Roaring Dragon"

The "Honji-Do" temple located in the "Nikko Toshogu" area of Nikko City, Japan, is famous for a strange acoustic phenomenon called the "Roaring Dragon." A dragon is painted on the ceiling of this building, and when hands are clapped under the head of the dragon, one can hear a strange fluttering echo. In this section, we present a challenging study that attempts to reproduce the "Roaring Dragon" phenomenon of the "Honji-Do" temple by combining the FDTD with the multichannel reproduction technique.

9.1.4.1 "Roaring Dragon" Phenomenon in "Honji-Do" Temple

The "Roaring Dragon" phenomenon in the "Honji-Do" temple is a fluttering echo that is caused by repeated reflections between the ceiling, which has very little curvature, and the flat floor. The temple was unfortunately destroyed by an accidental fire in 1961 and was rebuilt in 1969. In the reconstruction work, reproduction of the "Roaring Dragon" phenomenon was one of the most important items, and a one-fourth scale acoustic model experiment was conducted to study the cause of the acoustic phenomenon and to determine how it could be reconstructed. The following is a summary of the study:

1. The duration time of the fluttering echo becomes longer with increasing curvature of the ceiling.
2. The arch rise (the difference in height between the center of the ceiling and the edge) that best reproduces the "Roaring Dragon" phenomenon is 9 cm.
3. When hands are clapped under the head of the dragon, just to the side of directly below the center of the ceiling, the "Roaling Dragon" pulsates.

9.1.4.2 Three-Dimensional FDTD Simulation of "Roaring Dragon" Phenomenon

Three-dimensional FDTD simulation was conducted by modeling the room sound field in the "Honji-Do" temple. Figure 9.7 shows the plan of the "Honji-Do" temple.The source was positioned just below the painted dragon at a height of 1.2 m. The receiving point was positioned just above the source position at a height of 1.5 m. As the arch rise of the ceiling (the difference in height between the center of the ceiling and the edge) was assumed to be 9 cm. The spatial grid size and the time interval were set to be 0.02 m and 0.032 ms, respectively. The Type-II directivity factor, defined in Eq. (9.2), was selected for calculating the directional impulse responses.

9.1.4.3 Reproduction of "Roaring Dragon" Phenomenon by Six-Channel Sound Field Simulation

Figure 9.8 shows the system used to reproduce the "Roaring Dragon", based on a three-dimensional FDTD calculation and the six-channel sound field simulation technique [3]. The basic concept of this system is the same as that of the multichannel sound field simulation system described above. In this system, the sound of a handclap made at the center of the system was convolved in real time with the directional impulse responses at the receiving point, which was set in the sound

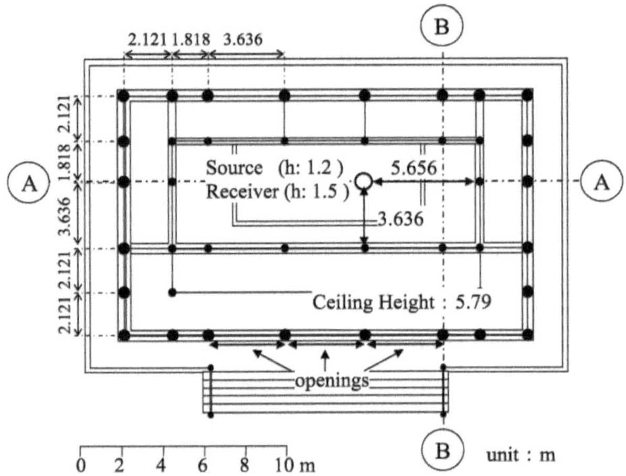

Fig. 9.7 Plan of the "Honji-Do" temple

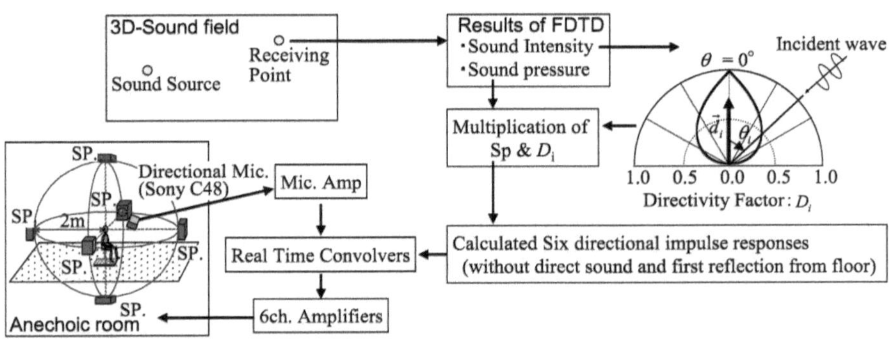

Fig. 9.8 Outline of the six-channel sound field simulation for reproduction of the "Roaring Dragon"

field of the model of the "Honji-Do" temple. The directional impulse responses were obtained by the FDTD calculation, and the resultant signals were reproduced through the loudspeakers set in an anechoic room. This system was originally developed to simulate the sound field on the stage of actual concert halls in an anechoic room to investigate the acoustic property for music players [4]. The technique has also been applied to the auralization of numerical simulation results. Figure 9.9 shows the directional impulse responses obtained by the FDTD calculations. In the figures, the front direction corresponds to the upward direction in Fig. 9.7. It can be seen that the fluttering echo persists for a long time in the directional impulse responses of Up and Down directions. This indicates that there were repeated strong reflections between the ceiling and the floor. Those who had visited "Honji-Do" temple and experienced the real "Roaring Dragon" within the previous year reported that the simulated fluttering echo could be perceived above their heads and that it pulsated as it decayed. Some subjects also commented that they perceived the duration time of

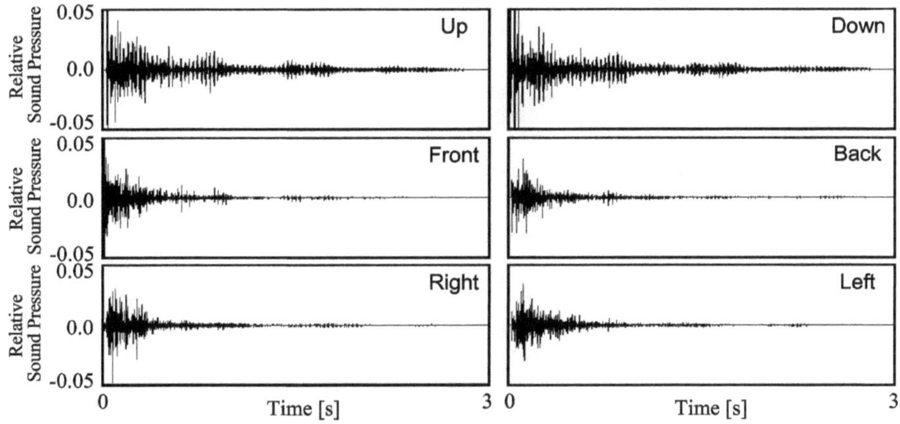

Fig. 9.9 Directional impulse responses at the receiving point (the sound pressure is normalized by the amplitude of the direct sound)

the simulated echo to be longer than that of the real "Roaring Dragon". One reason for this might be that there are many objects, such as ritual articles, which reflect or absorb the sound in the real "Honji-Do" temple. It may also be due to the low level of background noise in the anechoic room.

9.2 Noise Propagation

The frequency and time-transient characteristics of the leak sound transmitted into the residential buildings are strongly influenced by the sound insulation characteristics of the building façade. In this section, an auralization technique [5] in which the sound insulation characteristics of the façade can be realized through numerical simulation is described.

9.2.1 Auralization Method

Detailed scheme of the proposed auralization system shown in Fig. 9.10 is described in this section.

9.2.1.1 Recording of Vehicle Noise

A waveform of a pass-by noise of a vehicle was recorded by an omnidirectional microphone which was set at a point 7.5 m away from the running lane (waveform (a), shown in Fig. 9.10), and the waveform was divided into N sections with every

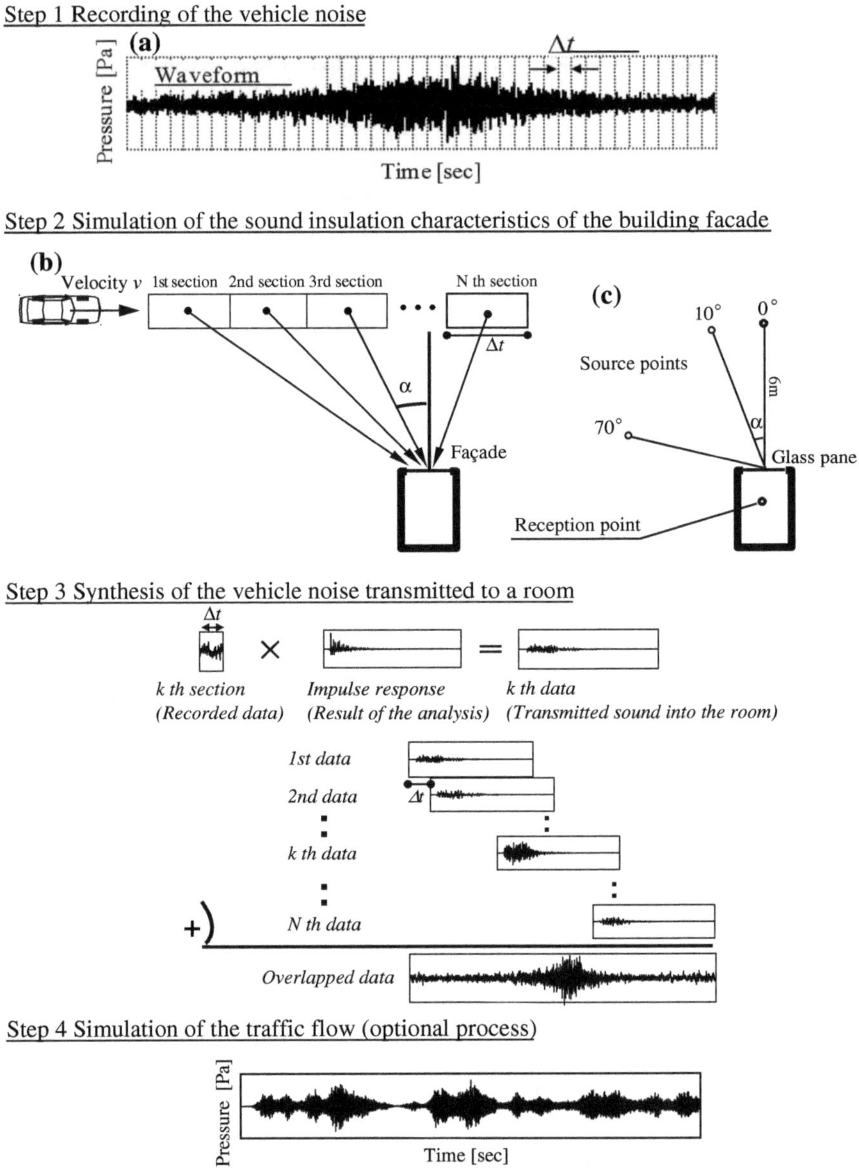

Fig. 9.10 Flow chart of the auralization scheme. **a** Waveform, **b** incident angle of the vehicle noise into the room, **c** arrangement of the numerical analysis (sound transmission through façade)

time interval, Δt. Signal processing on these divided N sections was carried out in the following steps.

9.2.1.2 Simulation of Sound Insulation Characteristics of Building Façade

As shown in Step 2 of Fig. 9.10, the incident angle, α, of the sound which propagates from the running vehicle to the building façade, is varying every moment and the sound insulation characteristics of building façade, especially glass pane, also vary in correspondence with the angle of sound incidence.

In order to simulate the directional characteristics of sound insulation of the building façade, vibroacoustic numerical analysis using FDTD method was applied. The plan of the sound field for the three-dimensional analysis is shown in Fig. 9.10c. Each of the three-dimensional sound field was analyzed, and the impulse response at the receiving point was calculated.

9.2.1.3 Synthesis of Vehicle Noise Transmitted into a Room

The incident angle, α, of the vehicle noise to the façade was estimated by the geometrical relationship between the positions of the vehicle and the building façade. The convolution of the kth vehicle noise whose incident angle to the building façade is α and the impulse response of the same incident angle, α, obtained by numerical analysis was performed for the data of N sections. The transmitted noise into the room was made by overlapping the convoluted data by shifting every data with interval, Δt as shown in Step 3 of Fig. 9.10.

9.2.1.4 Simulation of Traffic Flow

In order to simulate a road traffic noise with an arbitrary traffic volume, a pass-by noises of multiple automobiles were overlapped. An example of the simulated road traffic noise with a traffic volume, 1,500 vehicles/h, is shown in Step 4 of Fig. 9.10.

9.2.2 Simulation Results

In this study, a simulation of the vehicle noise at the point of reception in indoor spaces was performed using the proposed method. The assumed condition is shown in Fig. 9.11. The geometrical relationship between the running lane of the vehicle and the points of reception and the details of the reception room are shown in Fig. 9.11. It was assumed that a glass plate with dimensions of $1.8\,\mathrm{m}\ (W) \times 1.8\,\mathrm{m}\ (H)$ is set in the opening of a room whose dimensions are $2.7\,\mathrm{m}\ (W) \times 2.2\,\mathrm{m}\ (H) \times 3.6\,\mathrm{m}\ (D)$, as shown in the figure.

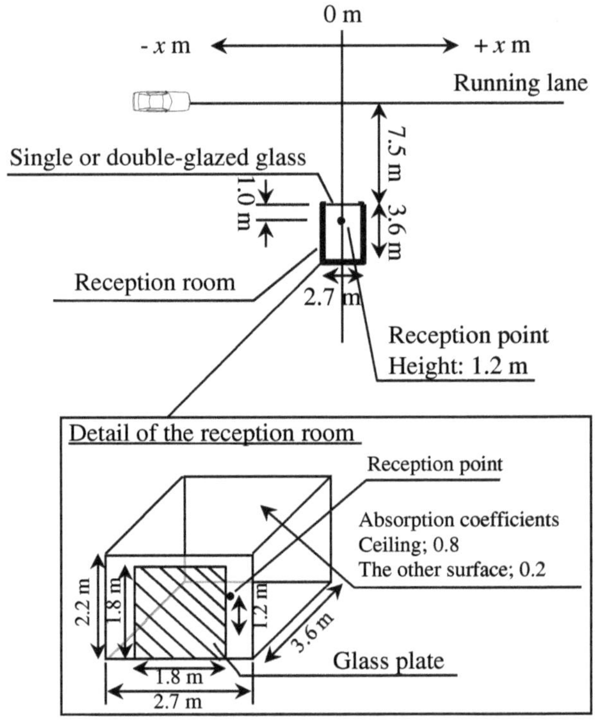

Fig. 9.11 Investigated conditions in the case study

The values of the parameters for the physical properties of the glass plate were set as follows: density, 2,500 kg/m^3; Young's modulus, 7.16×10^{10} N/m^2; and Poisson's ratio, 0.22. In addition, an elastically supported condition as described in Sect. 8.3.2 was applied in the vibration analysis. The absorption coefficient of the ceiling was set at 0.8, assuming a ceiling with absorption treatment, and that of the surfaces of the other walls were set at 0.2.

The investigated conditions are as follows: Case 0 (outdoor space), Case 1 (inside a room with a single glass plate with a thickness of 6 mm in the opening), Case 2 (inside a room with single glass plate of thickness 10 mm), and Case 3 (inside a room with double-glazed glass composed of two glass plates, each of thickness of 6 mm, separated by a 6-mm-thick layer of air).

Simulation results of the time-transient characteristics of sound pressure levels at the reception point caused by one pass-by vehicle is shown in each one-third Octave band in Fig. 9.12. In this graph, the timing at which the running vehicle reaches the 0 m position in Fig. 9.11 is set at 0 s. The sound pressure levels are calculated so that the sound exposure level of the pass-by sound is 75 dB. The speed of the vehicle is assumed to be 60 km/h, and the horizontal axis of the figure describes a relative time from the timing of the vehicle passing in front of the room.

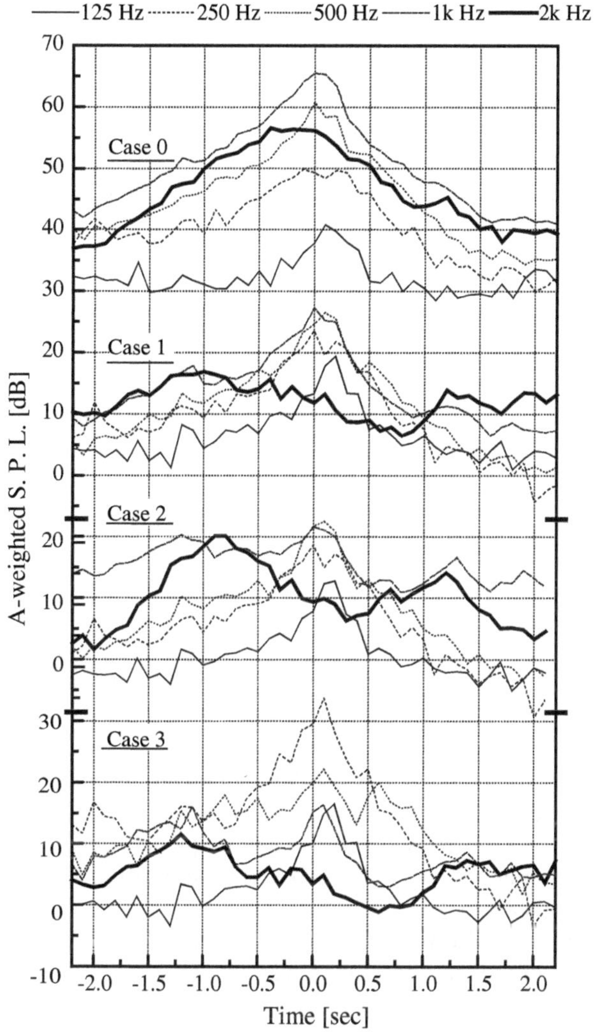

Fig. 9.12 Time-transient characteristics of the pass-by sound in 1/3. Oct. band frequency

In Case 0, the sound pressure level decreases as the position of the running vehicle moves further from the 0 m position in all frequency bands. However, in Case 1, the time-variant characteristics for 2 kHz have peaks at -1.2 s and $+1.2$ s, and the sound pressure level at 0 m is less than those at the peak positions. It is considered that a large quantity of sound energy is transmitted through the glass plate when the incident angle of sound to the plate is larger, in the 2 kHz band, which includes the coincidence cut-off frequency of a glass plate with a thickness of 6 mm.

In Case 2, the time-variant characteristics for 1 and 2 kHz have the same peak characteristics as those for 2 kHz in Case 1. It is also considered that this result is due

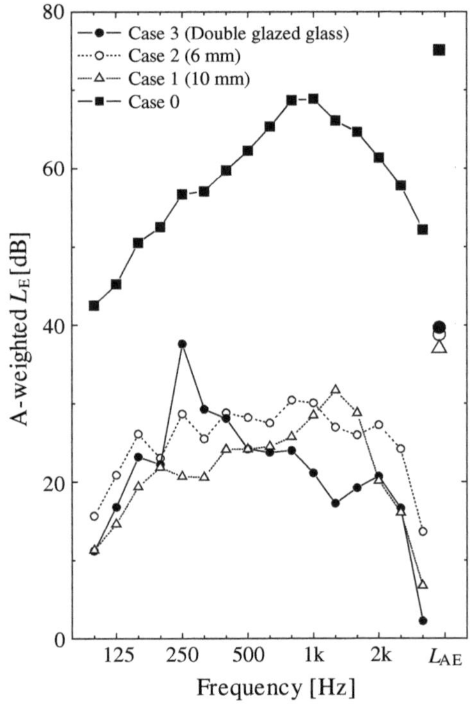

Fig. 9.13 Frequency characteristics of the transmitted sound into the room

to the coincidence phenomenon. In Case 3, the time-variant characteristics in 250 Hz have a large value, especially at around 0 s. The reason for this is that a double-glazed glass plate has a resonant frequency in the low frequency range, which is caused by the resonant phenomenon that characterizes its composition of mass (glass)-spring (air layer)-mass (glass).

Based on the obtained time-transient characteristics, the single event sound exposure levels, L_E, of the pass-by sounds are calculated, and the results are shown in Fig. 9.13. Comparing the conditions of Case 1 and 2, the sound energy level of Case 2 in 250 Hz has larger value than that of Case 1, and it is caused by the mass-spring-mass effect as described above.

9.3 Head-Related Transfer Functions

A wide variety of research fields are discussing the need to reproduce stereophonic sound fields. To do this, it is necessary to obtain head-related transfer functions (HRTFs), which are acoustic transfer functions between sound sources located around the human head and ear. Based on this, several measurements have been performed [6]. However, when the physical load of the subject is taken into account, it is extremely difficult to conduct the necessary procedures for obtaining HRTFs.

Attempts have therefore been made to obtain HRTFs by numerical analysis, resulting in a number of published reports.

Yet the numerical analysis also has some drawbacks. Huge calculation costs are required since data for the entire audible frequency range are required for HRTFs. For example, if analysis at 20 kHz is conducted using the boundary element method (BEM), the number of degrees of freedom (DOF) to represent a human head is around 100,000, and the required memory will exceed 160 GB. For this reason, upper limit frequencies have been limited to several kHz if the entire head is taken into account, or several assumptions have been made to conduct analysis up to 20 kHz (e.g., pinna is connected to the infinite baffle plane) [7].

Incidentally, in recent years, research has also been conducted on the implementation of sound field analysis using fast multipole algorithms with the three-dimensional BEM (see Sect. 4.3). Using this fast multipole BEM (FMBEM) it has been possible to calculate HRTFs for the entire audible frequency range within feasible memory sizes and calculation times [8]. This study outlines the results of these calculations.

9.3.1 Basic Examination

9.3.1.1 Checking Uniqueness of Solution

When analyzing the external field using the BEM, the unique solution cannot be obtained at the eigenfrequencies of an internal field whose boundary is the same as the target geometry. To avoid this, the proper formulation should be selected by comparing theoretical solutions with the solution from the BEM. Checked formulations are basic form (BF), normal derivative form (NDF) (see Sect. 4.1.1 for these formulations), Burton–Miller form (BMF) and dual form (DF) (see Sect. 4.2.1 for these formulations). A sphere 0.25 m in diameter is used as a simplified shape of a human head (Fig. 9.14). The angle θ is introduced to represent locations on the sphere, starting from 0° to 180°. The piston oscillation of 1 mm/s is defined in the range from $\theta = 0°$ to $\theta = 20°$. All other parts of the surface are taken to be rigid. In this case, the theoretical solution can be calculated with Eq. (9.5) [9]:

$$p(r,\theta) = \frac{j\rho c V_0}{2} \sum_{n=0}^{N} \left(P_{n-1}\cos\alpha - P_{n+1}\cos\alpha\right) \frac{h_n^{(1)}(kr)}{h_n^{(1)\prime}(ka)} P_n(\cos\theta), \qquad (9.5)$$

where j is the imaginary unit, ρ is the medium density, c is the sound speed, k is the wave number, V_0 is the vibration velocity, a is the sphere's radius, r is the distance between the sphere's center and the field point ($r > a$), P_n are the Legendre polynomials, $h_n^{(1)}$ are the spherical Hankel functions of the first kind, and $h_n^{(1)\prime}(x) = h_{n-1}^{(1)}(x) - ((n+1)/x)h_n^{(1)}(x)$. In this study, $\rho = 1.225\,\mathrm{kg/m^3}$, $c = 340\,\mathrm{m/s}$, $V_0 = 0.001\,\mathrm{m/s}$, $a = 0.125\,\mathrm{m}$, $\alpha = 20°$ are used and response at 20 kHz is calculated.

Fig. 9.14 Model used for checking uniqueness of the solution

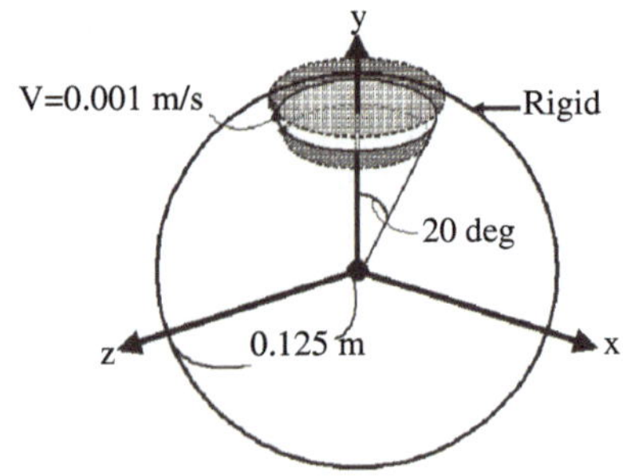

Fig. 9.15 Boundary element mesh (num. of elements: 31,200)

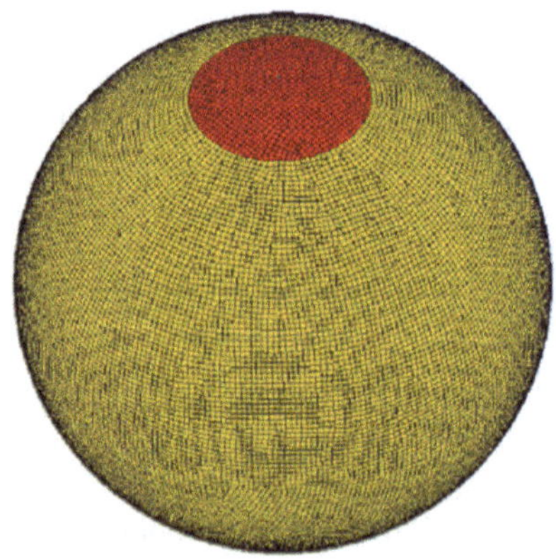

The mesh used for the boundary element model is shown in Fig. 9.15. The sphere is discretized with elements whose size is about one-sixth of a wavelength, so analysis can be performed up to 20 kHz and the number of elements is 31,200. The velocity B.C. ($V_0 = 0.001$) is defined as the part indicated with red in the figure. Only in the DF case, to avoid the fictitious eigenfrequency problem, the specific impedance of air is defined as impedance B.C.s to the negative side of all the elements. The ILUT(1^{-6},100) [10] was used as a preconditioning technique. The generalized minimal residual (GMRes) method was adopted as an iterative solver with the restart number being set to 2,528 for BF, NDF, and BMF cases, and 325 for the DF case, so 3 GB of memory is required. The machine used in this work is an IBM IntelliStation with the following specifications; CPU: AMD OpteronTM 2.79 GHz; OS: Windows XP Professional ×64 edition; memory: 8 GB.

Figure 9.16 shows the sound pressure distributions on the sphere as a function of the θ at 20 kHz. The results obtained by BF or NDF differ greatly from the theoretical solution, which is caused by the deteriorated uniqueness of the solution. With BMF, this problem does not occur, but analysis precision is poor, especially in the area of $\theta > 90°$. With DF, small differences from the theoretical solution were seen in the area of $\theta > 140°$, so we regard it as sufficiently accurate. Table 9.1 shows the

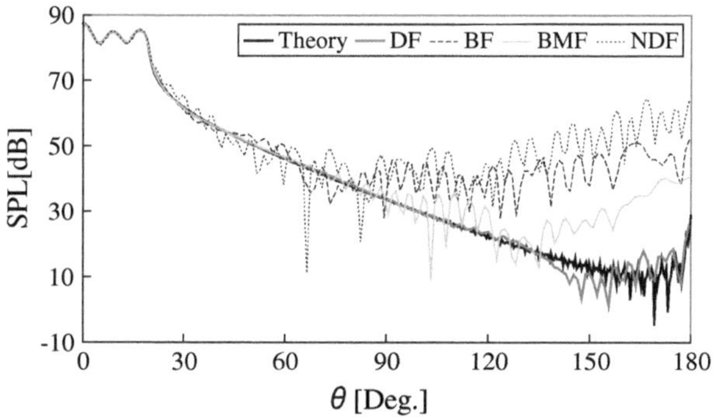

Fig. 9.16 Distribution of sound pressure on sphere

Table 9.1 Calculation cost

Formulation	BF	NDF	BMF	DF
Required memory (GB)	1.64	1.64	1.64	2.68
CPU time (min)	51.8	132.4	1432.7	23.6
Number of matrix-vector multiplications	65	166	1 836	18

amount of memory, CPU time, and number of iterations required for analysis. A larger amount of memory is required since the number of DOFs of the DF is twice that of other formulations. However, due to the smaller number of iterations, the analysis time is the shortest. DF is used for HRTF calculation because of its accuracy and short calculation time.

9.3.1.2 Checking Reciprocity

The reciprocity between the point sound sources and the field points is introduced to reduce the number of calculation cases. However, in the case of HRTF calculation using reciprocity, the distance between the point sound source and the boundary elements tends to be short, so these analysis precision goes down. Caution should therefore be paid to this point.

This study uses the same geometry as in the previous study. The field point position is defined 2 mm outside in the $\theta = 0°$ direction. The point sound source is defined at 1 m from the spherical center in the $\theta = 90°$ direction (Fig. 9.17). In analysis using reciprocity, the source and field point position are set reversed. The boundary element model is checked in two cases. Case 1 is where the surface is discretized uniformly with an element size enabling analysis of each frequency. Case 2 is where the portion of the surface that is assumed to be the range of the pinna (in this case $\theta < 20°$) is discretized with 1 mm (half the distance between the point sound source and the sphere) elements, the lower hemisphere ($90° < \theta < 180°$) is discretized

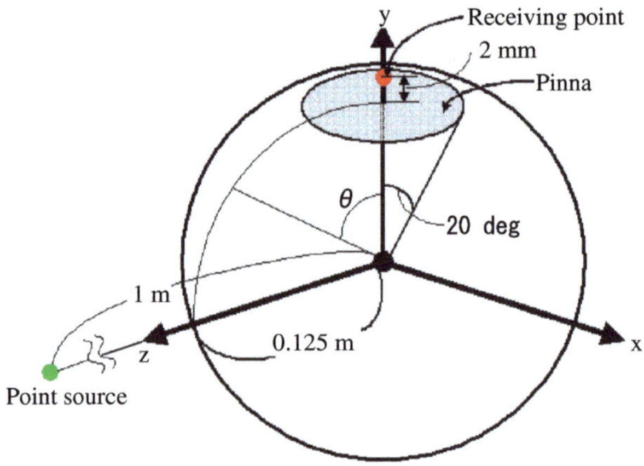

Fig. 9.17 Model used for reciprocity check

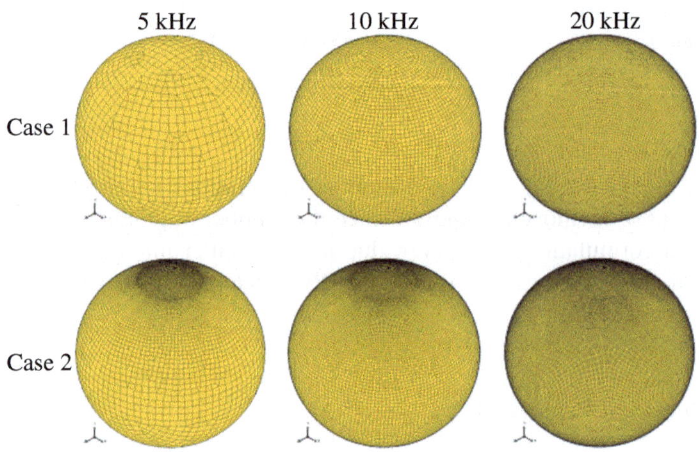

Fig. 9.18 Boundary element meshes used in checks of reciprocity

with the same condition as in Case 1 and the other part discretized to maintain the continuity of the elements. Figure 9.18 shows the meshes used in this check.

Table 9.2 shows sound pressure amplitude at the field point at 5, 10, and 20 kHz. The value indicated as an error is $(p_\mathrm{r} - p_\mathrm{n})/p_\mathrm{r}$. p_n is the pressure obtained by the model that does not consider reciprocity, and p_r is the pressure obtained by the model that does consider reciprocity. The reciprocity does not hold because errors in Case 1 are 24 %.

Table 9.2 Result of reciprocity checking

Frequency (kHz)	Case	Pressure in amplitude (Pa) Taking reciprocity into considerration		Relative error (%)
		No	Yes	
5	Case 1	1.160	0.874	24.66
	Case 2	1.180	1.169	0.932
10	Case 1	1.161	1.057	8.958
	Case 2	1.167	1.161	0.514
20	Case 1	1.139	1.127	1.053
	Case 2	1.138	1.138	0.000

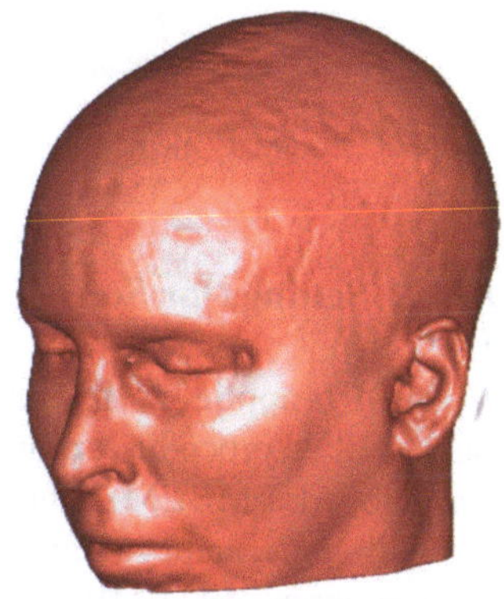

Fig. 9.19 Picture obtained by MRI

9.3.2 Simulation Results

9.3.2.1 Generation of Boundary Element Mesh

Individual geometry data is required to calculate the individual HRTF. To this end, the geometry used in the analysis is created from image data using magnetic resonance imaging (MRI). Using this geometry (Fig. 9.19), the surface is discretized.

By MRI, 108 pictures are taken with 1 mm clearance in the sagittal direction (parallel to the body axis z and separating the body into right and left). The format of these pictures in Digital Imaging and Communications in Medicine (DICOM), used as a format for medical pictures such as computed tomography (CT) and MRI, and those are read by a special program. Using this, the image information inside the eardrum is removed. The new image information is extracted to Standard Triangulated Language (STL), which is a data format to represent three-dimensional geometry as a cluster of small triangles. Finally, these data are used for the boundary element mesh. The generated boundary element mesh is shown in Fig. 9.20. Because the maximum element length is 2 mm (about one-sixth of a wavelength at 27 kHz when the sound speed is 340 m/s), the number of elements is 186,380.

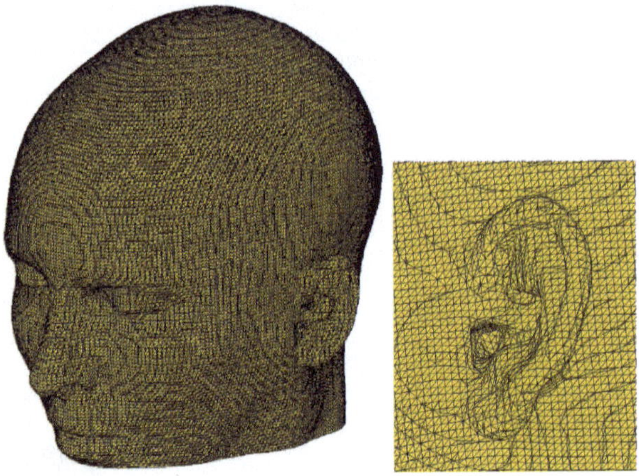

Fig. 9.20 Boundary element mesh used for HRTF calculation (num. of elements: 186,380)

9.3.2.2 Settings for Solver

When DF is applied for the calculation using the boundary element mesh obtained by the described procedure, more than 1 TB of memory is required if conventional BEM is employed. FMBEM is thereby employed. Moreover, in order to obtain HRTF for full audible frequency range with a single boundary element mesh, both FMBEM for low frequency (LF-FMBEM) and high frequency (HF-FMBEM) (see Sect. 4.3) are used. Detailed settings for LF-FMBEM and HF-FMBEM can be found in [11, 12]. The deepest cell level in the hierarchical cell structure L and the appropriate solver are selected so as to avoid exceeding the maximum memory requirement (in this case 16 GB). In this study, about 11.0 GB is required by LF-FMBEM below 6.2 kHz, and about 13.6 GB is required by HF-FMBEM for higher frequencies.

9.3.2.3 Approximation Using Rational Function

An approximated HRTF is evaluated from the response at the limited frequency by an interpolation with the aid of a rational function. The response at other frequencies can be obtained by directly evaluating this function. The calculation time was reduced because the approximated HRTF can be obtained from the limited frequency. The function is approximated using the following rational function:

$$f(x) = \frac{ax + b}{1 + cx}. \tag{9.6}$$

Coefficients a, b and c are calculated by the responses of the BEM model at three frequency steps: $x = x_0$ the center frequency, $x = x_{+1} = x_0 + \Delta x$ (where Δx is the difference in frequency between x_0 and the next calculated frequency to the center frequency), and $x = x_{-1} = x_0 - \Delta x$. The frequency x_0 is selected through the following procedure.

1. Two frequencies (the next to lowest frequency $[x_L]$ among the total frequency range and the frequency next to the highest one $[x_H]$) are selected as x_0.
2. For each x_0, calculate the response at three frequencies ($x_0 - \Delta x, x_0, x_0 + \Delta x$). Then with coefficients a, b, and c, the approximate functions Eq. (9.6) are defined. Those are called $f_{n,L}(x)$ and $f_{n,H}(x)$ later.
3. To define the switching frequency between $f_{n,L}(x)$ which is the approximate function for the lower frequency and $f_{n,H}(x)$, which is the approximate function for the higher frequency, explore the frequency x_k where the error defined below shows the minimum between x_L and x_H:

$$E = \frac{1}{p}\sum_{n=1}^{p}\left(\min\left[\frac{f_{n,H}(x_k) - f_{n,L}(x_k)}{f_{n,L}(x_k)}, \frac{f_{n,H}(x_k) - f_{n,L}(x_k)}{f_{n,H}(x_k)}\right]\right), \quad (9.7)$$

   where p is the number of field points.
4. If the error at x_k is lower than the specified value, $f_{n,L}(x)$ is selected as the approximation function below x_k and $f_{n,H}(x)$ is selected as the approximation function above x_k . In the case of greater than the specified value, first the center frequency x_M between x_L and x_H is calculated, next x_L and x_M are again set as a new variable group of x_L and x_H, and finally go back to 1. Similarly, x_M and x_H are set again as a new variable group and go to 1.

The approximation function for all frequency ranges is obtained by repeatedly executing the above procedure.

9.3.2.4 Other Settings

The actual calculation was conducted taking into account the reciprocity between the evaluation point and the sound source locations. Therefore the evaluation point and the source locations were switched, enabling the response calculation to be performed all at once.

The sound source is located 2 mm from the left eardrum (Fig. 9.22), and three field points are located as indicated by Fig. 9.21 in the horizontal plane, which includes the ear canal entrance. The sound speed is 340 m/s and medium density is 1.225 kg/m^3. Response is calculated by 50 Hz up to 22.05 kHz. The error indicated in Eq. (9.7) is 1.0. The Generalized Product Bi-Conjugate Gradient (GPBiCG) was adopted as an iterative solver. The machine used is a Dell Precision 690 with the following specifications; CPU: Intel R XeonR, X5355 processor, 2.66 GHz; OS: Windows XP Professional x64 edition; memory: 32.0 GB. The other settings are the same as in Sect. 9.3.1.

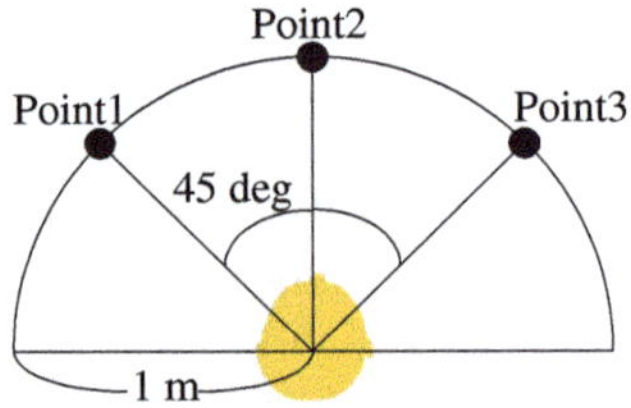

Fig. 9.21 Field point locations

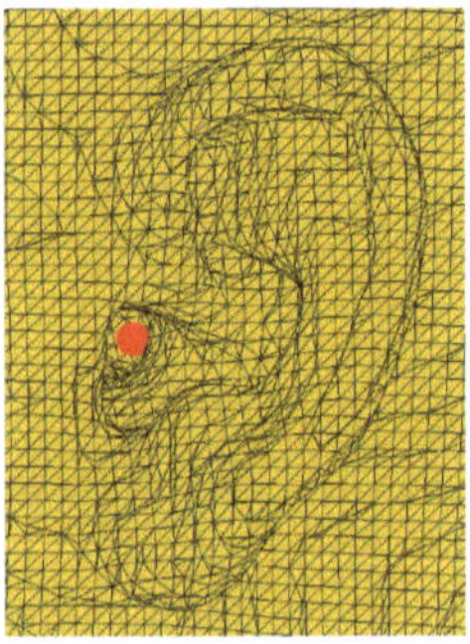

Fig. 9.22 Point source location (indicated by *circle* near entrance of ear canal)

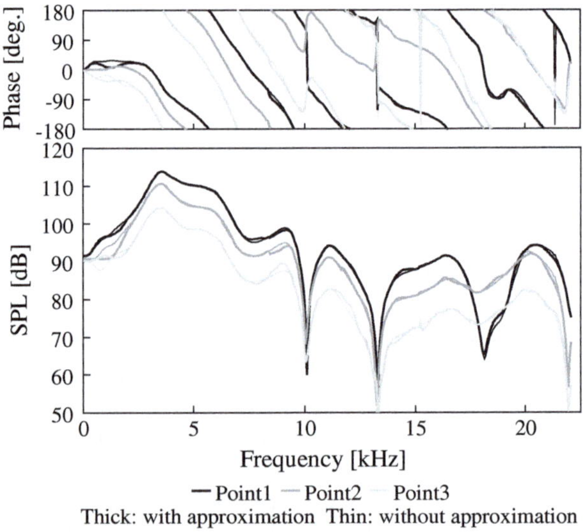

Fig. 9.23 HRTF: (*Top*) phase shift (value minus the change within distance between sound source and evaluation points), (*bottom*) sound pressure level (normalization of sound pressure at a point 1 m from sound source)

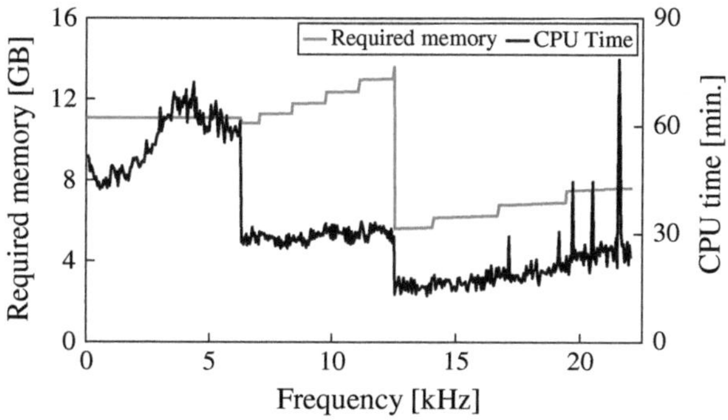

Fig. 9.24 Required amount of memory and calculation time

9.3.2.5 Results

The calculated HRTFs are shown in Fig. 9.23. In this figure, both HRTFs with and without rational function approximation are shown.

Figure 9.24 shows the required amount of memory and the calculation time of the case without rational function approximation. In this case it takes about 10.1 days to acquire the HRTFs for all frequency range, whereas by the case of using rational function approximation the total calculation time is reduced to 1.6 days because the number of calculation steps is reduced from 441 to 72.

References

1. T. Yokota, S. Sakamoto, H. Tachibana, Sound field simulation method by combining finite difference time domain calculation and multi-channel reproduction technique. Acoust. Sci. Tech. **25**, 15–23 (2004)
2. S. Yokoyama, K. Ueno, S. Sakamoto, H. Tachibana, 6-channel recording/reproduction system for 3-dimensional auralization of sound fields. Acoust. Sci. Tech. **23**, 97–102 (2002)
3. T. Yokota, S. Sakamoto, H. Tachibana, K. Ishii, Numerical analysis and auralization of the "ROARING DRAGON" phenomenon by the FDTD method. J. Environ. Eng. (Trans. AIJ) **629**, 849–856 (2008)
4. K. Ueno, H. Tachibana, Experimental study on the evaluation of stage acoustics by musicians using a 6-channel sound simulation system. Acoust. Sci. Tech. **24**, 130–138 (2003)
5. T. Asakura, T. Miyajima, S. Sakamoto, Prediction method for sound from passing vehicle transmitted through building facade. Appl. Acoust. **74**(5), 758–769 (2013)
6. B. Gardner, K. Martin. HRTF measurements of a KEMAR dummy-head microphone. Technical report, MIT Media Lab Perceptual Computing, 1994
7. Y. Kahana, P.A. Nelson, Boundary element simulations of the transfer function of human heads and baffled pinnae using accurate geometric models. J. Sound Vib. **300**, 552–579 (2007)
8. T. Masumoto, T. Oshima, Y. Yasuda, T. Sakuma, M. Kabuto, M. Akiyama. HRTF calculation in the full audible frequency range using FMBEM, in *Proceedings of the 19th International Congress on Acoustics*, Madrid, number COM-06-012, 2007 (In CD-ROM)
9. E.G. Williams, *Fourier acoustics: sound radiation and nearfield acoustical holography*, Chap.6.7 (Academic Press,1999)

10. Y. Saad, ILUT: a dual threshold incomplete LU factorization. Numer. Linear Algebra **1**, 387–402 (1994)
11. Y. Yasuda, T. Oshima, T. Sakuma, A. Gunawan, T. Masumoto, Fast multipole boundary element method for low-frequency acoustic problems based on a variety of formulations. J. Comput. Acoust. **18**, 363–395 (2010)
12. Y. Yasuda, T. Sakuma, Fast multipole boundary element method for large-scale steady-state sound field analysis. Part II: examination of numerical items. Acta Acust. United Ac. **89**(4), 28–38 (2003)

Index

T. Sakuma et al. (eds.), *Computational Simulation in Architectural, and Environmental Acoustics*, DOI: 10.1007/978-4-431-54454-8,

MIX
Papier aus verantwortungsvollen Quellen
Paper from responsible sources
FSC® C105338

If you have any concerns about our products,
you can contact us on
ProductSafety@springernature.com

In case Publisher is established outside the EU,
the EU authorized representative is:
Springer Nature Customer Service Center GmbH
Europaplatz 3, 69115 Heidelberg, Germany

Printed by Libri Plureos GmbH
in Hamburg, Germany